CARRIER TRANSPORT IN NANOSCALE MOS TRANSISTORS

CARRIER TRANSPORT IN NANOSCALE MOS TRANSISTORS

Hideaki Tsuchiya

Yoshinari Kamakura

This edition first published 2016

Registered Office
John Wiley & Sons Singapore Pte. Ltd., 1 Fusionopolis Walk, #07-01 Solaris South Tower, Singapore 138628.

For details of our global editorial offices, for customer services and for information about how to apply for permission to reuse the copyright material in this book please see our website at www.wiley.com.

Library of Congress Cataloging-in-Publication data applied for

ISBN: 9781118871669

A catalogue record for this book is available from the British Library.

Set in 10/12pt Times by SPi Global, Pondicherry, India

Printed and bound in Singapore by Markono Print Media Pte Ltd

1 2016

Contents

Preface

The device scaling concept, which can lead to increase in both switching speed and integrated density of MOSFETs with reasonable power consumption, has been the main guiding principle of the integrated device engineering over the past 40 years. It has been recognized, however, that conventional device scaling has confronted difficulties below the sub-100 nm regime, owing to several physical and essential limitations directly related to device miniaturization. As a consequence, new device technologies to overcome these difficulties are highly required. A group of these new device technologies, called technology boosters, include high-*k* gate stack technologies, high carrier mobility channels, ultrathin-body structures, multigate structures, metal source/drain, and novel operating principles. The basic purpose of these technologies are to boost or improve specific device parameters, such as carrier velocity, gate leakage current, short-channel effects, subthreshold slope, and so on.

Given the large number of technology options mentioned above, physically based device simulations will play an important role in developing the most promising strategies for forthcoming nanometer era. In particular, most of the device architecture and material options are expected to affect the performance of MOSFETs through the band structure, the electrostatics and the scattering rates of carriers in the channel region. Therefore, microscopic or atomistic modeling is necessary to obtain a physical insight and to develop a quantitative description of the carrier transport in ultrascaled MOSFETs. In this context, this book aims to offer a thorough explanation of carrier transport modeling of nanoscale MOSFETs, covering topics from the atomistic band structure calculation to the most recent challenges targeting beyond the end of the International Technology Roadmap for Semiconductors (ITRS). We also focus on the roles of phonon transport in ultrascaled MOSFETs, which are getting a lot more attention lately as major thermal management challenges on the LSI chip.

As for the modeling methodology, we have highlighted the multi-subband Monte Carlo method because of some distinct advantages compared to other methods. Specifically, it provides us with the ability to explore all transport regimes, including diffusive, quasi-ballistic and even quantum transport (by applying a Wigner Monte Carlo technique) regimes, and also introduces new scattering mechanisms without increasing its computational resources. The physical interpretation of calculated results is intuitively comprehensible, owing to its particle description of the carrier transport. The dynamical equation of the Wigner function (i.e. the Wigner transport equation) is very similar to the Boltzmann transport equation, except in the influence of the potential whose rapid space variations generate quantum mechanical effects. Furthermore, it coincides with the non-equilibrium Green's function formalism under a ballistic transport. We have illustrated the details of the Wigner Monte Carlo technique and its application to the quantum transport analysis of III-V MOSFETs in this book.

To go beyond the end of the ITRS roadmap, several alternative or innovative devices are being investigated, such as nanowires, carbon nanotubes, graphenes and tunnel-FETs. We have dealt with nanowires and some atomic layer 2-D materials related to graphene, and have discussed their performance potentials by comparisons with those of competitive MOSFETs composed of Si and III-V compound semiconductors.

This book was written for graduate students, engineers and scientists who are engaged in work on nanoscale electronic devices, and was designed to provide a deeper understanding of physical aspects of carrier transport in real electronic devices. Familiarity with quantum mechanics, basic semiconductor physics and electronics is assumed. After working through this book, students should be prepared to follow current device research, and to actively participate in developing future devices.

Hideaki Tsuchiya
Yoshinari Kamakura

Acknowledgements

The authors would like to warmly thank several colleagues of the Nanostructure Electronics Laboratory at Kobe University and the Integrated Functional Devices Laboratory at Osaka University. In particular, we wish to thank Tanroku Miyoshi, Matsuto Ogawa, Satofumi Souma, and Kenji Sasaoka, all at Kobe University (Japan), and to thank Nobuya Mori at Osaka University (Japan), Shigeyasu Uno at Ritsumeikan University (Japan), Masashi Uematsu and Kohei Itho at Keio University (Japan), all of whom are members of the joint research project of JST/CREST. We also wish to thank Gennady Mil'nikov, Hideki Minari, Yoshihiro Yamada, Shunsuke Koba, Kentaro Kukita, Indra Nur Adissilo, and Shiro Kaneko, together with many students who spent time in our laboratories.

The authors are also indebted to several people for their encouragement and support in writing this book and/or their kind comments and suggestions in the course of our work. In particular, we wish to thank Karl Hess and Umberto Ravaioli at University of Illinois at Urbana-Champaign (USA), Shin'ichi Takagi, Toshiro Hiramoto and Akira Toriumi at University of Tokyo (Japan), Ken Uchida at Keio University (Japan), Philippe Dollfus at University of Paris-Sud, Orsay (France), Kenji Natori at Tokyo Institute of Technology (Japan), Takanobu Watanabe at Waseda University (Japan), Nobuyuki Sano at University of Tsukuba (Japan), David K. Ferry at Arizona State University (USA), Tomislav Suligoj at University of Zagreb (Croatia), David Esseni at University of Udine (Italy), Ming-Jer Chen at National Chiao Tung University (Taiwan), and Björn Fischer at Infinion Technology (Germany). Finally, H.T. is especially thankful to my parents, Shiro and Satsuki, who encouraged me to undertake writing this book during the hospitalization for my anti-cancer drug treatment at Japanese Red Cross Kobe Hospital, and took care of my two sons, Seiji and Satoshi. H.T. dedicates this book to my late wife, Naomi, with love.

1

Emerging Technologies

1.1 Moore's Law and the Power Crisis

Figure 1.1 shows the famous Moore's law for a metal-oxide-semiconductor field-effect transistor (MOSFET) integrated in an electronic logic circuit, which illustrates the annual variations in the number of transistors and in transistor size in a simple way. Since large-scale integrated (LSI) circuit technology was invented in the 1960s, the progress of miniaturization techniques based on scaling law has achieved significant advancement in the electronics industry, up to the present date. However, from the year around 2005, the increase in power consumption of LSI circuits has become a major problem. To succeed in the scaling law, not only the geometrical dimensions of MOSFET, a basic building block of LSI circuit, but also their power supply voltage, are required to be scaled down simultaneously. However, the power supply voltage has ceased to fall, at around 1 V after 2005. There are various reasons for this – for example: to suppress characteristic variability among hundreds of millions of integrated MOSFETs; to cut wasteful power consumption in the off-state; to maintain high-speed performance, and so on. Consequently, LSI consumption power or, in terms of global influence, the total electrical power consumed by IT devices and systems all over the world, increases rapidly year by year.

The power consumption of a MOSFET is expressed by:

$$P = fC_{\mathrm{load}}V_{\mathrm{dd}}^2 + I_{\mathrm{off}}V_{\mathrm{dd}} \tag{1.1}$$

where f, C_{load}, V_{dd} and I_{off} represent the operating frequency, the load capacitance, the power-supply voltage, and the off-current, respectively. The first term on the right-hand side of Equation (1.1) corresponds to the power required to charge and discharge a MOS capacitor, (i.e., a consumed power at on-state), and the second term, consumed power at off-state. The ceasing to fall of V_{dd}, as mentioned above, has mainly induced the increase in consumed power at on-state. On the other hand, owing to the drain-induced barrier lowering (DIBL)

Carrier Transport in Nanoscale MOS Transistors, First Edition. Hideaki Tsuchiya and Yoshinari Kamakura.

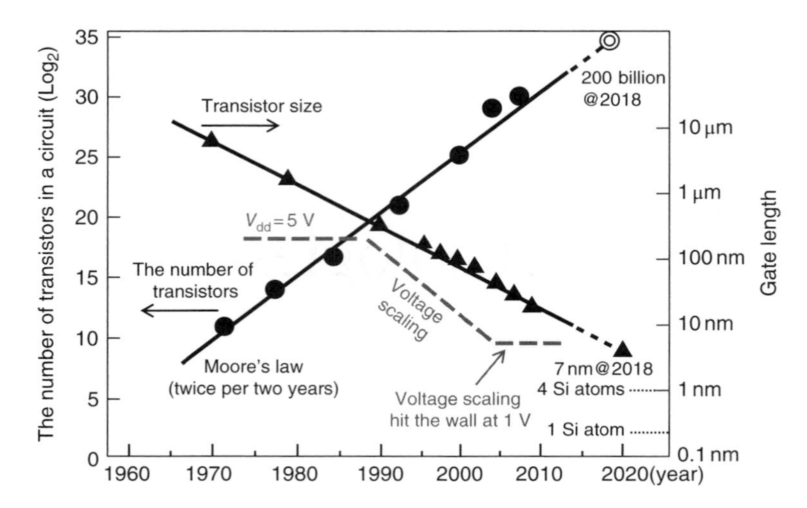

Figure 1.1 Moore's law for a MOSFET integrated in a LSI.

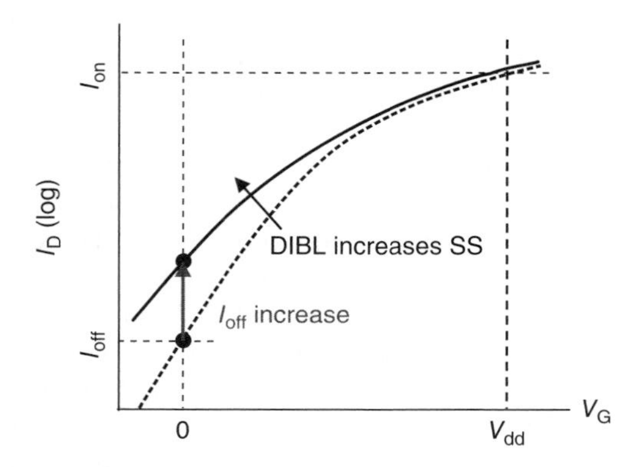

Figure 1.2 Influences of DIBL on $I_D - V_G$ characteristics. DIBL degrades the subthreshold slope (SS) and then causes an exponential I_{off} increase.

phenomenon, which is caused by reduction of the gate electrostatic control over the channel with decreasing the channel length, I_{off} is beginning to increase exponentially, as shown in Figure 1.2. This leads to a drastic increase in consumed power at off-state – which, for instance, decreases the battery life of mobile devices such as smartphones and wearable appliances.

1.2 Novel Device Architectures

To reduce the off-state power consumption, novel structure MOSFETs that possess better gate electrostatic control to suppress DIBL have received a lot of attention [1.1]. The representative new device structures are shown in Figure 1.3. In 2012, the Intel Corporation released an announcement stating that they were starting to manufacture central processing units (CPUs)

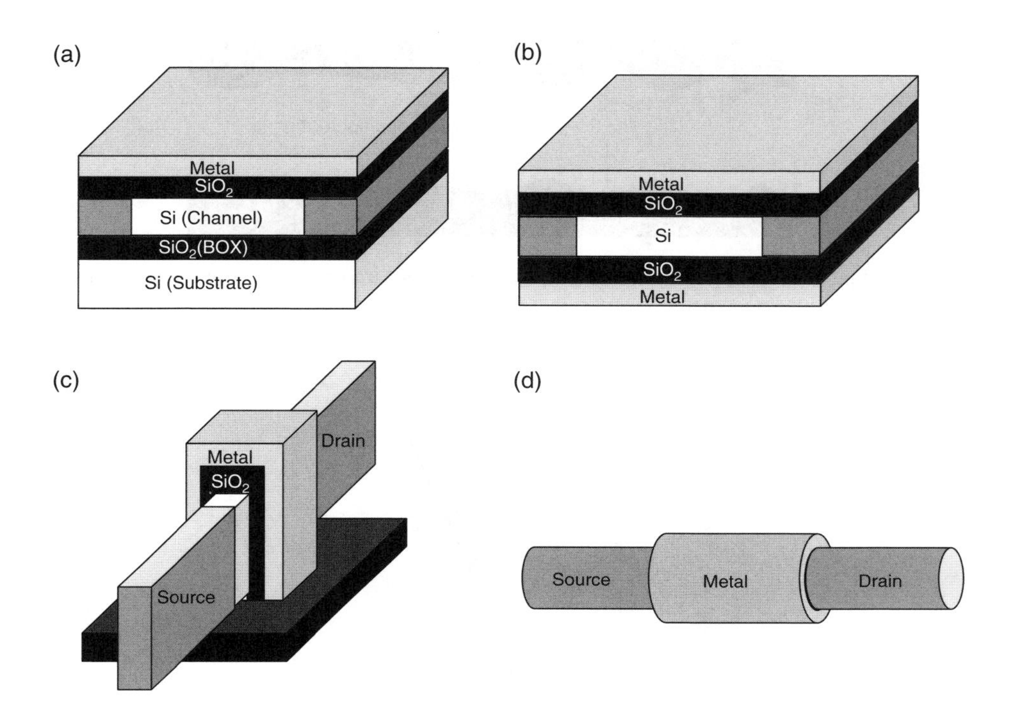

Figure 1.3 Representative new device structures. (a) ultrathin-body (UTB) silicon-on-insulator (SOI) structure; (b) double-gate (DG) structure; (c) Fin or trigate structure; and (d) gate-all-around (GAA) nanowire structure.

constructed from FinFETs [1.2]. This was a landmark in the electronic industry, because a three-dimensional transistor has been commercialized for the first time since the planar type MOS transistor was invented in 1960. A GAA nanowire MOSFET, shown in Figure 1.3(d), is considered one of the ultimate structures of FinFETs and, therefore, globally active and competitive research has been promoted.

As seen in Figure 1.3, these new structure MOSFETs have an ultrathin Si channel sandwiched in between gate oxides or insulators of substrate. In particular, Si channels in FinFET and GAA nanowire MOSFET are completely surrounded by oxides. As a result, the Si channel thickness T_{Si} fluctuates along a transport direction in atomic scale, as shown in Figure 1.4.

When T_{Si} is thinner than a spatial extent of carrier's wave function, the T_{Si} fluctuation produces spatial fluctuation of quantized sub-band along the transport direction, and thus leads to an additional scattering source for carriers. Consequently, the carrier mobility may seriously decrease in nanometer-scaled new structure MOSFETs. The influence of the T_{Si} fluctuation was first investigated by H. Sakaki *et al.* experimentally and theoretically for GaAs/AlAs quantum well structures [1.3]. They found that the electron mobility reduces in proportion to the sixth power of quantum well thickness, which shows that the interface fluctuation scattering is the dominant scattering mechanism in thin quantum well structures.

For SOI-MOSFETs, K. Uchida *et al.* experimentally demonstrated that the same channel thickness dependence as for Sakaki's result is obtained for T_{Si}s less than 3 nm, as shown in

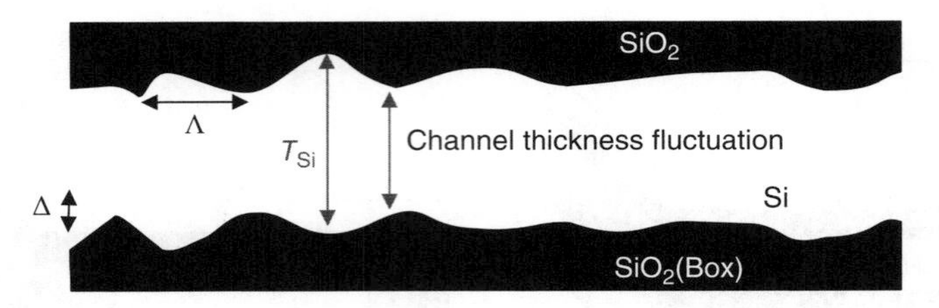

Figure 1.4 Spatial fluctuation of Si channel thickness along transport direction, which emerges in ultrathin Si films with a nanometer thickness.

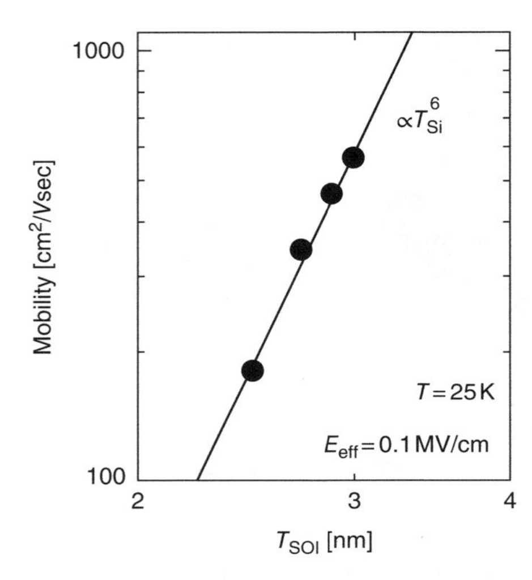

Figure 1.5 T_{Si} dependence of electron mobility at 25 K [1.4]. T_{Si}^{6} dependence is clearly observed for T_{Si}s less than 3 nm.

Figure 1.5 [1.4]. Therefore, there are growing concerns about the degradation of the on-state device performance in new structure MOSFETs with a nanometer channel thickness. However, the role of the T_{Si} fluctuation under a quasi-ballistic transport, where scattering events inside the channel decrease to several times, has not yet been fully understood. To deeply understand it, we need to develop a device simulation technique considering quantum confinement and scattering effects at the atomic level. We will describe such a challenge in Chapter 3.

In addition to the scattering by the T_{Si} fluctuation mentioned above, phonon scattering and impurity scattering also play an important role. In particular, intrinsic channels are likely adopted in novel structure MOSFETs and, thus, deep understanding of phonon scattering processes in ultrashort channel MOSFETs should be important. Carrier transport in this regime has been actively discussed in terms of the quasi-ballistic transport since K. Natori proposed the concept of ballistic MOSFET [1.5].

As for phonon scattering processes, interestingly, *inelastic phonon emission processes* can suppress carriers backscattering to the source and then promote ballistic transport, contrary to common sense, in the case of ultrashort-channel MOSFETs [1.6, 1.7]. This is considered to be due to the fact that once a carrier has lost its kinetic energy by a few multiples of k_BT (about 60 meV for silicon) via inelastic phonon emission processes, the carrier has little chance of returning to the source, due to the potential bottleneck barrier, and is eventually absorbed into the drain; thus, the ballisticity improves. We will discuss this subject in detail in the first half of Chapter 3.

The continued scaling of transistor dimensions and integrated density is causing major thermal management challenges on the LSI chip [1.8]. In particular, the novel structure MOSFETs have Si channels surrounded by the gate oxides and insulators, which have a lower thermal conductivity than Si [1.9]. Therefore, thermal energies generated in a device via optical phonon emission are readily accumulated inside the device, which might lead to degradation of the device performance. In Chapter 4, we will discuss phonon transport in Si nanostructures, to examine such a heat generation problem qualitatively.

1.3 High Mobility Channel Materials

The reduction of V_{dd} is essential to decrease on-state power consumption. Higher mobility channel materials can increase the on-current because the carrier's velocity becomes higher at the same V_{dd}, and thus they are expected to achieve equal or superior performance to Si MOSFETs under a lower V_{dd} operation [1.10], as shown in Figure 1.6.

The effective masses and mobilities of representative semiconductors are summarized in Table 1.1. Compared to Si, Ge has both a higher electron mobility and a higher hole mobility, while III-V compound semiconductors, that is, InP and $In_{0.53}Ga_{0.47}As$, have a significantly higher electron mobility. One of the important reminders is that the solid solubility of donors in III-V semiconductors is limited to less than, or comparable to, $2\times10^{19}\,cm^{-3}$ [1.11]. Consequently, III-V MOSFETs generally exhibit a higher parasitic resistance in source and drain electrodes than Si MOSFETs do [1.12–1.15]. This also may lead to "source starvation" [1.12, 1.13], which cannot maintain a large flow of ballistic carriers heading in the channel, owing to the insufficient impurity scattering in the lightly doped source. We will discuss this subject in the first half of Chapter 5.

The higher mobilities of III-V semiconductors are mainly due to their lighter effective masses. But then, a lighter effective mass carrier has a larger tunneling probability through a

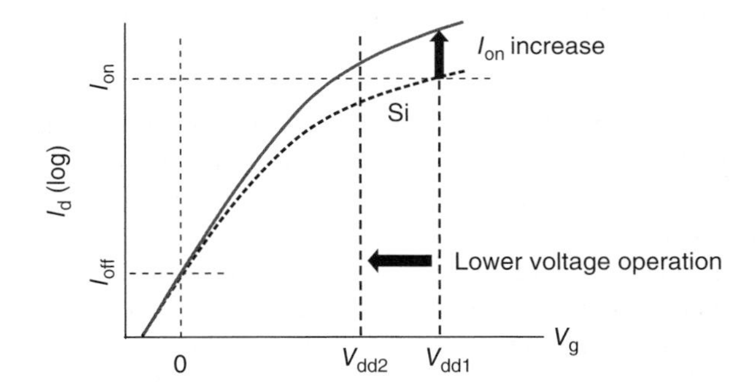

Figure 1.6 On-current increase due to high-mobility channel MOSFETs. They are expected to achieve a lower V_{dd} operation than conventional Si MOSFETs.

Table 1.1 Effective masses and mobilities of representative semiconductors.

Material		Si	Ge	InP	$In_{0.53}Ga_{0.47}As$
electron	mass m_e (m_0)	0.19/0.98 (m_t/m_l)	0.082/1.59 (m_t/m_l)	0.082	0.046
	mobility ($cm^2/V \cdot s$)	1600	3900	5400	25 000
hole	mass m_{hh}/m_{lh} (m_0)	0.49/0.16	0.28/0.044	0.45/0.12	0.51/0.22
	mobility ($cm^2/V \cdot s$)	430	1900	200	450

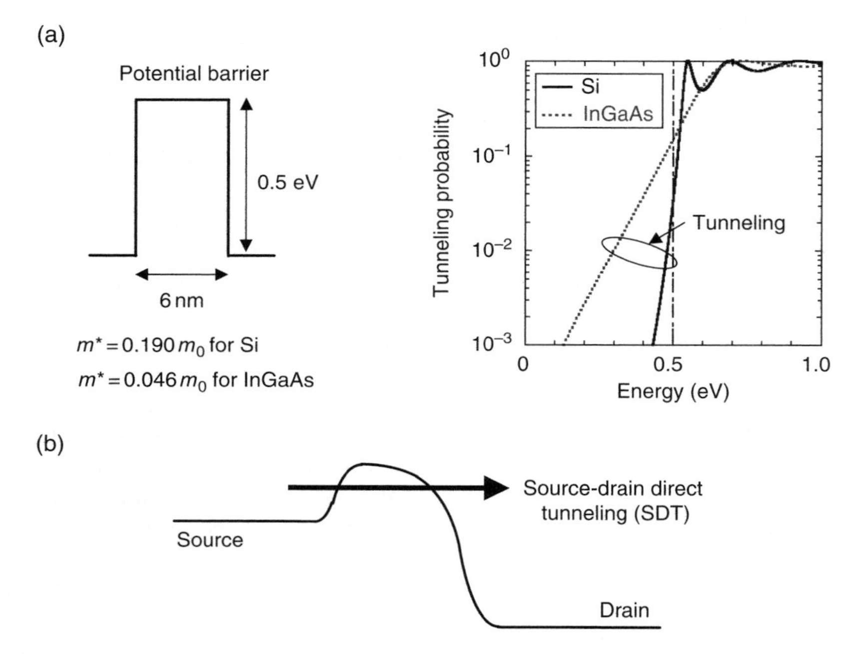

Figure 1.7 (a) Tunneling probabilities calculated for Si and $In_{0.53}Ga_{0.47}As$ through a single potential barrier and (b) schematic of source-drain direct tunneling (SDT) at off-state.

finite potential barrier. Figure 1.7(a) shows the tunneling probabilities calculated for Si and $In_{0.53}Ga_{0.47}As$ through the potential barrier with 0.5 eV height and 6 nm width, which supposes an off-state of a sub-10 nm MOSFET. The effective masses were given as $m^* = 0.19\,m_0$ for Si and $0.046\,m_0$ for $In_{0.53}Ga_{0.47}As$.

It is found that $In_{0.53}Ga_{0.47}As$ exhibits several orders of magnitude larger tunneling probability than Si. This phenomenon leads to a tunneling leakage current between source and drain electrodes at off-state, as shown in Figure 1.7(b). Therefore, this is called "source-drain direct tunneling (SDT)." SDT might be a major obstacle in downscaling III-V MOSFETs into the deca-nanometer or nanometer scale [1.16, 1.17]. We will discuss this subject in the second half of Chapter 5.

1.4 Two-Dimensional (2-D) Materials

Graphene, a one-atom-thick carbon sheet arranged in a honeycomb lattice [1.18], is known to exhibit the highest electron mobility of all presently-known materials and, therefore, its application to high-speed electronic devices is strongly anticipated. However, since graphene has no band gap, the electrical conduction cannot be fully switched off by tuning the gate voltage, which is necessary for digital applications. To open a band gap, several methods have been proposed, as shown in Figure 1.8. Graphene nanoribbon (Figure 1.8(a)) uses quantum confinement effect in its transverse direction, while bilayer graphene (Figure 1.8(b)) introduces symmetry breaking between two carbon layers via a vertical electric field or interaction between a graphene layer and its substrate.

Although these methods actually open a band gap in graphene, the characteristic linear dispersion relation is distorted and, furthermore, an effective mass appears in graphene nanoribbons. Accordingly, accurate consideration of the band structure is important in order to assess the device characteristics of semiconducting graphene devices.

Practical application of graphene devices require a reliable substrate, but the mobility in graphene on SiO_2 substrates is limited to 25 000 cm²/Vs [1.19–1.21]. The reason for this mobility reduction on SiO_2 substrates is considered to be the additional scattering mechanisms induced by the substrate, such as charged impurities, polar and non-polar surface optical phonons in the SiO_2, and substrate surface roughness. On the other hand, a drastic improvement of the mobility to 140 000 cm²/Vs near the charge neutrality point was reported using a hexagonal boron nitride (h-BN) substrate [1.22]. Hence, h-BN substrate is expected to be suitable for graphene electronic devices. For practical design and analysis of graphene devices, we will need to consider the scattering from the substrates precisely.

The discovery of graphene, and the succeeding tremendous advancement in this field of research, have promoted the search for similar two-dimensional (2-D) materials composed of other group IV elements. The silicon or germanium equivalents of graphene are called silicene and germanene. A 2-D silicene has been successfully fabricated on (0001)-oriented thin films of zirconium diboride (ZrB_2), that were grown epitaxially on Si (111) wafers [1.23]. Theoretical simulations showed that silicene and germanene have no band gap, similar to graphene. On the other hand, 2-D materials such as silicane, germanane [1.24], MoS_2 [1.25, 1.26] and black phosphorus [1.27, 1.28] exhibit a band gap larger than about 1 eV, though they are monolayer-thick materials. These 2-D materials with a sufficiently large band gap may have advantages in the application to LSI devices over graphene-based materials. We will discuss the electronic properties of 2-D materials and their performance potentials as an FET channel in Chapter 7.

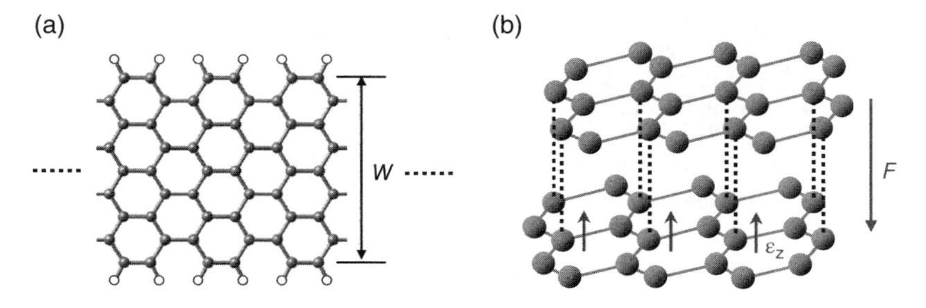

Figure 1.8 (a) Graphene nanoribbon; and (b) bilayer graphene to open a band gap.

1.5 Atomistic Modeling

As described in Section 1.2, Si UTB-SOI MOSFET, shown in Figure 1.3 (a) has better gate electrostatic control over the channel than conventional Si bulk MOSFET, and thus is expected to be immune to short-channel effects such as DIBL, and threshold voltage lowering with decreasing the channel length. Experimentally, extremely-scaled SOI-MOSFETs with Si channel thickness less than 1 nm have been fabricated, as shown in Figure 1.9(a) [1.29].

Currently, hydrogen termination of the channel is used in device modeling, as a compromise between efficiency and accuracy. However, in such atomic-scale dimensions, not only the quantum confinement effects, but also the roles of interfaces between the Si channel and the SiO_2 oxides, will be important to achieve good agreement with experimental results. Thus, state-of-the-art *ab initio* simulation techniques, such as a density-functional first-principles method [1.30], or a density-functional tight-binding method [1.31], where practical atomic structures are assumed for SiO_2 layers, as shown in Figure 1.9(b), have been applied to reveal large quantitative differences, in comparison with simulations of H-terminated Si film.

Furthermore, considering the sub-10 nm technology node, GAA nanowire MOSFET with ideal gate electrostatic controllability is attracting a lot of attention. Figure 1.10 shows the schematic of GAA nanowire MOSFET constructed from a diamond crystal, where a square-shaped cross-section is assumed. In nanowire MOSFETs with the channel length shorter than 10 nm, its cross-sectional dimensions are supposed to be less than 5 nm. Hence, the number of atoms in the cross-section becomes countable, as shown in Figure 1.10(a), (b) and (c).

Accordingly, the electronic states significantly depend on geometrical parameters such as wire orientation and cross-sectional dimension [1.32]. Moreover, not only electrons, but also phonons, are spatially confined to a nanometer scale [1.33, 1.34]. To deeply understand their behaviors, we need an atomistic device simulation technique based on a first-principles approach [1.35, 1.36], or a tight-binding approach [1.37, 1.38], if necessary coupled with phonon band structure calculation and electron-phonon interaction modeling [1.33, 1.34]. We will describe such a challenge in Chapters 2 and 6.

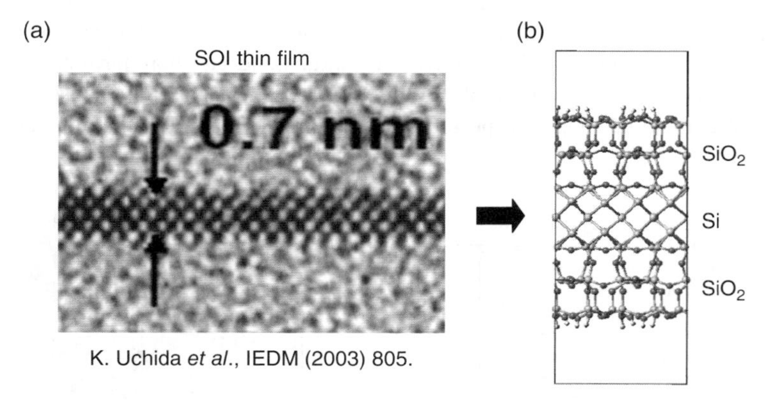

Figure 1.9 (a) Experimentally fabricated SOI-MOSFET with Si channel thickness of 0.7 nm, which corresponds to the 5-atomic-layer thickness of Si atoms [1.29]. (b) Example of SOI atomic structure used for *ab initio* simulations.

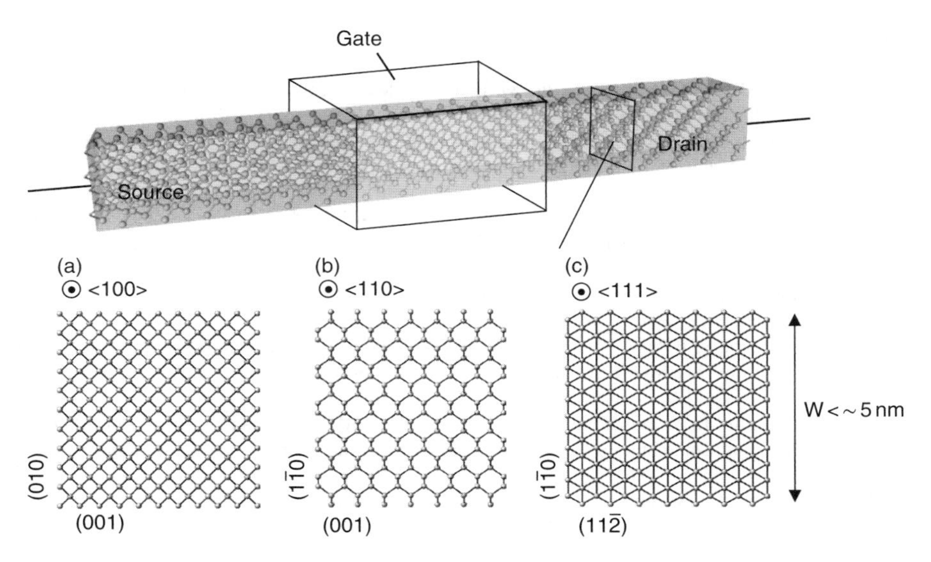

Figure 1.10 Schematic of GAA nanowire MOSFET constructed from a diamond crystal. As shown in the lower figures, when the cross-sectional dimensions are equal to or less than 5 nm, the number of atoms in the cross-section becomes countable, where the wire direction is (a) <100>-; (b) <110>-; and (c) <111>-orientations.

In summary, the introduction of new device structures and new channel materials are expected to improve the device performance of MOSFETs, without depending on conventional geometrical scaling. On the other hand, they can exhibit quite different features from conventional MOSFETs, because they are "new" technologies. Most of them are caused by quantum mechanical properties of carriers. In a sense, breaking the miniaturization limit of Si MOSFETs relies on the well-managed control of quantum mechanical effects and, therefore, the role of quantum mechanical simulation and atomistic analysis techniques will become more and more important than ever.

References

[1.1] *International Technology Roadmap for Semiconductors* [http://www.itrs.net/].

[1.2] C. Auth, C. Allen, A. Blattner, D. Bergstrom, M. Brazier *et al.* (2012). A 22 nm high performance and low-power CMOS technology featuring fully-depleted tri-gate transistors, self-aligned contacts and high density MIM capacitors. *2012 Symposium on VLSI Technology*, 131–132.

[1.3] H. Sakaki, T. Noda, K. Hirakawa, M. Tanaka and T. Matsusue (Dec. 1987). Interface roughness scattering in GaAs/AlAs quantum wells. *Applied Physics Letters* **51**(23), 1934–1936.

[1.4] K. Uchida, H. Watanabe, A. Kinoshita, J. Koga, T. Numata and S. Takagi (2002). Experimental study on carrier transport mechanism in ultrathin-body SOI n- and p-MOSFETs with SOI thickness less than 5 nm. *IEDM Technical Digest* 47–50.

[1.5] K. Natori (Oct. 1994). Ballistic metal-oxide-semiconductor field effect transistor. *Journal of Applied Physics* **76**(8), 4879–4890.

[1.6] K. Natori (July 2008). Ballistic/quasi-ballistic transport in nanoscale transistor. *Applied Surface Science* **254**(19), 6194–6198.

[1.7] H. Tsuchiya and S. Takagi (Sep. 2008). Influence of elastic and inelastic phonon scattering on the drive current of quasi-ballistic MOSFETs. *IEEE Transactions on Electron Devices* **55**(9), 2397–2402.

[1.8] J.A. Rowlette and K.E. Goodson (Jan. 2008). Fully coupled nonequilibrium electron-phonon transport in nanometer-scale silicon FETs. *IEEE Transactions on Electron Devices* **55**(1), 220–232.

[1.9] Y. Kamakura, I. N. Adisusilo, K. Kukita, G. Wakimura, S. Koba, H. Tsuchiya and N. Mori (2014). Coupled Monte Carlo simulation of transient electron-phonon transport in small FETs. *IEDM Technical Digest*, 176–179.

[1.10] S. Takagi, R. Zhang, J. Suh, S.-H. Kim, M. Yokoyama, K. Nishi and M. Takenaka (May 2015). III-V/Ge channel MOS device technologies in nano CMOS era, *Japanese Journal of Applied Physics* **54**, p. 06FA01.

[1.11] M.V. Fischetti and S.E. Laux (Mar. 1991). Monte Carlo simulation of transport in technologically significant semiconductors of the diamond and zinc-blende structure – Part II: Submicrometer MOSFET's. *IEEE Transactions on Electron Devices* **38**(3), 650–660.

[1.12] M.V. Fischetti, L. Wang, B. Yu, C. Sachs, P.M. Asbeck, Y. Taur and M. Rodwell (2007). Simulation of electron transport in high-mobility MOSFETs: Density of states bottleneck and source starvation. *IEDM Technical Digest*, 109–112.

[1.13] M.V. Fischetti, S. Jin, T.-W. Tang, P. Asbeck, Y. Taur, S.E. Laux, M. Rodwell and N. Sano (Jun. 2009). Scaling MOSFETs to 10 nm: Coulomb effects, source starvation and virtual source model. *Journal of Computational Electronics* **8**(2), 60–77.

[1.14] T. Mori, Y. Azuma, H. Tsuchiya and T. Miyoshi (Mar. 2008). Comparative study on drive current of III-V semiconductor, Ge and Si channel n-MOSFETs based on quantum-corrected Monte Carlo simulation. *IEEE Transactions on Nanotechnology* **7**(2), 237–241.

[1.15] H. Tsuchiya, A. Maenaka, T. Mori and Y. Azuma (Apr. 2010). Role of carrier transport in source and drain electrodes of high-mobility MOSFETs. *IEEE Electron Device Letters* **31**(4), 365–367.

[1.16] S.R. Mehrotra, S.G. Kim, T. Kubis, M. Povolotskyi, M.S. Lundstrom and G. Klimeck (July 2013). Engineering nanowire n-MOSFETs at $Lg < 8$ nm. *IEEE Transactions on Electron Devices* **60**(7), 2171–2177.

[1.17] S. Koba, M. Ohmori, Y. Maegawa, H. Tsuchiya, Y. Kamakura, N. Mori and M. Ogawa (Feb. 2014). Channel length scaling limits of III-V channel MOSFETs governed by source-drain direct tunneling. *Japanese Journal of Applied Physics* **53**, 04EC10.

[1.18] K.S. Novoselov, A.K. Geim, S.V. Morozov, D. Jiang, Y. Zhang, S.V. Dubonos, I.V. Grigorieva and A. A. Firsov (Oct. 2004). Electric field effect in atomically thin carbon films, *Science* **306**, 666–669.

[1.19] Y.-W. Tan, Y. Zhang, K. Bolotin, Y. Zhao, S. Adam, E.H. Hwang, S. Das Sarma, H.L. Stormer and P. Kim (Dec. 2007). Measurement of scattering rate and minimum conductivity in graphene. *Physical Review Letters* **99**(24), p. 246803.

[1.20] S. Cho and M.S. Fuhrer (Feb. 2008). Charge transport and inhomogeneity near the minimum conductivity point in graphene. *Physical Review B* **77**(8), p. 081402(R).

[1.21] J. Yan and M.S. Fuhrer (Nov. 2011). Correlated charged impurity scattering in graphene. *Physical Review Letters* **107**(20), p. 206601.

[1.22] C.R. Dean, A.F. Young, I. Meric, C. Lee, L. Wang, S. Sorgenfrei, K. Watanabe, T. Taniguchi, P. Kim, K.L. Shepard and J. Hone (Oct. 2010). Boron nitride substrates for high-quality graphene electronics. *Nature Nanotechnology* **5**, 722–726.

[1.23] A. Fleurence, R. Friedlein, T. Ozaki, H. Kawai, Y. Wang and Y. Yamada-Takamura (June 2012). Experimental evidence for epitaxial silicene on diboride thin films. *Physical Review Letters* **108**(24), p. 245501.

[1.24] E. Bianco, S. Butler, S. Jiang, O.D. Restrepo, W. Windl and J.E. Goldberger (Mar. 2013). Stability and exfoliation of germanane: A germanium graphane analogue. *ACS Nano* **7**(5), 4414–4421.

[1.25] B. Radisavljevic, A. Radenovic, J. Brivio, V. Giacometti and A. Kis (Mar. 2011). Single-layer MoS_2 transistors. *Nature Nanotechnology* **6**, 147–150.

[1.26] J. Kang, W. Liu and K. Banerjee (Mar. 2014). High-performance MoS2 transistors with low-resistance molybdenum contacts. *Applied Physics Letters* **104**(9), p. 093106.

[1.27] S.P. Koenig, R.A. Doganov, H. Schmidt, A.H. Castro Neto and B. Özyilmaz (Mar. 2014). Electric field effect in ultrathin black phosphorus. *Applied Physics Letters* **104**(10), p. 103106.

[1.28] A.N. Rudenko and M.I. Katsnelson (May 2014). Quasiparticle band structure and tight-binding model for single- and bilayer black phosphorus. *Physical Review B* **89**(20), p. 201408(R).

[1.29] K. Uchida, J. Koga and S. Takagi (2003). Experimental study on carrier transport mechanisms in double- and single-gate ultrathin-body MOSFETs – Coulomb scattering, volume inversion and δT_{SOI}-induced scattering. *IEDM Technical Digest*, 805–808.

[1.30] T. Hara, Y. Yamada, T. Maegawa and H. Tsuchiya (2008). Atomistic study on electronic properties of nanoscale SOI channels. *Journal of Physics: Conference Series* **109**, 012012.
[1.31] S. Markov, B. Aradi, C.-Y. Yam, H. Xie, T. Frauenheim and G. Chen (Mar. 2015). Atomic level modeling of extremely thin silicon-on-insulator MOSFETs including the silicon dioxide: Electronic structure. *IEEE Transactions on Electron Devices* **62**(3), 696–704.
[1.32] N. Neophytou and H. Kosina (Aug. 2011). Atomistic simulations of low-field mobility in Si nanowires: Influence of confinement and orientation. *Physical Review B* **84**(8), p. 085313.
[1.33] W. Zhang, C. Delerue, Y.-M. Niquet, G. Allan and E. Wang (Sep. 2010). Atomistic modeling of electron-phonon coupling and transport properties in *n*-type [110] silicon nanowires. *Physical Review B* **82**(11), p. 115319.
[1.34] Y. Yamada, H. Tsuchiya and M. Ogawa (Mar. 2012). Atomistic modeling of electron-phonon interaction and electron mobility in Si nanowires. *Journal of Applied Physics* **111**(6), p. 063720.
[1.35] W. Kohn and L.J. Sham (Nov. 1965). Self-consistent equations including exchange and correlation effects. *Physical Review* **140**(4A), A1133–A1138.
[1.36] G. Kresse and J. Furthmüller (Oct. 1996). Efficient iterative schemes for *ab initio* total-energy calculation using a plane-wave basis set. *Physical Review B* **54**(16), 11169–11186.
[1.37] T.B. Boykin, G. Klimeck and F. Oyafuso (Mar. 2004). Valence band effective-mass expressions in the $sp^3d^5s^*$ empirical tight-binding model applied to a Si and Ge parametrization. *Physical Review B* **69**(11), p. 115201.
[1.38] J.-M. Jancu, R. Scholz, F. Beltram and F. Bassani (Mar. 1998). Empirical $spds^*$ tight-binding calculation for cubic semiconductors: General method and material parameters, *Physical Review B* **57**(11), 6493–6507.

2

First-principles Calculations for Si Nanostructures

A knowledge of the band structure of a material is the first necessary step to understanding and predicting its electronic properties. With the recent advancement in semiconductor micro-fabrication technologies, novel device architectures, coupled with nano-structures such as ultrathin films and nanowires, have been proposed. The band structure plays a fundamental role in determining both the electrostatics and the dynamics of carriers in those nano-structures. In this chapter, the band structures computed for Si nano-structures using first-principle density-functional theory are presented, and atomistic effects in Si nano-structures are discussed.

2.1 Band Structure Calculations

2.1.1 Si Ultrathin-body Structures

With aggressively downscaling to a nanometer scale of the VLSI technology, ultrathin-body (UTB) structures of Si, such as silicon-on-insulator (SOI) and FinFETs, are getting a lot of attention because of their superior immunity to short channel effects. Experimentally, extremely-scaled SOI-MOSFETs with Si channel thickness less than 1 nm were fabricated as shown in Figure 2.1, and fundamental device operation was successfully reported [2.1]. In such atomic-scale dimensions, not only the quantum confinement effects, but also the roles of interfaces between Si channel and gate oxides will be important. We here present our atomistic investigation on electronic properties of SOI channels, based on a first-principles simulation.

Figure 2.2 shows the atomic structures used in the Si-UTB simulations, where we employed three types of spatial confinement [2.2]. The SiO_2 layers were assumed to be crystalline and placed onto the Si (001) surfaces without any defects – that is, the Si/SiO_2 interfaces are geometrically abrupt. To apply the supercell technique, vacuum layers with a sufficient thickness are included above and below the structures. We refer to (a) and (b) as the silicon-on-insulator (SOI) model, and (c) as the H-terminated model, respectively.

Carrier Transport in Nanoscale MOS Transistors, First Edition. Hideaki Tsuchiya and Yoshinari Kamakura.

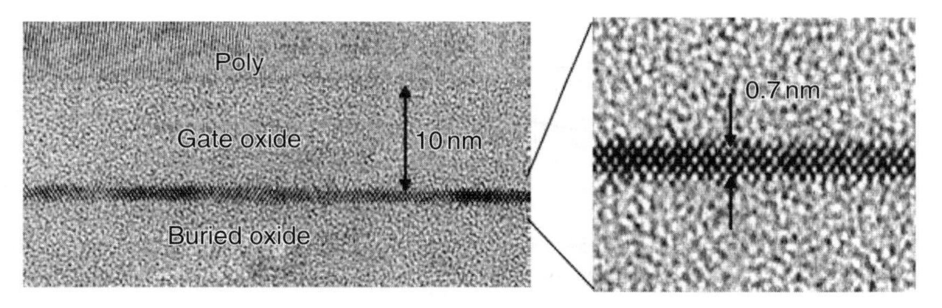

Figure 2.1 Cross-sectional TEM photograph of 0.7 nm SOI MOSFET [2.1]. The thickness of 0.7 nm corresponds to the 5-atomic-layer thickness of Si atoms.

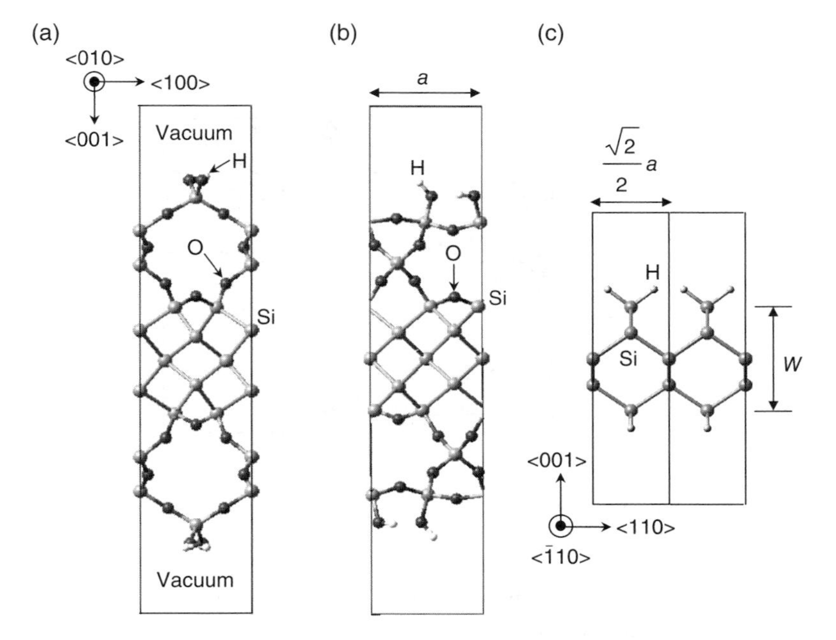

Figure 2.2 Atomic structures used in the Si-UTB simulations, where we employed three types of spatial confinement by using (a) β-cristobalite SiO_2 layers; (b) α-quartz SiO_2 layers; and (c) hydrogen-termination of surface dangling bonds.

The band structure calculations in this section were performed by using a first-principles simulation package based on the density-functional theory (DFT), VASP, where the electron-electron exchange and correlation interactions were treated within the generalized gradient approximation (GGA). The Kohn-Sham equation was solved by using a plane-wave basis set; based on the ultra-soft pseudopotentials and projector-augmented-wave method [2.3]. To obtain stabilized structures, we performed structure optimization procedures after placing all atoms in a unit cell, where a conjugate gradients minimization method was employed to relax all atomic coordinates and cell shape and size via total energy minimization.

First, Figure 2.3 shows the electronic bandstructures along the Γ to X direction computed for the β-cristobalite SOI model.

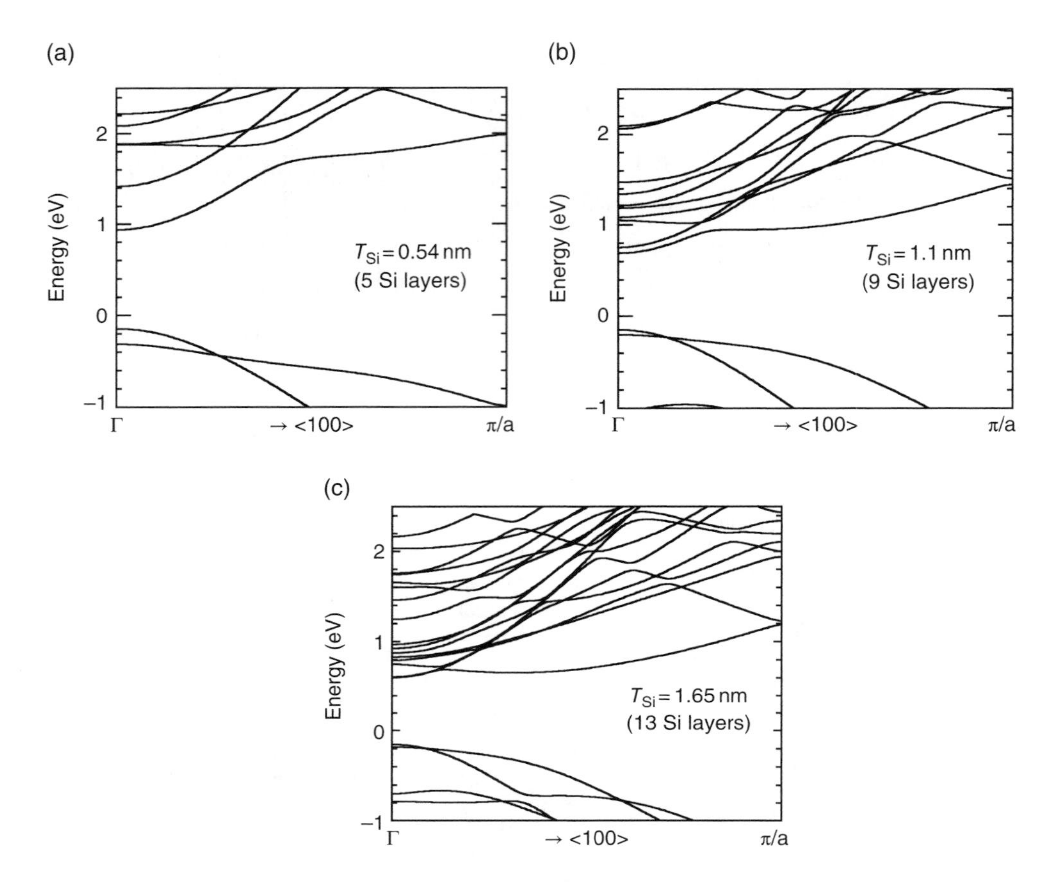

Figure 2.3 Electronic band structures computed for β-cristobalite SOI model, where Si thickness T_{Si} is given at (a) 0.54; (b) 1.1; and (c) 1.65 nm.

First, it is found that the Si 2-D bandstructure confined in the <001>direction becomes direct. This is explained as follows [2.2]: In the Si 2-D channels, the six equivalent valleys at the bulk conduction band minimum are folded onto quantized-k_z plane due to the quantum confinement effect as shown in Figure 2.4(b). Those minimum energies are determined by their quantized energy levels, and thus the valleys 1 and 2 located at the Γ point become the lowest valleys, because of a larger confinement effective mass of m_l. Note that the width of the first Brillouin zone (FBZ) for the SOI model is half of that for the bulk Si, as shown in the solid square of Figure 2.4(b), so the valleys 3–6 are again folded into the halved FBZ in the Γ-X direction. Therefore, another conduction band minimum appears away from the Γ point, as seen in Figure 2.3(c). Here, it is interesting to note that the energies of the two conduction band minima in Figure 2.3(c) are almost identical, since the quantization has little impact on the bandstructure for the thicker SOI channel.

Next, Figure 2.5 shows the bandstructures computed for β-cristobalite SOI, α-quartz SOI and H-terminated models [2.4], where the Si channels consist of five Si-atomic layers with 0.54 nm thickness. They all have a conduction band minimum at the Γ point, as explained in the previous paragraph. Here, note that the width of FBZ for the SOI model is half of that for the H-terminated model, since the size of the unit cell is larger in the SOI model, as shown in Figures 2.2 (a) and (b).

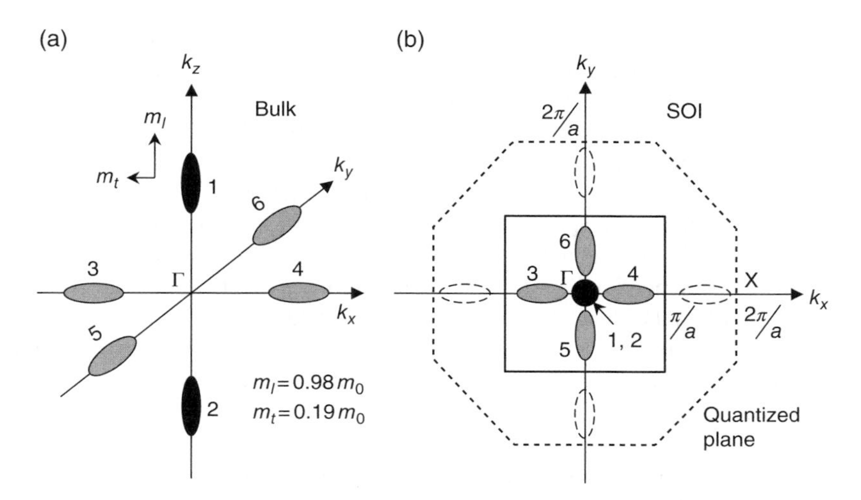

Figure 2.4 Schematic diagrams of conduction band valleys for: (a) bulk Si; and (b) SOI channel. In (b), the solid square and the dashed octagon represent the first Brillouin zones for SOI channel and bulk Si, respectively.

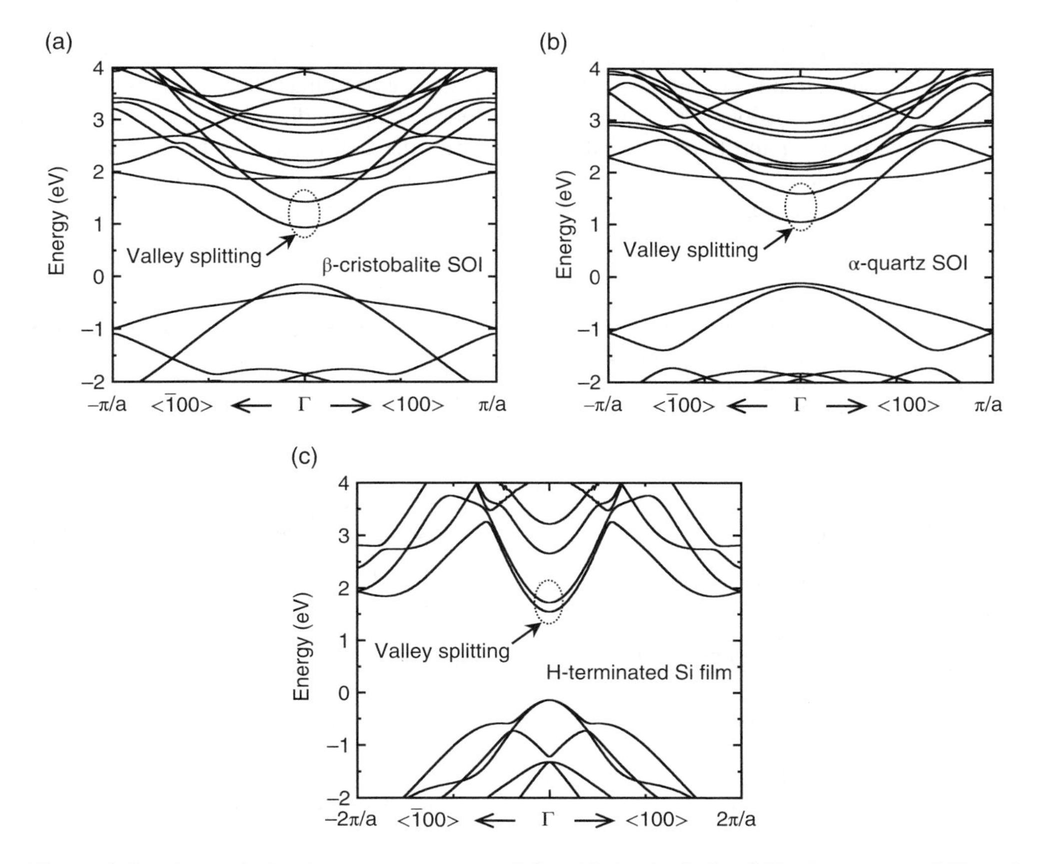

Figure 2.5 Electronic band structures computed for: (a) β-cristobalite SOI; (b) α-quartz SOI; and (c) H-terminated models, where Si thickness T_{Si} is given at 0.54 nm.

We further notice that the two equivalent valleys projected at the Γ point are split by the interactions [2.5, 2.6]. Such a valley splitting is known to affect carrier transport, because the density-of-states at conduction band minimum decreases [2.7–2.9]. Here, it is noteworthy that the two SOI models predict significantly larger valley splitting than the H-terminated model.

To examine the differences in their valley splitting behaviors, we calculated the valley splitting energies for the three confinement models as a function of the Si layer thickness, as shown in Figure 2.6(a), where dependence of the valley splitting energy on applied electrostatic field is not considered in this study. It is found that the two SOI models exhibit nearly identical splitting energies, and predict significantly larger splitting energy than the H-terminated model, especially when the Si layer thickness becomes smaller than 1 nm. On the other hand, as shown in Figure 2.6(b), the effective mass of electron parallel to <110>direction at the conduction band minimum is approximately equal to the bulk transverse effective mass, m_t, even for sub-1 nm thickness.

Next, Figure 2.7 shows the T_{Si} dependences of bandgap energy, where the experimental data [2.10] are also plotted. Note that a bandgap energy discrepancy (0.54 eV) between the density-functional theory and the experimental data for bulk Si is added to the computed results of Figure 2.7. It is found that the experimental bandgap energies are systematically lower than those of the H-terminated Si film, and the bandgap energies predicted by the two SOI models are in good agreement with the experimental data [2.11]. As pointed out in [2.10], this is considered to be due to an influence of transition regions formed at the Si/SiO_2 interfaces.

To investigate the details of such transition regions, we plot in Figure 2.8 the spatial variations in the conduction band minimum (CBM) and valence band maximum (VBM) energies, and also the charge density profiles at the Γ point along the normal direction for β -cristobalite oxide, α-quartz oxide and H-terminated models. Here, the CBM and VBM variations were obtained by calculating the local density-of-states for each atom. It is found that electronic transition regions are formed on the SiO_2 layers, and the CBM varies rather gradually in the

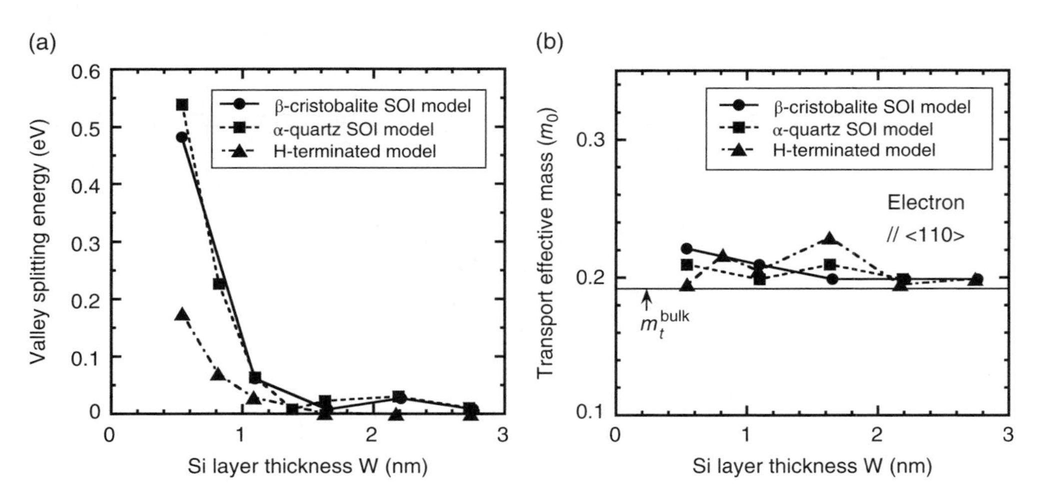

Figure 2.6 (a) Valley splitting energies; and (b) electron effective masses parallel to the <110>direction at conduction band minimum for the three confinement models, which is calculated as a function of Si layer thickness. The horizontal line in (b) represents the bulk transverse effective mass m_t.

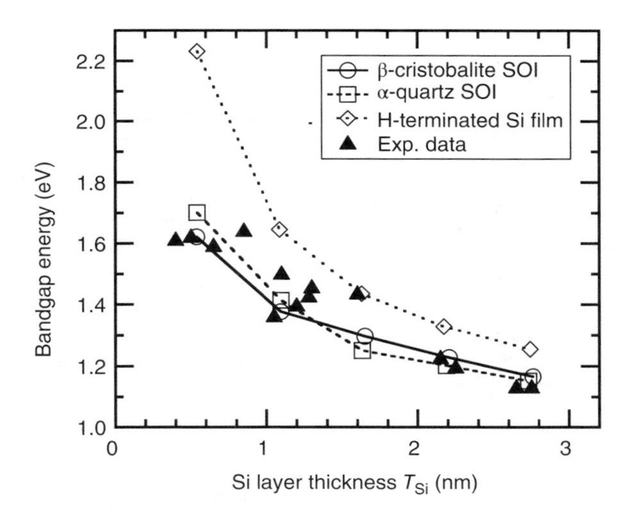

Figure 2.7 T_{Si} dependences of bandgap energy for β-cristobalite SOI, α-quartz SOI and H-terminated Si film. Note that the bandgap energy discrepancy (0.54 eV) between the density-functional theory and the experiment for bulk Si is corrected.

transition regions [2.12]. As a result, the charge densities, which correspond to the wavefunctions, spread deeply into the SiO_2 layers, as shown in Figures 2.8 (a) and (b), while they exponentially decay in the case of the H-terminated model shown in Figure 2.8(c). Consequently, an effective thickness of the silicon quantum well becomes larger in the SOI models and, thus, the bandgap energy decreases, as shown in Figure 2.7. In summary, the spatial confinement using the SiO_2 layers affects valley splitting and bandgap energy, but not the effective mass of electron.

2.1.2 Si Nanowires

It is well-known that Si nanowire (SiNW) MOSFETs should offer better electrostatic gate control than planar Si MOSFETs, by employing the GAA configurations shown in Figure 1.3(d) or Figure 1.10. Furthermore, its device performance should exhibit a strong orientational dependence via the band structure effects [2.4, 2.13]. In this section, we investigate the electronic properties of <110>- and <100>-oriented SiNWs, by using VASP within the GGA electron-electron exchange and correlation interactions, because they exhibit better transport performances than other orientations [2.13], and we compare them with those of Si-UTB structures.

In ref. [2.4], we have reported that the H-terminated model describes well the quantum confinement effects on ballistic $I_D - V_G$ characteristics, at least under low-bias condition, for the Si-UTB structures. Hence, we employ the hydrogen-termination of dangling bonds at the surfaces of SiNWs in this section. Figure 2.9 shows the atomic structures of SiNWs with two orientations, (<110>and <100>), where the square-shaped cross-section is used, and confinement directions are indicated in each figure. Note that all surface dangling bonds are terminated by hydrogen atoms. In the <110>-oriented SiNWs, a characteristic wire width is defined as

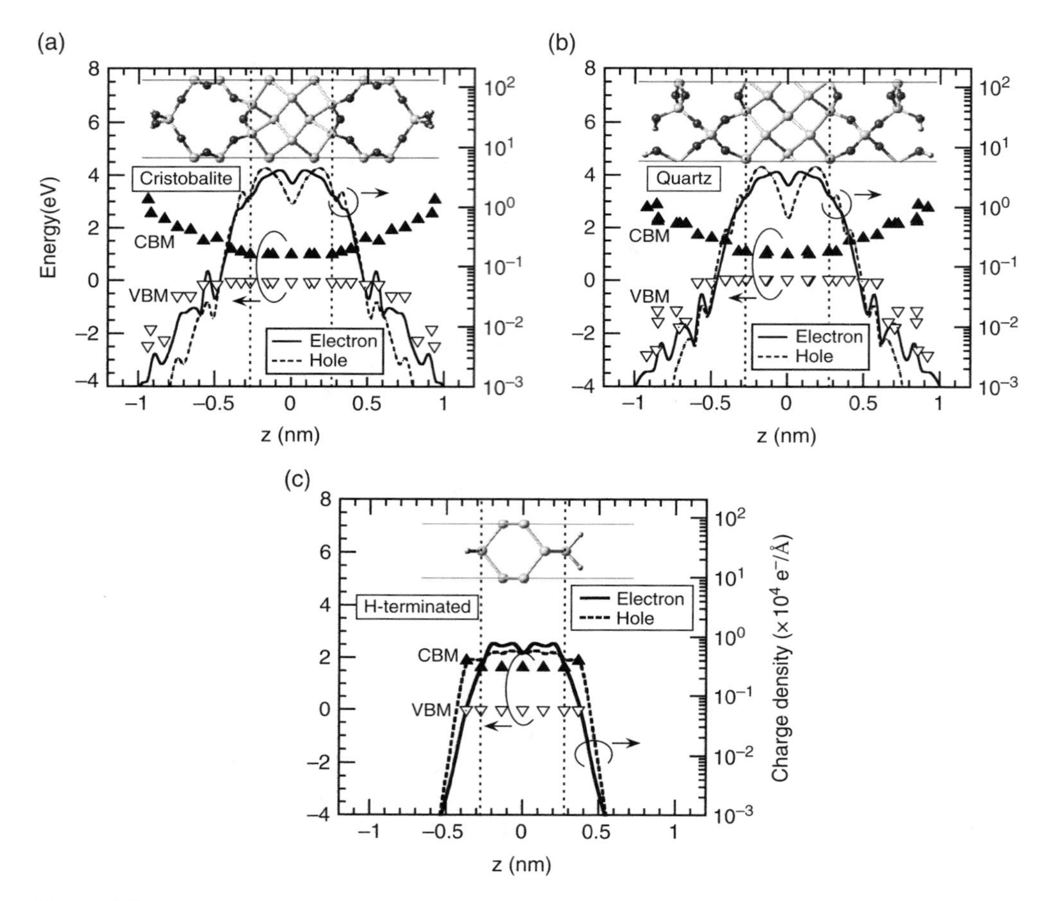

Figure 2.8 Spatial variations in conduction band minimum (CBM) and valence band maximum (VBM) energies and charge density profiles at Γ point along normal direction for: (a) β-cristobalite oxide; (b) α-quartz oxide; and (c) H-terminated models. The results for the five Si-layer models are shown here.

$W \equiv \sqrt{W_1 \times W_2}$, because the two directions have slightly different dimensions. All atoms, including the hydrogen atoms, have been relaxed by performing the structural optimization procedures.

Figure 2.10 shows the band structures computed for <110>- and <100>-SiNWs with about 1 nm wire width. They also have a conduction band minimum at the Γ point [2.4, 2.13, 2.14]. It is found that the valley splitting at the Γ point is significantly larger in the <110>-SiNW than in the <100>-SiNW, which is identical with the previous tight-binding results reported in [2.13].

Figure 2.11(a) shows how this effect varies with the spatial confinement, together with the Si-UTB results. Valley splitting in the <110>-SiNWs can reach up to several hundreds meV for sizes smaller than 2 nm, which is much larger than the room temperature thermal energy ($k_BT = 26$ meV) and, thus, they are expected to have a significant influence on the transport properties. Furthermore, Figure 2.11(b) shows the effective masses at the conduction band

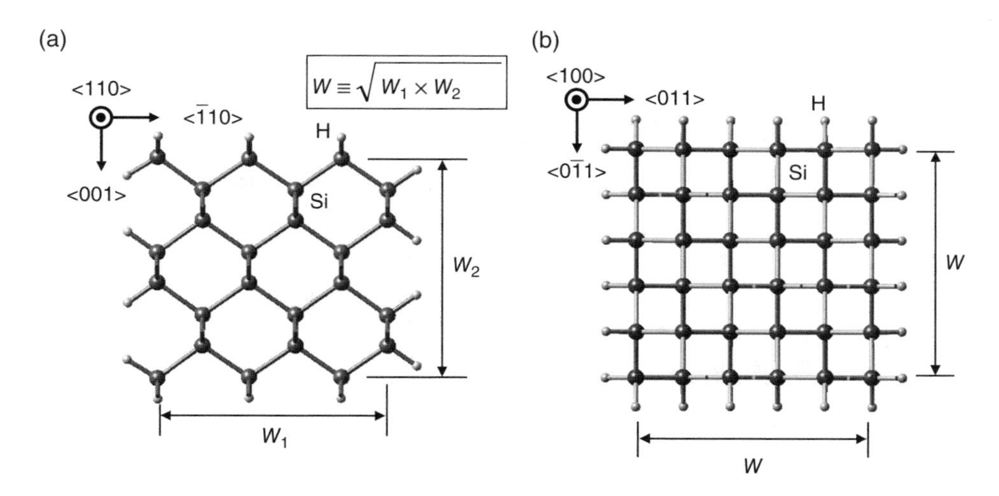

Figure 2.9 Atomic structures used in the SiNW simulations, where we use square-shaped cross-section and two different orientations: (a) <110>; and (b) <100>. Note that all surface dangling bonds are terminated by hydrogen atoms.

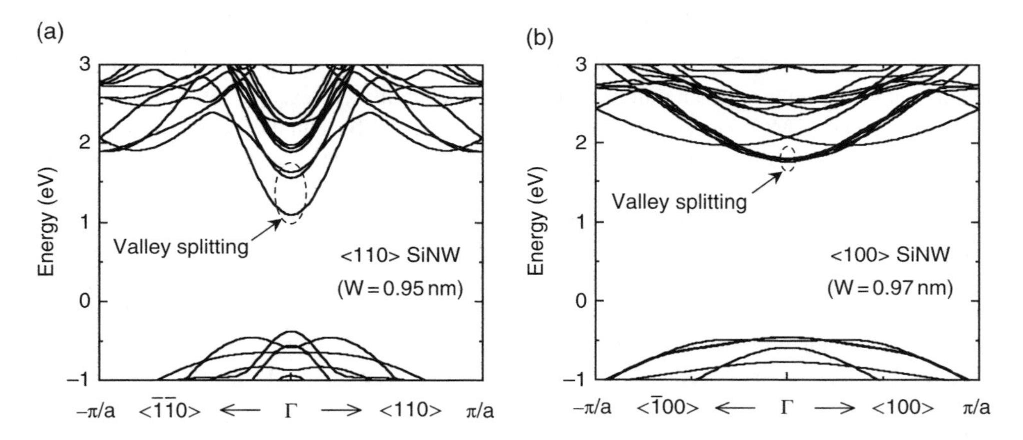

Figure 2.10 Band structures computed for (a) <110>- and (b) <100>-SiNWs with about 1 nm wire width.

minimum. It is found that the <110>-SiNW effective mass is smaller than the bulk transverse effective mass for sizes smaller than 3 nm, while the <100>-SiNW effective mass increases with increase in quantization.

The above effective mass variations in the SNWs are also identical with the previous tight-binding results [2.13], and therefore they can be explained using the anisotropy and the non-parabolicity in the Si conduction band Brillouin zone [2.14]. As later discussed in Chapter 5, the effective mass variations mentioned above basically govern the device performance of SiNW-MOSFETs.

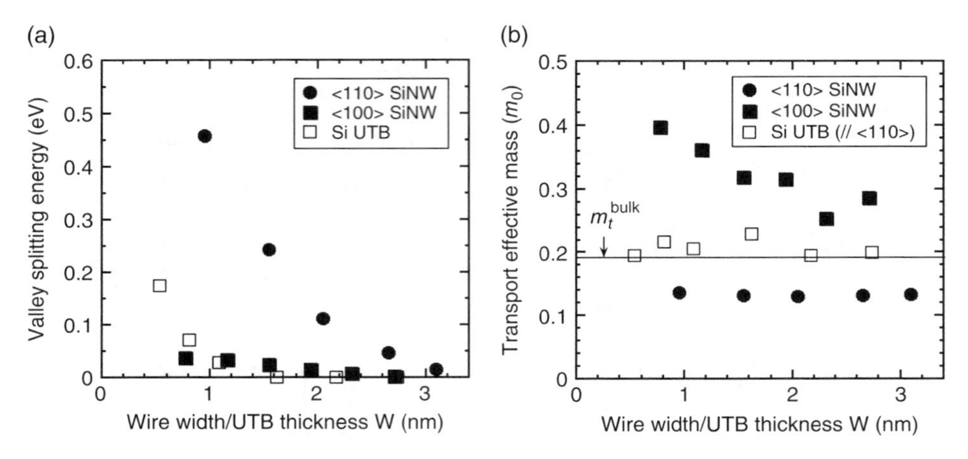

Figure 2.11 (a) Valley splitting energies; and (b) electron effective masses at conduction band minimum for <110>- and <100>-oriented SiNWs. The results for Si-UTB are also plotted. In (b), the horizontal solid line represents the bulk transverse effective mass of Si.

2.1.3 Strain Effects on Band Structures: From Bulk to Nanowire

As the size of Si-MOSFETs shrinks down to the nanometer scale, new device technologies, such as mobility enhanced channels and multi-gate architectures, are strongly needed to realize advanced CMOS devices. Among these, strained-Si channel engineering, which includes the choices of surface orientations, channel directions, strain configurations and channel materials, is recently becoming more important [2.15]. In particular, uniaxial strain can be more effective than biaxial strain in increasing the current drive of short-channel devices, because of the effective mass reduction of holes [2.16, 2.17] and even electrons [2.18]. On the other hand, device structures with new gate configurations are preferred to provide better electrostatic control than conventional planar structures. Thus, UTB channels and SiNW channels with multi-gate or GAA structures are highly expected.

So far, the effective mass reduction of carriers by applying uniaxial strain has been experimentally demonstrated for bulk- [2.16–2.18] and ultrathin-body SOI MOSFETs [2.18]. Understanding the physical mechanisms for such an effective mass reduction in both p- and n-channel MOSFETs has progressed on the basis of theoretical computational studies for bulk-Si MOSFETs [2.16–2.20]. For ultrathin-body structures, there are few theoretical reports discussing strain effects on the effective mass [2.21], to the best of our knowledge. However, to further scale down MOSFETs with desired higher performance, an introduction of the strain technology to Si nanostructure channels will be explored in future.

In this section, we describe strain effects in Si nanostructures – that is, two-dimensional (2-D) thin film and one-dimensional (1-D) nanowire, including bulk material. Under extreme scaling of MOSFETs, the number of atoms in the cross-section becomes countable, and the consideration of crystal symmetry and bond orientation, in addition to quantum mechanical confinement, is important. Thus, we employed a first-principles method to identify the band structure parameters that greatly influence carrier transport properties [2.14]. For SiNWs, strain-induced modulation of band gap and effective mass has been recently reported, based on similar first-principles calculations, for various growth directions and cross-sectional shapes and sizes [2.22–2.24]. In this section, we intend to highlight the unique strain properties of SiNWs by comparing with those of bulk and 2-D thin film.

2.1.3.1 Bulk Si

First, we present variations in energy splitting of conduction band edge and effective mass of electrons, by applying biaxial and uniaxial strains to bulk Si. Figure 2.12 shows the lattice models used in the calculation, where uniaxial strain is parallel to <100>and <110>directions as shown in Figs. 2.12 (b) and (c), respectively. The plane and directions parallel to strain are uniformly deformed from –1% to 1%, and the lattice constants perpendicular to the strain direction are carefully relaxed. All calculations were performed by using a first-principles simulation package, *VASP*, where the electron-electron exchange and correlation interactions were treated within the generalized gradient approximation (GGA) [2.3].

Figs. 2.13 (a)-(c) show the calculated energy splitting of the conduction band edge, ΔE_C, under biaxial, uniaxial <100>and uniaxial <110>strains, respectively, where energy of the conduction band edge at unstrained condition was set to be zero. The insets show the schematic diagrams of six conduction band valleys on a Si (001) surface, numbered from 1 to 6, and also strain directions. As is well known, the strains cause band splitting of the 6-fold degenerate

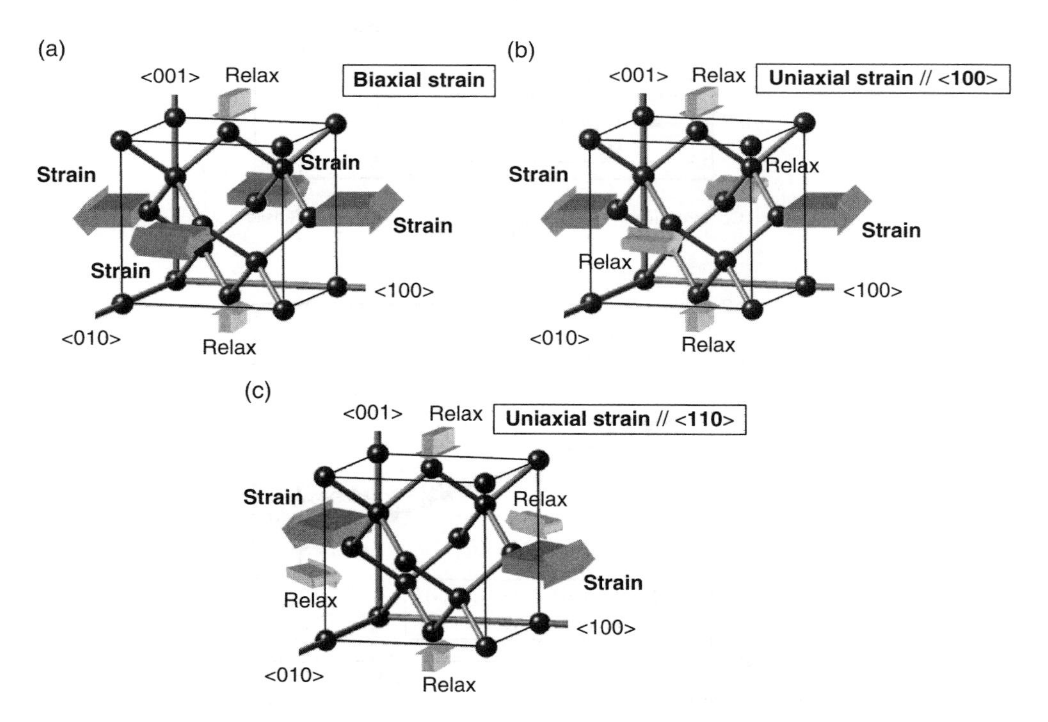

Figure 2.12 Lattice models used in the calculation for bulk Si, where: (a) biaxial; (b) uniaxial <100>; and (c) uniaxial <110>strains. The plane and directions parallel to strain are uniformly deformed from –1% to 1%, and the lattice constant perpendicular to the strain direction is carefully relaxed.

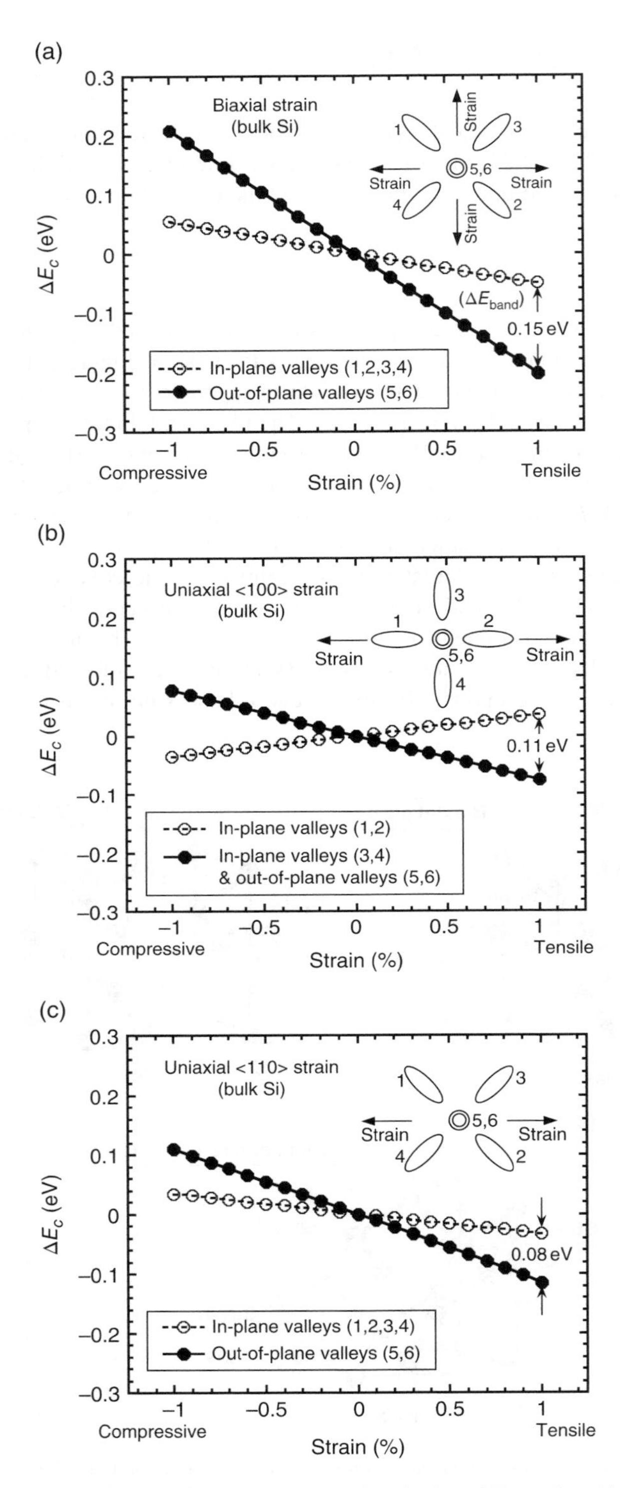

Figure 2.13 Calculated energy splitting of conduction band edge, ΔE_C, under: (a) biaxial; (b) uniaxial <100>; and (c) uniaxial <110> strains, where the energy of conduction band edge at unstrained condition was set to be zero. The insets show the schematic diagrams of six conduction band valleys on a Si (001) surface, and strain directions.

valleys into the 2-fold (5, 6) and the 4-fold (1, 2, 3, 4) valleys, as shown in Figure 2.13, the amount of which is the largest for biaxial strain and is the smallest for uniaxial <110>strain.

The present results are consistent with the previous results, based on the empirical non-local pseudopotential method by Uchida *et al.* [2.18] and Ungersboeck *et al.* [2.20]. Figure 2.14 shows the in-plane effective masses computed for the same strain conditions as in Figure 2.13. As Uchida *et al.* first demonstrated [2.18], the m_T reduction and enhancement are observed by applying uniaxial <110>strain, as shown in Figure 2.14(c).

For comparison purposes, the computed m_T values are compared with those of Uchida *et al.* [2.18] and Ungersboeck *et al.* [2.20], as shown in Figure 2.15. It is found that the trend of effective mass modulation due to uniaxial <110>strain is the same for both methods, though the present first-principles method predicts smaller variation in m_T than the empirical non-local pseudopotential method. Therefore, the first-principles approach might underestimate the performance enhancement of the strained-MOSFETs, and the details are under investigation. In the following sections, we examine the effects of uniaxial <110>strain on Si nanostructures.

2.1.3.2 Si Thin Film

Next, we present the calculated results for Si thin film. Figure 2.16 shows the unit cells used in the calculation, where we only considered uniaxial strain parallel to <110>direction. Dangling bonds of surface Sis are terminated with hydrogen atoms, and vacuum layers with 0.4 nm thick are added above and below the structure.

More realistic models sandwiched between two SiO_2 layers [2.2] will be applicable to band structure calculations under strain. Here, note that the axes in the plane perpendicular to the confinement direction are <110>and <1$\bar{1}$0> in Figure 2.16. The Si thin film consists of five Si atomic layers, which corresponds to Si layer thickness of 0.54 nm. Therefore, electrons confined in the Si thin film are considered to be an ideal 2-D electron gas. Such an extremely scaled SOI-MOSFET, with five Si atomic layers, was successfully fabricated, as presented in 2.1.1 [2.1]. In this section, we investigate strain effects on such ideal 2-D electron gas.

Figure 2.17 shows the dispersion curves computed for the Si thin films, where uniaxial <110>strain is (a) 1% compressive, (b) 0% and (c) 1% tensile. Here, note that for all three dispersions, the energy reference of the vertical axis was set to be the Fermi energy obtained in the case of unstrained Si thin film, to examine quantitative energy splitting of the conduction band edge due to strain, as discussed later. From Figure 2.17, they have a conduction band minimum at the Γ point, which results from the *k*-space projection of the two ellipsoidal bands (5, 6) onto the <001>plane of quantization [2.13].

There are four more valleys residing off- Γ states (two in the positive and two in the negative k_x axis), that result from the four off-plane ellipsoidal bands (1, 2, 3, 4). The former Γ valleys appear at lower energies because of their heavy quantization mass, and have lighter transport mass. On the other hand, the latter off-Γ valleys appear at higher energies because of the lighter quantization mass, and have heavier transport mass. In Figure 2.17, we again note that the two-fold Γ valleys are split by the interactions between the two equivalent valleys.

It is also found from Figure 2.17 that uniaxial <110>strain causes band splitting between the Γ and off- Γ valleys in Si thin film. We plotted the calculated energy splitting of the conduction band edge, ΔE_C, as shown in Figure 2.18, where the off- Γ valleys have much higher energies, due to quantum confinement. We can see that the ground-state Γ valley moves up and down due to uniaxial <110>strain, while the off- Γ four-fold valleys are almost constant.

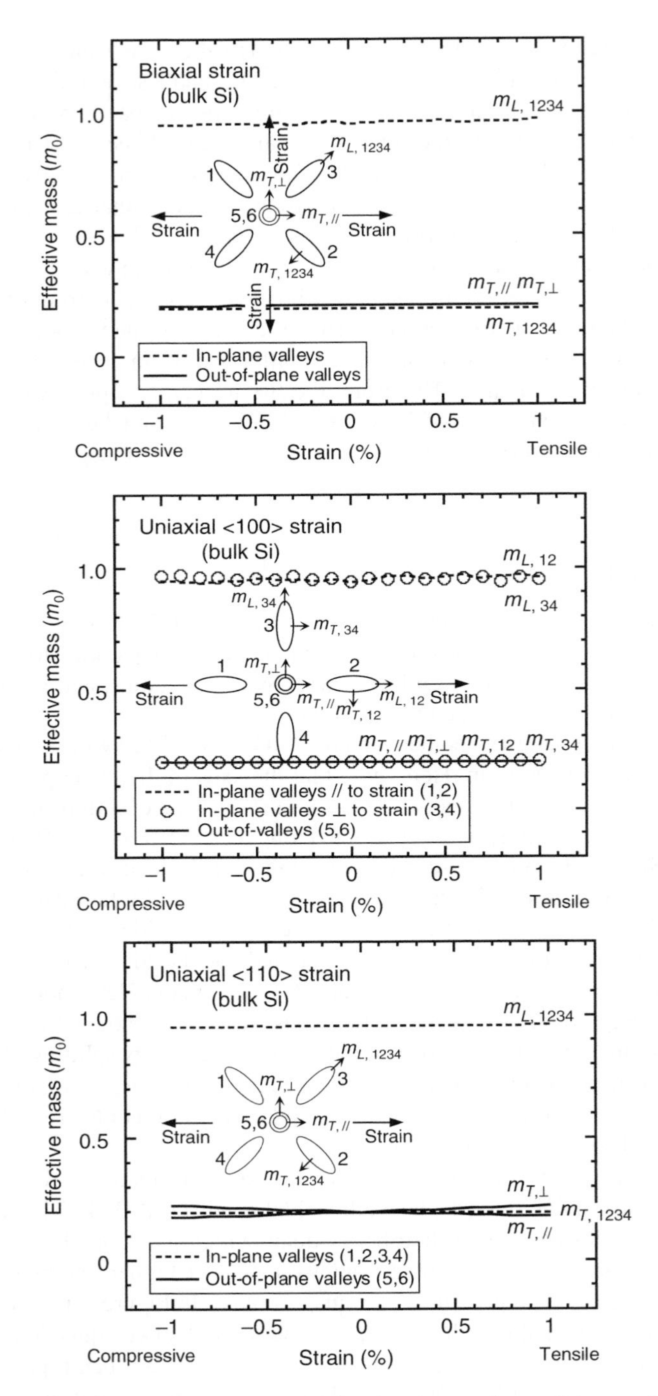

Figure 2.14 In-plane effective masses computed for the same strain conditions as in Figure 2.13. The m_T reduction and enhancement are observed by applying uniaxial <110>strain.

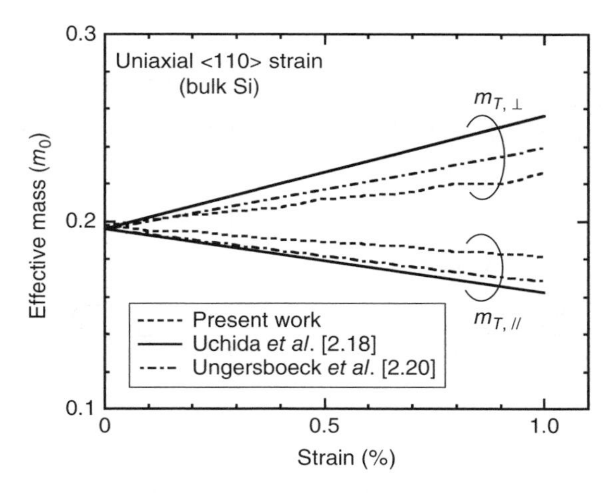

Figure 2.15 Comparison of computed m_T values with results from the empirical non-local pseudopotential method [2.18, 2.20].

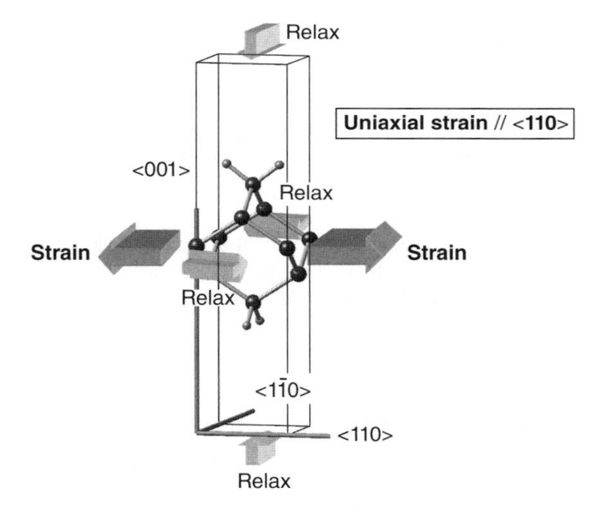

Figure 2.16 Unit cells used in the calculation for Si thin film, where we considered uniaxial strain parallel to <110>direction. Dangling bonds of surface Sis are terminated with hydrogen atoms, and vacuum layers are included above and below the structures. Note that the axes in plane perpendicular to confinement direction is <110>and <1$\bar{1}$0>. The Si thin film consists of five Si atomic layers (T_{Si}=0.54 nm).

Furthermore, Figure 2.19 shows the in-plane effective masses of the ground-state Γ valley. It is found that the m_T reduction and enhancement are expected to occur, even in uniaxial <110>-strained Si thin film. Indeed, the electron mobility enhancement and reduction under uniaxial <110>strain has been observed experimentally in such ultrathin-body MOSFETs [2.18], which is attributed to the m_T change due to the uniaxial strain. Based on the above results, Si thin film is found to have similar strain effects as in bulk Si under uniaxial <110>strain.

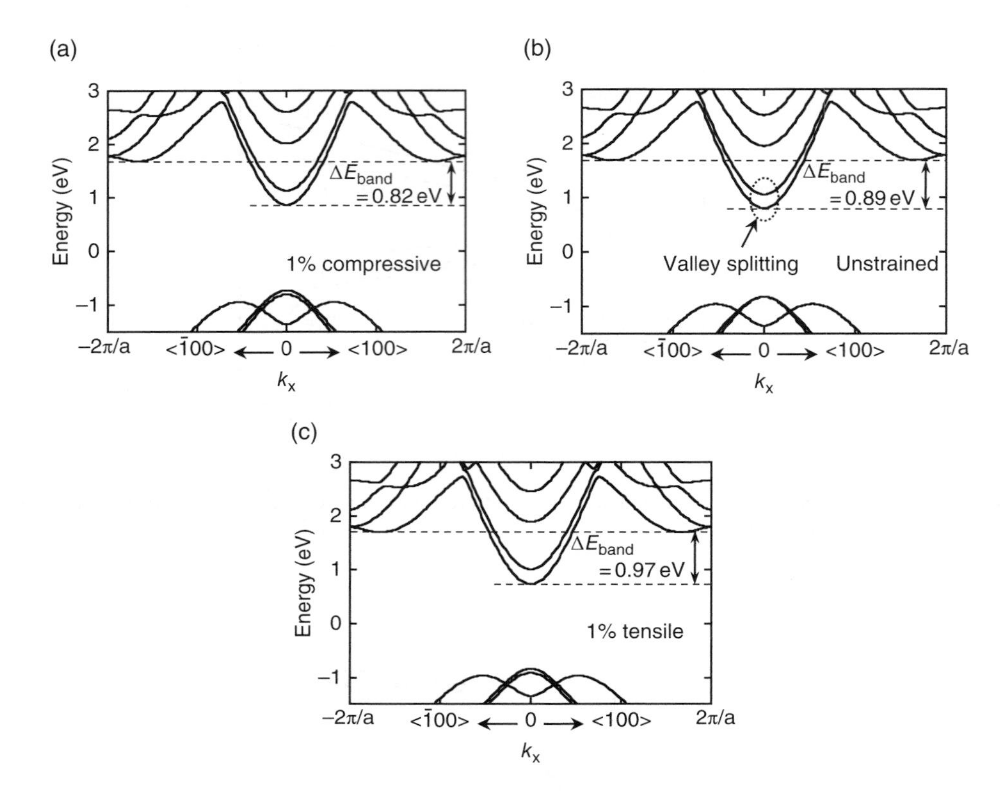

Figure 2.17 Dispersion curves computed for Si thin film, where uniaxial <110>strain is: (a) 1 % compressive; (b) 0 %; and (c) 1 % tensile. Here, the energy reference of the vertical axis was set to be the Fermi energy, in the case of unstrained Si thin film.

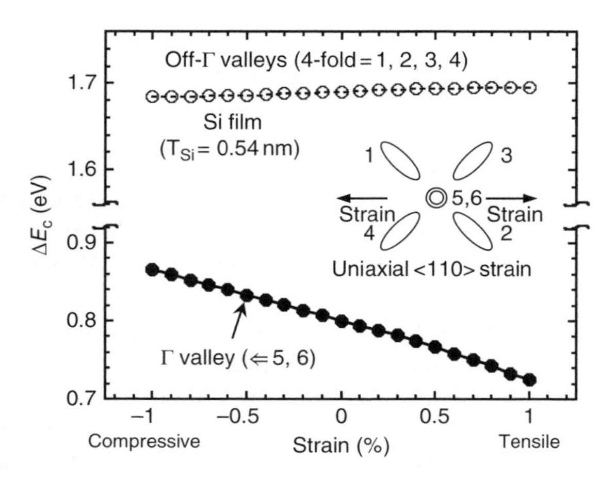

Figure 2.18 Calculated energy splitting of Si thin film under uniaxial <110>strain, where the off-Γ valleys have much higher energies due to quantum confinement.

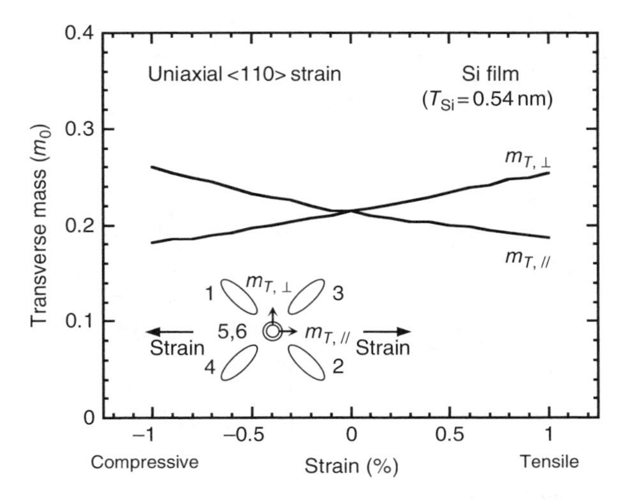

Figure 2.19 In-plane effective masses of ground-state Γ valley estimated for Si thin film under uniaxial <110> strain. The m_T reduction and enhancement are expected to occur, even in Si thin film.

2.1.3.3 SiNW

Finally, we present the calculated results for SiNW. Figure 2.20 shows the simulation model of SiNW and its unit cell. As in the Si thin film, dangling bonds of surface Sis are terminated with hydrogen atoms, and the nanowire is surrounded by vacuum layers with 0.3 nm thick. Quantum confinement directions are <001> and $\langle 1\bar{1}0\rangle$, and their dimensions are 0.94 nm and 0.77 nm, respectively. Therefore, electrons in the nanowire are considered to be an ideal 1-D electron gas.

Relaxed nanowire is 3 % stretched along the axis, even in the unstrained condition, which agrees well with the previously reported theoretical results [2.23, 2.25]. Then, uniaxial compressive and tensile strains were applied along the <110> direction.

Figure 2.21 shows the dispersion curves computed for the <110>-oriented SiNWs under 1% compressive, 0% and 1% tensile uniaxial strains, where the energy reference of the vertical axis for all the three dispersions is the Fermi energy in the case of unstrained Si nanowire.

Overall features, such as conduction band minimum and valley splitting at the Γ point, are similar to the Si thin film, but the four-fold off- Γ valleys are found to be split by the valley interactions in the SiNW, which results from the *k*-space projection of the two ellipsoidal bands (1, 4) or (2, 3) onto the $\langle 1\bar{1}0\rangle$ plane of quantization [2.13]. Since the valley splitting depends on quantization, the splitting energies at the Γ point in Figure 2.21 are somewhat larger than those in Figure 2.17 for Si thin film.

Next, Figure 2.22 shows the calculated energy splitting of the conduction band edge, ΔE_C. The strain dependence of the conduction band edge is quite similar to the Si thin film, though the off- Γ valleys indicate an opposite strain dependence from Figure 2.18.

The strain dependence of off-Γ valleys in Figure 2.22 is different from the previous study [2.23, 2.24]. We think this is because we set the energy reference of all the dispersion curves at the Fermi energy in the case of unstrained SiNW. Indeed, when the energy offset is not corrected, the same strain dependence of off-Γ valleys as in [2.23, 2.24] is obtained. Figure 2.23

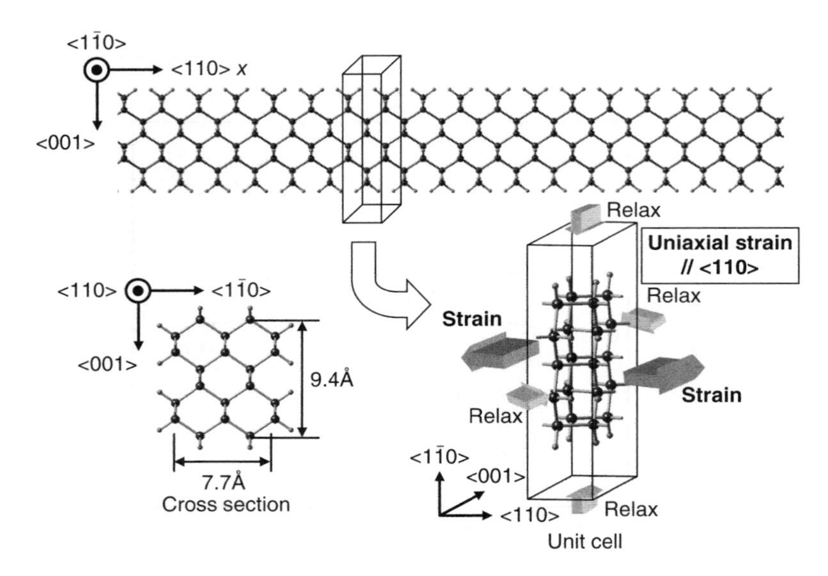

Figure 2.20 Simulation model of SiNW and its unit cell. Quantum confinement directions are <001> and $\langle 1\bar{1}0 \rangle$. The nanowire direction was chosen at <110>, because a higher current capability compared to other directions, as <100> and <111> is expected [2.13]. Uniaxial strain was applied along <110> direction.

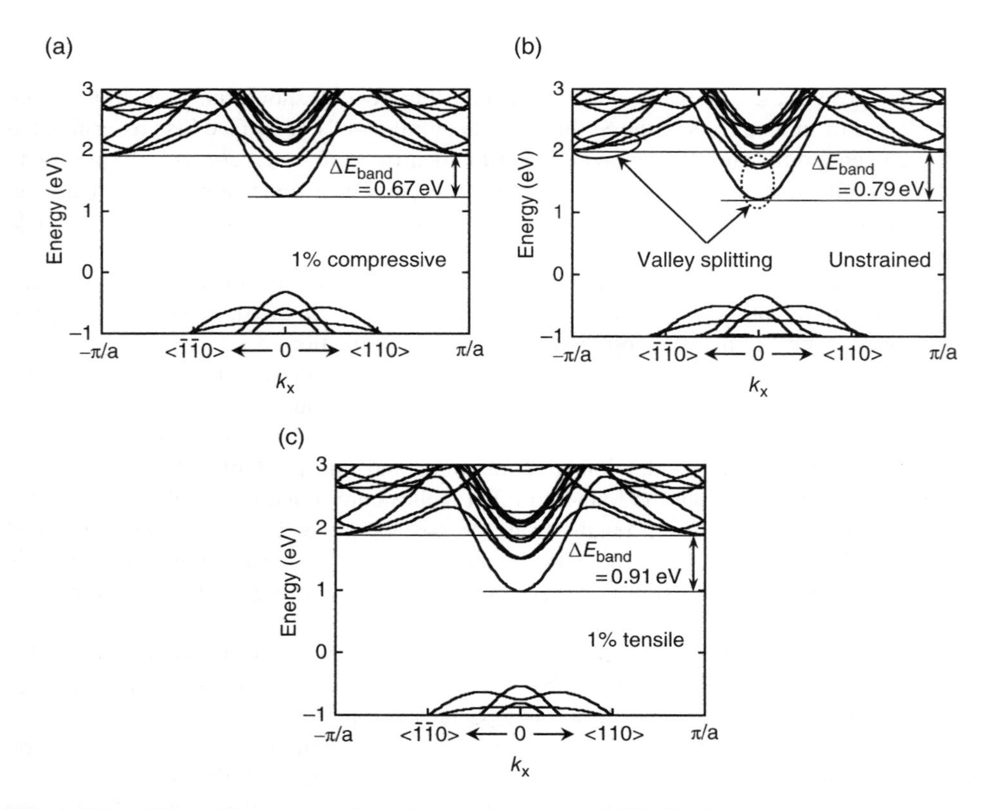

Figure 2.21 Dispersion curves computed for <110>-oriented SiNW, where uniaxial strain is: (a) 1 % compressive; (b) 0 %; and (c) 1 % tensile, where the energy reference of the vertical axis is the Fermi energy in the case of unstrained SiNW.

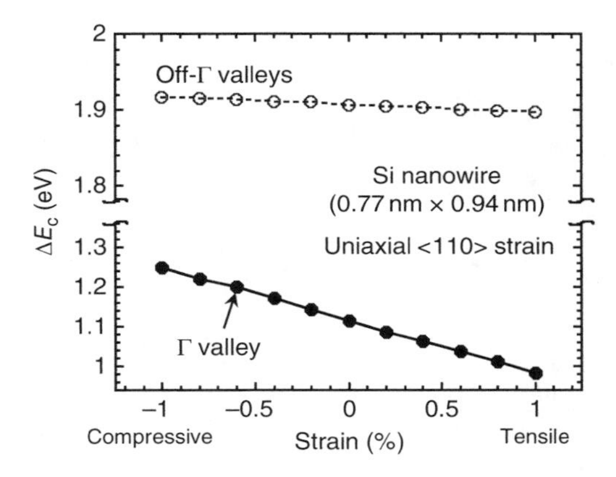

Figure 2.22 Calculated energy splitting of SiNW under uniaxial <110>strain.

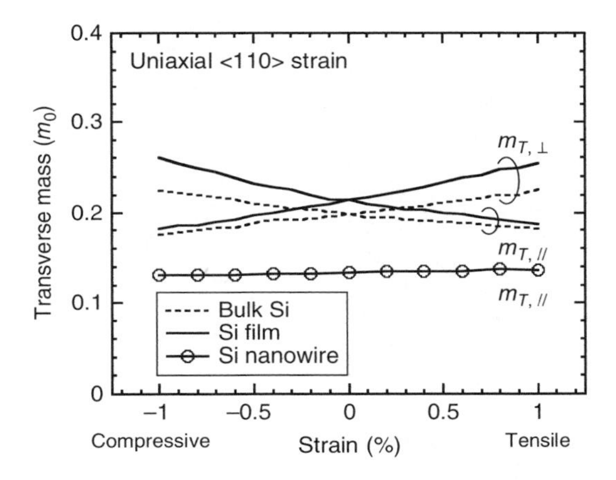

Figure 2.23 In-plane effective mass of ground-state Γ valley estimated for SiNW under uniaxial <110>strain. Results for bulk and thin film are also plotted for comparison.

shows the in-plane effective mass of the ground-state Γ valley, which is also compared with bulk and thin film. It is found that the m_T reduction due to uniaxial <110>tensile strain is not observed, but a smaller m_T is predicted in the SiNW. The present results are consistent with those of [2.22, 2.23].

The estimated transport mass is $\approx 0.13\,m_0$, independently of strain, which is smaller by 32% than the bulk m_T. Recently, such a smaller mass has been reported for unstrained SiNWs, based on the empirical tight-binding band calculation, and its physical mechanism was clearly explained in terms of semi-analytical construction of the nanowire dispersions by using the anisotropy and nonparabolicity in the Si conduction band Brillouin zone [2.13]. We apply that argument to explain why the effective mass of SiNW is independent of the strain, contrary to that of thin film as follows.

Figure 2.24 shows the energy contours in the k_x–k_y plane near the conduction band minimum of Si thin films under 1% compressive, 0% and 1% tensile uniaxial strains, where

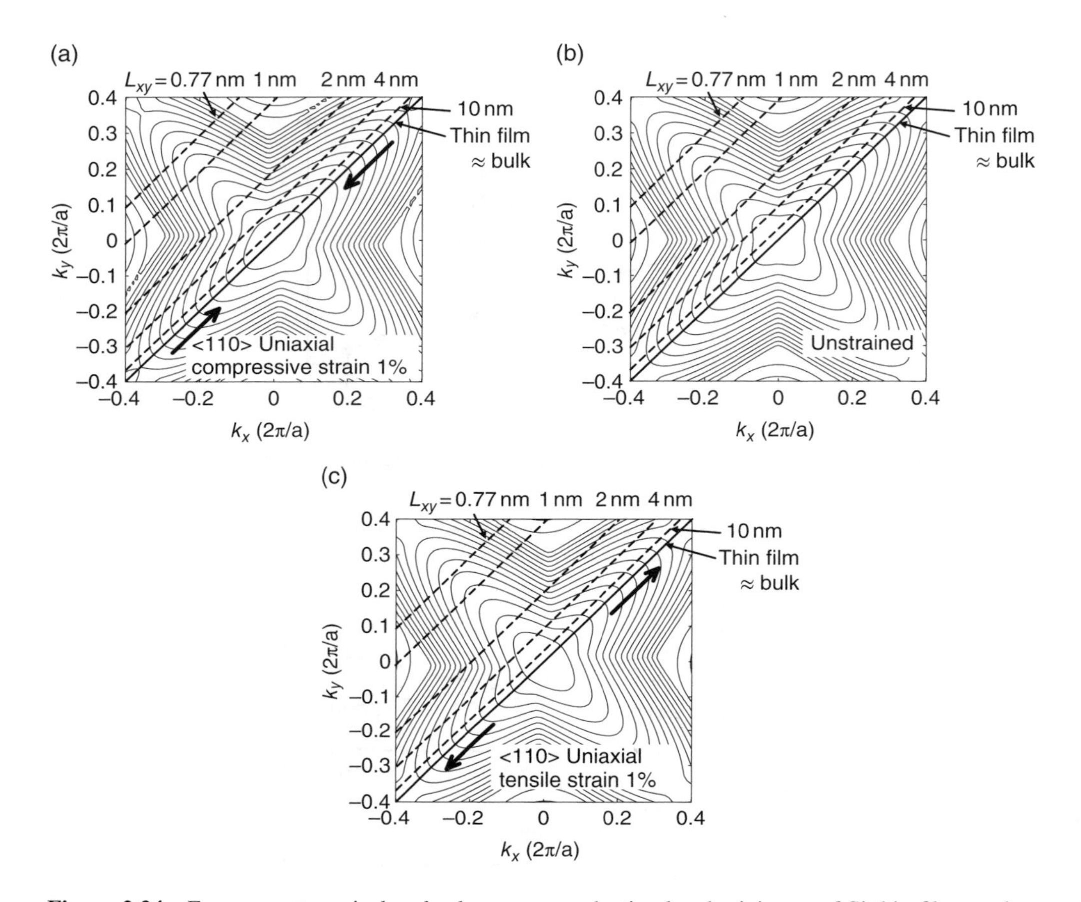

Figure 2.24 Energy contours in $k_x - k_y$ plane near conduction band minimum of Si thin films under: (a) 1% compressive; (b) 0%; and (c) 1% tensile uniaxial strains, where the confinement direction is <001>and the thickness L_z is taken at 0.94 nm. The dashed lines that cross at 45° represent an extra quantization in the $\langle 1\bar{1}0 \rangle$-direction in the present nanowire. A line for L_{xy}=0.77 nm corresponds to the present nanowire quantization, and lines for L_{xy} = 1 nm, 2 nm, 4 nm, 10 nm and ∞ (thin film limit) are also drawn for comparison.

the confinement direction is <001>and the thickness L_z is taken at 0.94 nm. Note that an extra quantization in the $\langle 1\bar{1}0 \rangle$-direction takes place in the present nanowire, as indicated by the dashed lines that cross Figure 2.24 at 45°, where "cut" through the energy contours along those lines semi-analytically forms 1-D dispersion curves of the nanowire [2.13]. In Figure 2.24, a line for L_{xy} = 0.77 nm, which corresponds to the present nanowire quantization, and also lines for L_{xy} = 1 nm, 2 nm, 4 nm, 10 nm and ∞ (thin film limit) are drawn for comparison.

It is found that higher k_{xy} region has larger curvature of the dispersion, and is immune to uniaxial <110>strain. As a result, the effective mass of the nanowire becomes smaller than those of bulk and thin film, and also changes little by uniaxial <110>strains, as found in

Figure 2.23. However, as the structure becomes thicker in the $\langle 1\bar{1}0 \rangle$-direction, the quantization line moves toward that of thin film limit, and its effective mass increases to the bulk transverse mass and a response to the strain is also restored. Recently, in [2.26], it has been reported that the transition from the bulk-like properties to the 1-D properties mentioned above actually occurs when the effective size of the nanowire becomes smaller than 2–4 nm. At least, the $\langle 110 \rangle$-oriented Si nanowire with nanoscale cross-section is found to have unique strain properties, unlike bulk and thin film.

2.2 Tunneling Current Calculations Through Si/SiO_2/Si Structures

Today's VLSI technology requires an atomic-scale understanding of physical phenomena arising from the miniaturization of Si-MOSFETs. One of the most important technological roadblocks for further miniaturization is the high-quantum mechanical tunneling current through the gate oxide, which greatly increases VLSI power consumption. Therefore, an accurate modeling of tunneling current through the ultrathin SiO_2 gate oxide [2.27, 2.28] and an alternative high-*k* gate oxide [2.29], is currently of utmost importance for device design.

So far, an atomistic formalism based on the empirical tight-binding (TB) method, embedded in a transfer matrix scheme [2.27], has been applied to calculate tunneling current through three structural models of SiO_2 layers, which are β-cristobalite, β-quartz and β-tridymite [2.28]. The TB approach showed that the β-cristobalite model gives the largest tunneling current for a wide range of oxide thickness, from sub-1 nm to 5 nm. Consequently, the best agreement with experimental results [2.30] was obtained for the β-cristobalite model. On the other hand, the β-cristobalite phase on Si (001) surface has been reported to be unstable, because of a large lattice mismatch at the Si/SiO_2 interface, and it transforms into a highly distorted structure [2.12]. In this section, we examine the influence of the microscopic oxide structure on gate tunneling current properties, by using a fully first-principles approach [2.31].

In the present approach, tunneling probabilities through SiO_2 layers are calculated using the linear combination of atomic orbitals basis ATK code [2.32, 2.33], which is based on the density-functional theory (DFT) and the non-equilibrium Green's function (NEGF) method. Although the DFT theory is known to underestimate bandgaps significantly, the conduction band dispersions are reproduced accurately and, thus, transport parameters such as the conduction band effective mass are reliable. To overcome the bandgap shortcoming, new density functional methods have been presented in the literature. These include: "sX-LDA", employing screened nonlocal exchange and LDA correction potentials within DFT [2.34]; "EXX method", allowing the determination of the exact Kohn-Sham exchange potential [2.35, 2.36]; and "SIC method", introducing self-interaction corrections [2.37]. Furthermore, a first-principles many-body theory of quasi-particle energies, using the GW approximation [2.38, 2.39], whose bandgaps generally agree much better with experiment than the standard DFT method, has also been presented.

Such an exact description of excited states in solids is a challenging and important problem in solid-state physics, but it is still difficult to apply these methods to transport problems in nanoscaled structures, due to the requirement of huge computational resources. Accordingly, we employed the standard DFT method with the generalized gradient approximation (GGA) exchange-correlation functionals, to investigate the gate-tunneling properties of SiO_2 gate oxides in this study.

2.2.1 Atomic Models of Si (001)/SiO_2/Si (001) Structures

Following Ref. [2.28], we adopted three crystalline models of SiO_2 gate oxide, as shown in Figure 2.25.

In this study, we attempted to connect the SiO_2 layers to Si (001) surface by rotating the SiO_2 crystals, so that the lattice constant closest to 5.43 Å of bulk Si crystal can be obtained. As a consequence, the interlayer SiO_2 crystals should be distorted in shape to be different from their equilibrium structures as follows. For β-quartz and β-cristobalite, they need to be biaxially tensiled in the plane perpendicular to <001> direction, and compressed in <001> direction, and vice versa for β-tridymite.

We first obtain such strained SiO_2 crystals by applying structural optimization procedures, where a conjugate gradients minimization method was employed to relax all atomic coordinates and cell shape and size. Then, Si and O atoms at the Si/SiO_2 interfaces are bonded so as not to show any defects, as shown in Figure 2.25. Here, we adopted a simpler interface structure for the β-cristobalite model than that in [2.28], in order for the width of the transition region to be minimized. In practice, the transition region of the β-cristobalite model was constructed in the same way as in the β-quartz model. Subsequently, the structural optimization procedures were again applied for all atoms in the SiO_2 layer and Si atoms neighboring the interfaces. Si atoms locating away from the interfaces were fixed at the equilibrium positions.

For simplicity, the structural optimization procedures were performed only for the zero bias condition, so a structural change due to the applied bias voltage is not considered here. In addition, a periodical boundary condition was employed at the ends of the left and right boundaries. The oxide thickness T_{SiO_2} is defined as a distance between Si atoms residing at the two surfaces.

As described in the introduction, tunneling probabilities were first calculated by using the linear combination of atomic orbitals basis ATK code [2.32, 2.33], where we used the GGA

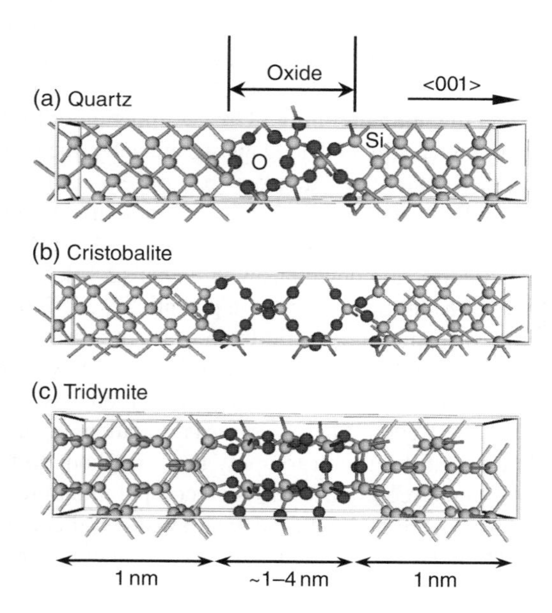

Figure 2.25 Three atomic models of Si (001)/SiO_2/Si (001) structures used in the simulation, where: (a) β-quartz; (b) β-cristobalite; and (c) β-tridymite SiO_2 models.

exchange-correlation functionals of Perdew, Burke and Ernzerhof, and core electrons of atoms are represented by norm-conserving pseudopotentials with the Troullier-Martins parameterization. Next, tunneling current densities were estimated by using the Laudauer-Büttiker formulation, with the calculated tunneling probabilities and the Fermi-Dirac distribution functions of n+-Si and p-Si electrodes. The Fermi energies are given as $E_F \cong 150\,\text{meV}$ above the conduction band edge in n+-Si electrode and $E_F \cong 240\,\text{meV}$ above the valence band edge in p-Si electrode, which correspond to background charge concentrations of $3\times10^{20}\,\text{cm}^{-3}$ and $1\times10^{15}\,\text{cm}^{-3}$, respectively [2.28]. Furthermore, only the negative gate bias is considered in the calculations to meet the same condition in [2.28].

2.2.2 *Current-voltage Characteristics*

First, Figure 2.26 shows the oxide thickness dependences of current-voltage characteristics computed for β-quartz, β-cristobalite and β-tridymite. As expected, the tunneling current increases exponentially as T_{SiO_2} decreases.

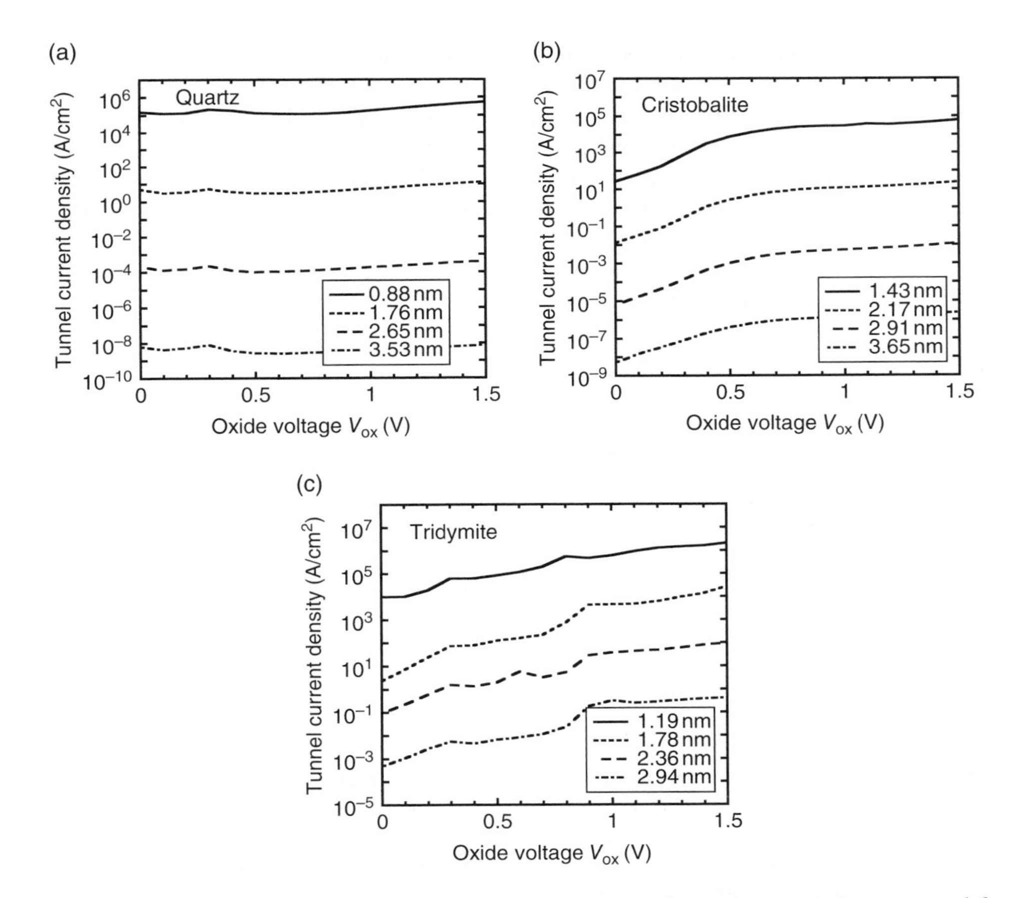

Figure 2.26 Oxide thickness (T_{SiO_2}) dependences of current-voltage characteristics computed for (a) β-quartz, (b) β-cristobalite and (c) β-tridymite.

Here, it should be noted that tunneling current is almost constant with the oxide voltage in the β-quartz model, whose behavior is obviously different from the experimental results [2.30]. There may be interesting physics behind this, but it is still puzzling and is under investigation. On the other hand, the β-cristobalite and β-tridymite models show the increase in the tunneling current with the oxide voltage, which have similar current-voltage characteristics to the experimental one, though numerical fluctuations are observed in the β-tridymite model. A reason for the fluctuations may be due to some technical difficulties in optimizing such a large atomic system. In other words, since the β-tridymite model has the highest atomic density among the three SiO_2 models, and there exists a large compressive strain in the plane perpendicular to the current direction, it is likely hard to obtain a sufficiently relaxed $Si/SiO_2/Si$ structure.

As discussed in [2.28], the tunneling currents under negative gate bias consist of three main contributions, which are schematically depicted in Figure 2.27. One is direct tunneling of conduction band electrons from the gate to the substrate (*C-C*), which gives the most important contribution. The second is due to valence band holes from the substrate to the gate (*V-V*), and the third one is due to electrons from the valence band of the gate to the conduction band of the substrate (*V-C*), which emerges as the oxide voltage increases larger than the Si bandgap energy, as shown in Figure 2.27(c), where the bandgap energy of bulk Si was estimated to be 0.7 eV from our bandstructure calculations – rather smaller than 1.2 eV of [2.28] due to the DFT/GGA formalism.

These three components are separately plotted in Figure 2.28 for the three SiO_2 models, where the oxide thicknesses used in the calculation are indicated in each figure. The *V-V* component (i.e. hole current) is negligible, due to higher Si/SiO_2 valence band offset [2.28]. The *C-C* component is dominant as expected, but the *V-C* component becomes comparable to the *C-C* contribution for $V_{ox} > 0.7$ V. As shown in Figure 2.28(a), the *V-C* contribution in the β-quartz model is excessive, so the β-quartz model indeed has different tunneling properties from the other models. Incidentally, the *C-C* and *V-V* currents flow at $V_{ox} = 0$ V, because the Fermi energies between n^+-Si and p-Si electrodes are different, as shown in Figure 2.27(a).

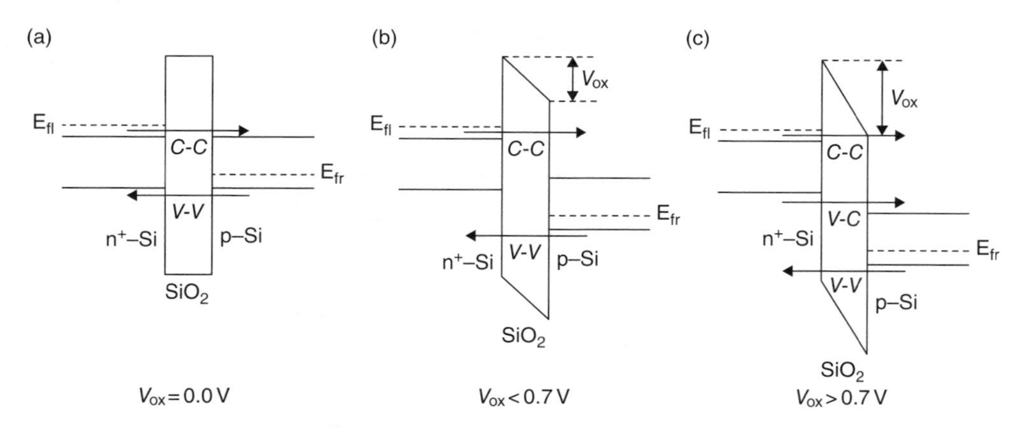

Figure 2.27 Conduction- and valence- band diagrams of n^+-Si/SiO_2/p-Si structures for: (a) $V_{ox} = 0$ V; (b) $V_{ox} < 0.7$ V; and (c) $V_{ox} > 0.7$ V. *C-C* represents the direct tunneling of conduction band electrons from gate to substrate, *V-V* the tunneling due to valence band holes from substrate to gate, and *V-C* the tunneling due to electrons from valence band of gate to conduction band of substrate, which emerges as oxide voltage V_{ox} increases larger than Si bandgap energy (0.7 eV).

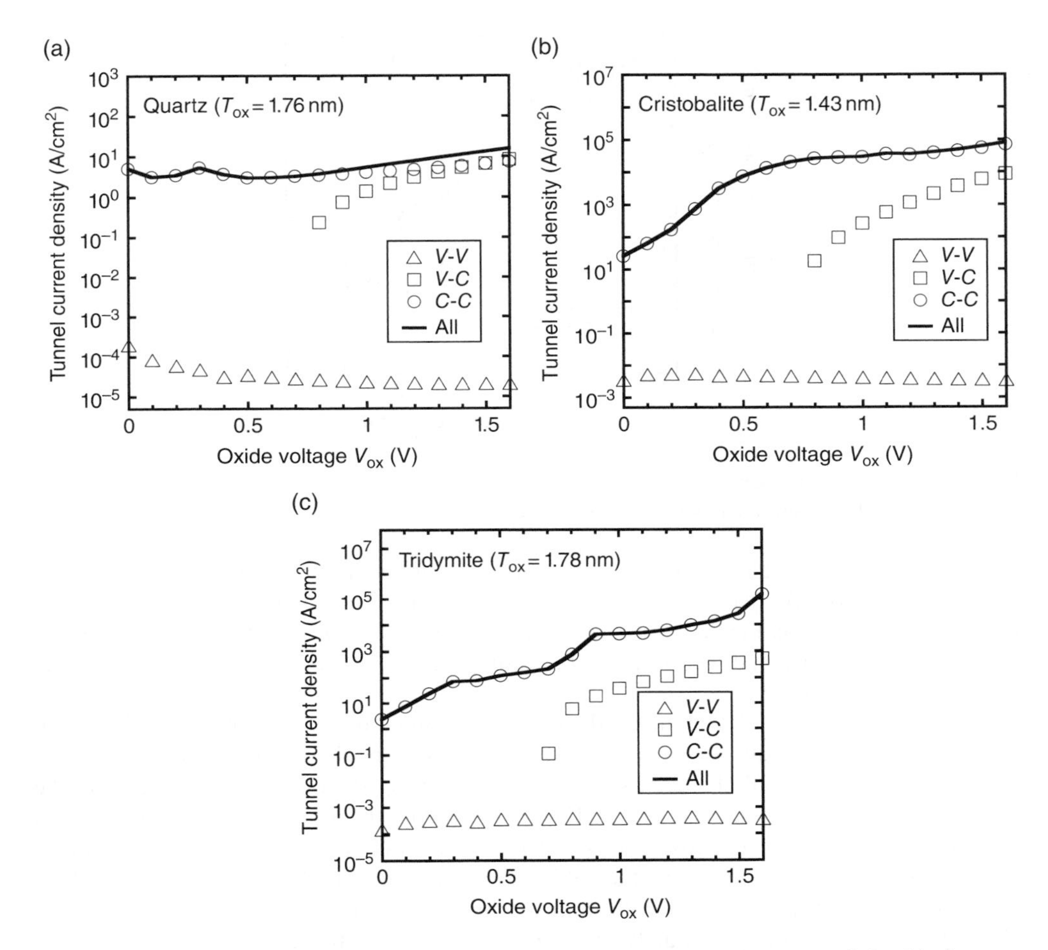

Figure 2.28 Three tunneling current components, *C-C*, *V-V*, and *V-C*, computed for (a) β-quartz, (b) β-cristobalite and (c) β-tridymite. The oxide thicknesses used in the calculation are indicated in each figure.

2.2.3 SiO_2 Thickness Dependences

Next, Figure 2.29 shows the tunneling current density as a function of the oxide thickness at V_{ox} = 0.7 V, where the tunneling current densities are determined by the *C-C* component, as shown in Figure 2.28.

Similar oxide thickness dependences of the tunneling current density have been also reported in [2.40]. It is found that the β-cristobalite and β-tridymite models have similar current-thickness properties, while the β-quartz model gives at least two orders of magnitude lower current density. The present results are basically consistent with the previous results based on the TB formalism, where the lower tunneling current in the β-quartz model was plausibly explained, using its largest conduction band effective mass [2.28]. Therefore, we further examined the band structure parameters for bulk SiO_2 crystals.

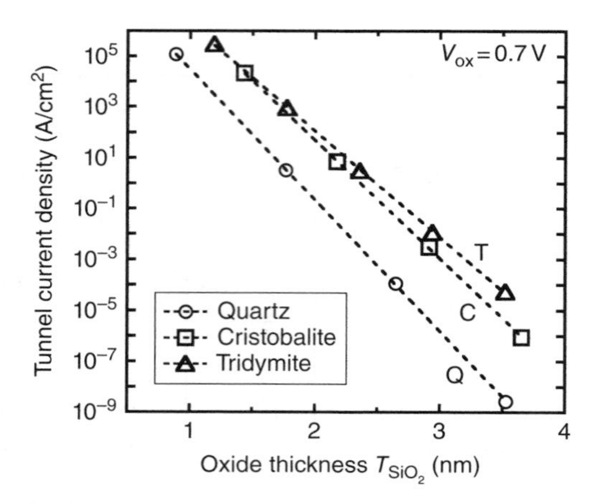

Figure 2.29 Oxide thickness dependences of tunneling current density at V_{ox} = 0.7 V, where the tunneling current is determined by the *C-C* component.

Figure 2.30 shows the computed band structures along the tunneling current direction <001> for β-quartz, β-cristobalite and β-tridymite crystals. The length of the first-Brillouin zone is different among the three SiO_2 crystals, because each SiO_2 crystal has a different unit cell size along the current direction as $a_q = 0.88$ nm (β-quartz), $a_c = 0.74$ nm (β-cristobalite) and $a_t = 0.56$ nm (β-tridymite).

It is found in Figure 2.30 that the conduction band minimum is located at the Γ point ($k = 0$) and the valence band width is very narrow, which had been already reported by several authors [2.41, 2.42]. Such overall characteristics are similar among the three SiO_2 crystals, but the detailed band structure parameters at the Γ point depend on the SiO_2 crystals, as summarized in Table 2.1, where the effective masses at the conduction band minimum and bandgap energies are compared. Although the DFT/GGA formalism underestimates the bandgap energies considerably, compared with the experimental one (≈9.0 V), the conduction band dispersions are reproduced accurately, and the estimated conduction band masses are reliable. According to Table 2.1, the largest values in both the conduction band masses and bandgap energies are obtained for the bulk β-quartz crystal, while the smallest values are obtained for the bulk β-tridymite crystal. These results are in reasonable agreement with the magnitude of tunneling current densities shown in Figure 2.29.

To understand correlations between the magnitude relation of the tunneling current densities in Figure 2.29 and the band structure parameters of the SiO_2 crystals in Table 2.1, we calculated the current-oxide thickness dependences corresponding to the *C-C* component by using the bandstructure parameters of Table 2.1, based on the conventional transfer-matrix method. In the transfer-matrix calculation, the conduction band discontinuity energy ΔE_C was simply given by $\Delta E_C = [E_G^{ox}(\text{DFT}) - E_G^{Si}(\text{DFT})]/2$, where $E_G^{ox}(\text{DFT})$ and $E_G^{Si}(\text{DFT})$ are bandgap energies calculated for the bulk SiO_2 and Si crystals based on the present DFT approach. The results are plotted in Figure 2.31, which represents that the same magnitude relation among the three oxide structures as those in Figure 2.29 is obtained, though difference between the β-cristobalite and β-tridymite models is estimated to be larger.

To further analyze if the difference in the tunneling current densities is more due to the different effective masses or is due to the different bandgaps, we also calculated the current-oxide

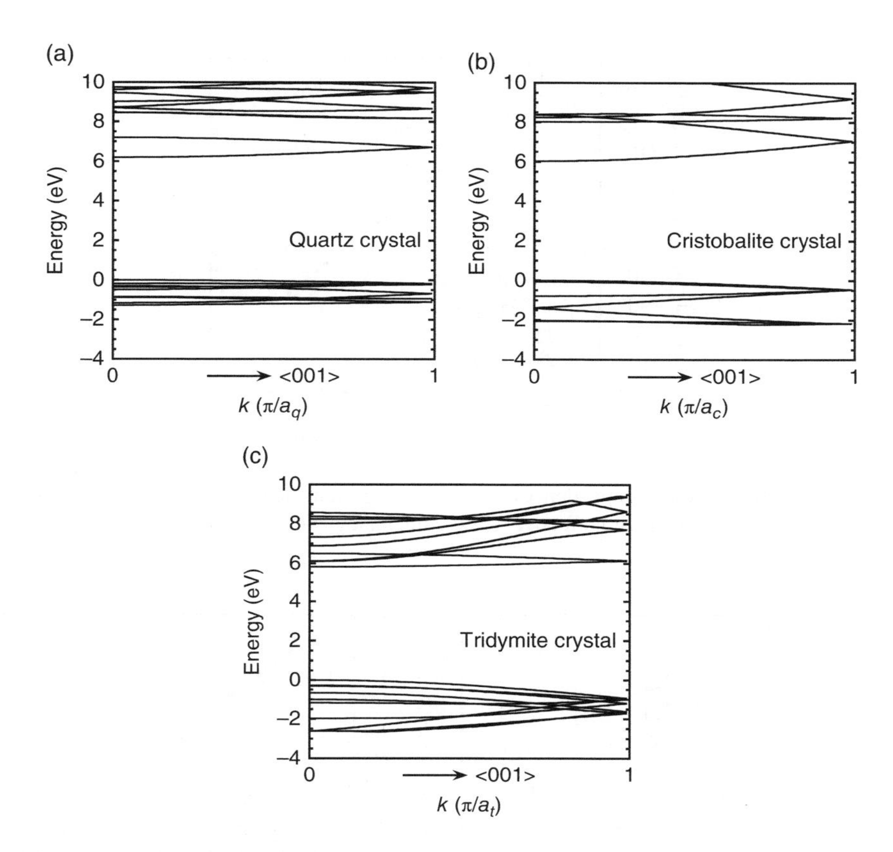

Figure 2.30 Computed band structures along tunneling current direction <001> for: (a) bulk β-quartz; (b) bulk β-cristobalite; and (c) bulk β-tridymite crystals, where SiO_2 lattice constants were taken as the same values as in Si/SiO_2/Si structures shown in Figure 2.25. Each crystal has a different unit cell size, as $a_q = 0.88$ nm, $a_c = 0.74$ nm and $a_t = 0.56$ nm for the β-quartz, β-cristobalite and β-tridymite crystals, respectively.

Table 2.1 Conduction band effective masses and bandgap energies extracted from the band structure calculations for the bulk SiO_2 crystals.

	Quartz	Cristobalite	Tridymite
Conduction band mass (m_0)	0.77	0.65	0.30
Bandgap energy (eV)	6.19	6.03	5.82

thickness dependences by altering only effective mass or bandgap energy, as shown in Figure 2.32, where bandgap energy and effective mass are fixed at the values of the β-quartz crystal, respectively.

From Figures 2.31 and 2.32, we can conclude that the tunneling current densities are more sensitive to the difference in the conduction band effective masses than that in the bandgaps. In addition, the small discrepancy between the β-cristobalite and β-tridymite models observed

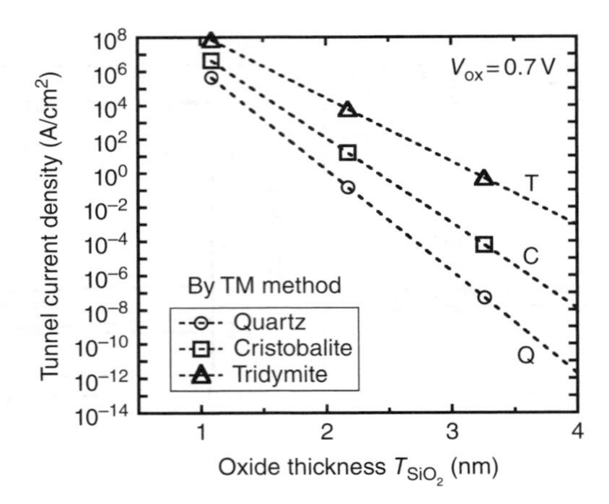

Figure 2.31 Oxide thickness dependences of tunneling current density at $V_{ox}=0.7\,V$, computed by using a transfer-matrix method with effective masses and bandgap energies of Table 2.1.

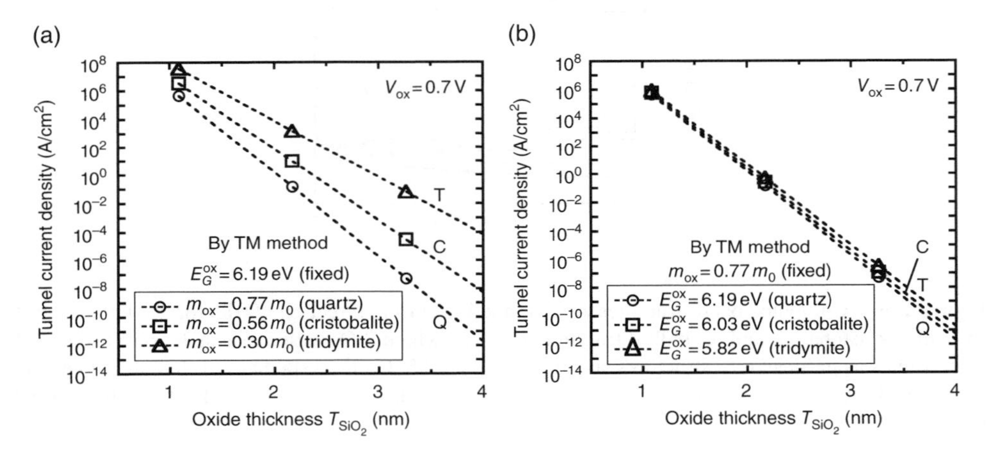

Figure 2.32 Oxide thickness dependences of tunneling current density at $V_{ox}=0.7\,V$, computed by using the transfer-matrix method with altered only (a) effective mass; and (b) bandgap energy, where (a) bandgap energy; and (b) effective mass are fixed at the value of β-quartz crystal, respectively.

in Figure 2.29 suggests a possibility of reconstruction of the SiO_2 atomic structures in the presence of the Si/SiO_2 interfaces. Nevertheless, the above results mean that the electronic properties of bulk SiO_2 crystals can still be important for the tunneling current behaviors, even in the nanoscaled SiO_2 gate oxides.

References

[2.1] K. Uchida, J. Koga and S. Takagi (2003). Experimental study on carrier transport mechanisms in double- and single-gate ultrathin-body MOSFETs – Coulomb scattering, volume inversion and δT_{SOI}-induced scattering. *IEDM Technical Digest*, 805–808.

[2.2] T. Hara, Y. Yamada, T. Maegawa and H. Tsuchiya (2008). Atomistic study on electronic properties of nanoscale SOI channels. *Journal of Physics: Conference Series* **109**, 012012.

[2.3] G. Kresse and J. Furthmüller (Oct. 1996). Efficient iterative schemes for ab initio total-energy calculation using a plane-wave basis set. *Physical Review B* **54**(16), 11169–11186.

[2.4] H. Tsuchiya, H. Ando, S. Sawamoto, T. Maegawa, T. Hara, H. Yao and M. Ogawa (Feb. 2010). Comparisons of performance potentials of silicon nanowire and graphene nanoribbon MOSFETs considering first-principles bandstructure effects. *IEEE Transactions on Electron Devices* **57**(2), 406–414.

[2.5] T. Ando, A.B. Fowler and F. Stern (Apr. 1982). Electronic properties of two-dimensional systems. *Reviews of Modern Physics* **54**(2), 437–672.

[2.6] A. Rahman, G. Klimeck, M. Lundstrom, T. B. Boykin and N. Vagidov (Apr. 2005). Atomistic approach for nanoscale devices at the scaling limit and beyond – Valley splitting in Si. *Japanese Journal of Applied Physics* **44**(4B), 2187–2190.

[2.7] J. Wang, A. Rahman, A. Ghosh, G. Klimeck and M. Lundstrom (Jul. 2005). On the validity of the parabolic effective-mass approximation for the *I-V* calculation of silicon nanowire transistors. *IEEE Transactions on Electron Devices* **52**(7), 1589–1595.

[2.8] Y. Liu, N. Neophytou, T. Low, G. Klimeck and M. Lundstrom (Mar. 2008). A tight-binding study of the ballistic injection velocity for ultrathin-body SOI MOSFETs. *IEEE Transactions on Electron Devices* **55**(3), 866–871.

[2.9] H. Scheel, S. Reich and C. Thomsen (Oct. 2005). Electronic band structure of high-index silicon nanowires. *Physica Status Solidi (B)* **242**(12), 2474–2479.

[2.10] Z.H. Lu and D. Grozea (2002). Crystalline Si/SiO_2 quantum wells. *Applied Physics Letters* **80**(2), 255–257.

[2.11] S. Markov, B. Aradi, C.-Y. Yam, H. Xie, T. Frauenheim and G. Chen (Mar. 2015). Atomic level modeling of extremely thin silicon-on-insulator MOSFETs including the silicon dioxide: Electronic structure. *IEEE Transactions on Electron Devices* **62**(3), 696–704.

[2.12] T. Yamasaki, C. Kaneta, T. Uchiyama, T. Uda and K. Terakura (Mar. 2001). Geometric and electronic structures of SiO_2/Si(001) interfaces. *Physical Review B* **63**, 115314.

[2.13] N. Neophytou, A. Paul, M. Lundstrom and G. Klimeck (June 2008). Bandstructure effects in silicon nanowire electron transport. *IEEE Transactions on Electron Devices* **55**(6), 1286–1297.

[2.14] T. Maegawa, T. Yamauchi, T. Hara, H. Tsuchiya and M. Ogawa (Apr. 2009). Strain effects on electronic bandstructures in nanoscaled silicon: From bulk to nanowire. *IEEE Transactions on Electron Devices* **56**(4), 553–559.

[2.15] S. Takagi, T. Irisawa, T. Tezuka, T. Numata, S. Nakaharai, N. Hirashita, Y. Moriyama, K. Usuda, E. Toyoda, S. Dissanayake, M. Shichijo, R. Nakane, S. Sugahara, M. Takenaka and N. Sugiyama (Jan. 2008). Carrier-transport-enhanced channel CMOS for improved power consumption and performance. *IEEE Transactions on Electron Devices* **55**(1), 21–39.

[2.16] S.E. Thompson, G. Sun, K. Wu, J. Lim and T. Nishida (2004). Key differences for process-induced uniaxial vs. substrate-induced biaxial stressed Si and Ge channel MOSFETs. *IEDM Technical Digest*, 221–224.

[2.17] L. Shifren, X. Wang, P. Matagne, B. Obradovic, C. Auth, S. Cea, T. Ghani, J. He, T. Hoffman, R. Kotlyar, Z. Ma, K. Mistry, R. Nagisetty, R. Shaheed, M. Stettler, C. Weber and M.D. Giles (Dec. 2004). Drive current enhancement in p-type metal-oxide-semiconductor field-effect transistors under shear uniaxial stress. *Applied Physics Letters* **85**(25), 6188–6190.

[2.18] K. Uchida, T. Krishnamohan, K.C. Saraswat and Y. Nishi (2005). Physical mechanisms of electron mobility enhancement in uniaxial stressed MOSFETs and impact of uniaxial stress engineering in ballistic regime. *IEDM Technical Digest*, 135–138.

[2.19] E.X. Wang, P. Matagne, L. Shifren, B. Obradovic, R. Kotlyar, S. Cea, M. Stettle and M.D. Giles (Aug. 2006). Physics of hole transport in strained silicon MOSFET inversion layers. *IEEE Transactions on Electron Devices* **53**(8), 1840–1851.

[2.20] E. Ungersboeck, S. Dhar, G. Karlowatz, H. Kosina and S. Selberherr (2007). Physical modeling of electron mobility enhancement for arbitrarily strained silicon. *Journal of Computational Electronics* **6**, 55–58.

[2.21] J. Yamauchi (Feb. 2008). Effective mass anomalies in strained-Si thin films and crystals. *IEEE Electron Device Letters* **29**(2), 186–188.

[2.22] D. Shiri, Y. Kong, A. Buin and M.P. Anantram (2008). Strain induced change of bandgap and effective mass in silicon nanowires. *Applied Physics Letters* **93**(7), 073114.

[2.23] P.W. Leu, A. Svizhenko and K. Cho (2008). *Ab initio* calculations of the mechanical and electronic properties of strained Si nanowires. *Physical Review B* **77**(23), 235305.

[2.24] K.-H. Hong, J. Kim, S.-H. Lee and J.K. Shin (2008). Strain-driven electronic band structure modulation of Si nanowires. *Nano Letters* **8**(5), 1335–1340.

[2.25] S. You, M. Gao and Y. Wang (Sep. 2007). Band structure of surface terminated silicon nanowire. *Extended Abstracts of International Conference on Solid State Devices and Materials (SSDM)*, Tsukuba, 698–699.

[2.26] L. Zhang, H. Lou, J. He and M. Chan (Nov. 2011). Uniaxial strain effects on electron ballistic transport in gate-all-around silicon nanowire MOSFETs. *IEEE Transactions on Electron Devices* **58**(11), 3829–3836.

[2.27] M. Städele, B.R. Tuttle and K. Hess (Jan. 2001). Tunneling through ultrathin SiO_2 gate oxides from microscopic models. *Journal of Applied Physics* **89**(1), 348–363.

[2.28] F. Sacconi, A.D. Carlo, P. Lugli, M. Städele and J-M. Jancu (May 2004). Full band approach to tunneling in MOS structures. *IEEE Transactions on Electron Devices* **51**(5), 741–748.

[2.29] F. Sacconi, J-M. Jancu, M. Povolotskyi and A.D. Carlo (Dec. 2007). Full-band tunneling in high-*k* oxide MOS structures. *IEEE Transactions on Electron Devices* **54**(12), 3168–3176.

[2.30] Khairurrijal, W. Mizubayashi, S. Miyazaki and M. Hirose (Mar. 2000). Analytic model of direct tunnel current through ultrathin gate oxides, *Journal of Applied Physics* **87**(6), 3000–3005.

[2.31] Y. Yamada, H. Tsuchiya and M. Ogawa (Apr. 2009). A first principles study on tunneling current through $Si/SiO_2/Si$ structures. *Journal of Applied Physics* **105**, 083702.

[2.32] J.M. Soler, E. Artacho, J.D. Gale, A. García, J. Junquera, P. Ordejón and D. Sánchez-Portal (Mar. 2002). The SIESTA method for *ab initio* order-*N* materials simulation. *Journal of Physics: Condensed Matter* **14**(11), 2745–2766.

[2.33] M. Brandbyge, J-L. Mozos, P. Ordejón, J. Taylor and K. Stokbro (Mar. 2002). Density-functional method for nonequilibrium electron transport. *Physical Review B* **65**, 165401.

[2.34] A. Seidl, A. Görling, P. Vogl and J.A. Majewski (Feb. 1996). Generalized Kohn-Sham schemes and the band-gap problem. *Physical Review B* **53**(7), 3764–3774.

[2.35] T. Kotani and H. Akai (Dec. 1995). Exact exchange potential band-structure calculations for simple metals: Li, Na, K, Rb and Ca. *Physical Review B* **52**(24), 17153–17157.

[2.36] M. Städele, J.A. Majewski, P. Vogl and A. Görling (Sep. 1997). Exact Kohn-Sham exchange potential in semiconductors. *Physical Review Letters* **79**(11), 2089–2092.

[2.37] J.P. Perdew and A. Zunger (May 1981). Self-interaction correction to density-functional approximations for many-electron systems. *Physical Review B* **23**(10), 5048–5079.

[2.38] L. Hedin (Aug. 1965). New method for calculating the one-particle Green's function with application to the electron-gas problem. *Physical Review* **139**(3A), A796–A823.

[2.39] M.S. Hybertsen and S.G. Louie (Oct. 1986). Electron correlation in semiconductors and insulators: Band gaps and quasiparticle energies. *Physical Review B* **34**(8), 5390–5413.

[2.40] E. Nadimi, P. Plänitz, R. Öttking, K. Wieczorek and C. Radehaus (Mar. 2010). First principle calculation of the leakage current through SiO_2 and SiO_xN_y gate dielectrics in MOSFETs. *IEEE Transactions on Electron Devices* **57**(3), 690–695.

[2.41] E. Gnani, S. Reggiani, R. Colle and M. Rudan (Oct. 2000). Band-structure calculations of SiO_2 by means of Hartree-Fock and density-functional techniques. *IEEE Transactions on Electron Devices* **47**(10), 1795–1803.

[2.42] Y. Xu and W.Y. Ching (Nov. 1991). Electronic and optical properties of all polymorphic forms of silicon dioxide. *Physical Review B* **44**(20), 11048–11059.

3

Quasi-ballistic Transport in Si Nanoscale MOSFETs

To achieve a continuous improvement in drive current of Si-MOSFETs, as the International Technology Roadmap for Semiconductors (ITRS) requires, carrier velocity enhancement utilizing ballistic transport, combined with the use of high-mobility channel materials and novel device architectures, is considered to be necessary. With the significant advances in lithography technology, channel length of Si-MOSFET continues to shrink rapidly down to a sub-10 nm regime, and the nanoscale channel lengths open up a possibility to realize a quasi-ballistic operation of MOSFETs. The device physics of the quasi-ballistic MOSFETs has been studied extensively by analytical and compact models, but a much more detailed treatment of dominant scattering processes, and consideration of multi-dimensional quantum transport, will be indispensable to explore an ultimate device performance of Si-MOSFETs. In this chapter, we present quasi-ballistic transport properties in several Si nanoscale MOSFETs, clarified by using semi-classical Monte Carlo simulations.

3.1 A Picture of Quasi-ballistic Transport Simulated using Quantum-corrected Monte Carlo Simulation

A schematic diagram representing how to determine a current drive in the quasi-ballistic model is shown in Figure 3.1. The source is treated as a reservoir of thermal carriers, which injects carriers from source to forward channel, with an injection velocity of v_{inj}.

When scattering exists in the channel, an average carrier velocity at the potential maximum point in the channel (bottleneck point) v_s decreases from v_{inj} and is expressed as follows [3.1]:

$$v_{\mathrm{s}} = v_{\mathrm{inj}} \times \frac{1-R}{1+R\left(v_{\mathrm{inj}} / v_{\mathrm{back}}\right)} \tag{3.1}$$

Carrier Transport in Nanoscale MOS Transistors, First Edition. Hideaki Tsuchiya and Yoshinari Kamakura.

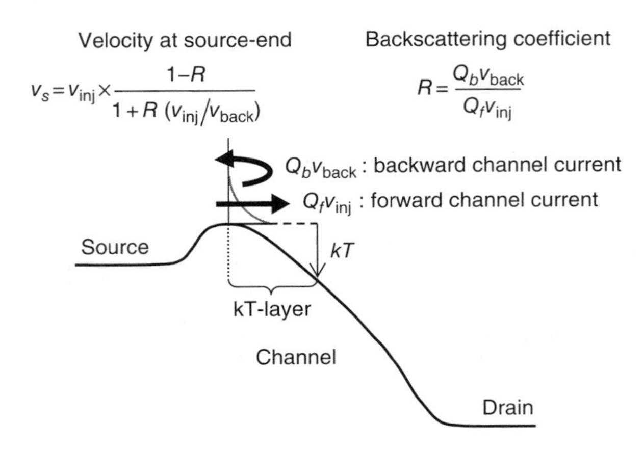

Figure 3.1 Schematic diagram representing how to determine current drive in quasi-ballistic model. Definitions of quasi-ballistic transport parameters are also given. Backscattering coefficient R is defined as the ratio between the backward and forward channel currents.

where v_{back} is a backward channel velocity and R a backscattering coefficient, defined as the ratio between the backward and forward channel currents given by $R=Q_b v_{back}/Q_f v_{inj}$, where Q_f and Q_b are the forward and backward channel charge densities at the bottleneck point, respectively. Equation (3.1) is derived from the relations $Qv_s=Q_f v_{inj} - Q_b v_{back}$ and $Q=Q_f - Q_b$. In the ballistic limit ($R=0$), $v_s=v_{inj}$. To explore an ultimate device performance of MOSFETs, a physical mechanism creating v_{inj} and roles of backscattering in Equation (3.1) need to be fully understood. In this section, we discuss these issues, based upon a quantum-corrected Monte Carlo (MC) device simulation considering quantum confinement effect in the inversion layer, and propose a picture of quasi-ballistic transport in nanoscale MOSFETs [3.2].

3.1.1 Device Structure and Simulation Method

The device structure used in the simulation is shown in Figure 3.2. Since the suppression of short-channel effects is crucial in a precise analysis of nanoscale MOSFETs, we adopted a double-gate (DG) and ultrathin-body (UTB) structure. The Si body thickness T_{Si} is set as 3 nm, where most of the electrons propagate in the two-fold valleys with a higher mobility along the channel (valleys of 1 and 2 in Figure 3.2(b)), and phonon scattering between the sub-bands is forbidden [3.3]. Furthermore, the extremely thin gate oxide of $T_{ox}=0.5$ nm is adopted, which is also effective in eliminating the short-channel effects. For simplicity, the gate tunneling effect is neglected in this study. The channel length L_{ch} is varied from 50 to 8 nm, where we confirmed that the short-channel effects are sufficiently suppressed until $L_{ch}=8$ nm. The donor concentration in the source and drain regions is $N_D=1\times10^{20}\,\text{cm}^{-3}$, and the channel region is assumed to be undoped. The width of the device is fixed at 100 nm, and the potential and carrier density in the z-axis are assumed to be uniform.

The electrical characteristics were computed using the MC device simulator, with quantum mechanical correction of potential developed at Kobe University. In this method, electrons are treated as three-dimensional (3-D) particles, and the equations of motion for the particles in free flight are given as [3.2, 3.4–3.7].

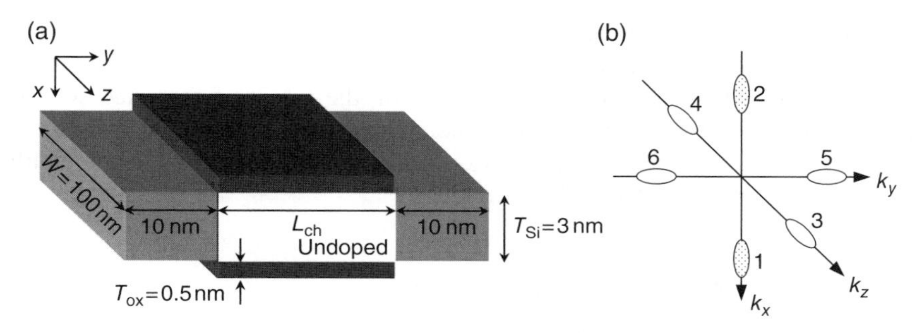

Figure 3.2 (a) DG-MOSFET used in the simulation. $T_{Si} = 3$ nm and $T_{ox} = 0.5$ nm. The donor concentration in the source and drain regions is $N_D = 1 \times 10^{20}\,\mathrm{cm}^{-3}$, and the channel region is assumed to be undoped. A metal gate is employed. (b) Equi-energy surfaces of Si conduction band, where the numbers (1, 2, …, 6) denotes valley index ν.

$$\frac{d\mathbf{r}}{dt} = \mathbf{v} \tag{3.2}$$

$$\frac{d\mathbf{k}}{dt} = -\frac{1}{\hbar}\nabla_{\mathbf{r}}\left(U + U_{\nu}^{\mathrm{QC}}\right) \tag{3.3}$$

where $\hbar$ is the Dirac constant ($= h/2\pi$). The velocity equation (3.2) is the same as the one used in the standard MC technique, but the force equation (3.3) is modified so that particles evolve under the enforcement, not only by the classical built-in potential U, but also by the quantum correction of potential U_{ν}^{QC}. This approximately represents the lowest-order quantized sub-band, given by:

$$U_{\nu}^{\mathrm{QC}} = -\frac{\hbar^2}{12m_x^{\nu}}\frac{\partial^2 \ln(n_{\nu})}{\partial x^2} - \frac{\hbar^2}{12m_y^{\nu}}\frac{\partial^2 \ln(n_{\nu})}{\partial y^2} \tag{3.4}$$

where the index $\nu = (1, 2, \ldots, 6)$ denotes the six equivalent valleys of the Si conduction band as shown in Figure 3.2(b). n_{ν} represents the carrier density, and m_x and m_y are the effective masses of the ellipsoidal bandstructure in the x- and y-directions, respectively, with $m_l = 0.98\,m_0$ (longitudinal effective mass) and $m_t = 0.19\,m_0$ (transverse effective mass), where m_0 is the free electron rest mass.

To prevent a divergence of U_{ν}^{QC}, the carrier density n_{ν} is time-averaged for several 10 ps during the MC simulation. In addition, we adopted a smoothing technique for the quantum potential $U + U_{\nu}^{\mathrm{QC}}$ by using the nearest neighbor spatial points. Furthermore, we should pay attention if we apply Equation (3.4) to carrier transport simulation at the Si/SiO_2 interface. The carrier density at the interface usually becomes very small due to a large potential barrier and, thus, the second-order space derivative of n_{ν} tends to diversify at the interface.

To overcome this problem, we introduce a coupled method with an effective potential approach [3.8, 3.9]. In the coupled method, the quantum correction of potential U_{ν}^{QC} is used only in the Si region, while the effective potential is employed in the SiO_2 region [3.6]. In the calculation of transition rates of scattering processes, the quantum correction of potential U_{ν}^{QC}

should be taken into account in the evaluation of carrier's total energy. Poisson's equation is self-consistently solved in the x- and y-directions, with 2-D carrier density obtained from the quantum-corrected MC simulation. The validity of the quantum-corrected MC method has been demonstrated by comparing with more accurate quantum models, such as one-dimensional self-consistent Shrödinger-Poisson solutions [3.4, 3.5] and the non-equilibrium Green's function approach [3.10].

3.1.2 Scattering Rates for 3-D Electron Gas

The scattering processes considered are ionized impurity (II) scattering, intravalley acoustic phonon (AP) and intervalley phonon (IP) scatterings, including *f*- and *g*-phonons as shown in Figure 3.3, surface roughness (SR) scattering, and electron-electron scatterings, including electron-plasmon and short-range Coulomb interactions.

The II scattering, which is considered only in the source and drain regions, is treated by using the Brooks-Herring model, in which the scattering rate is represented as a function the electron wavenumber **k**:

$$W(\mathbf{k}) = \frac{2\pi N_I Z^2 e^4 N(E_{\mathbf{k}})}{\hbar\varepsilon^2} \frac{1}{q_D^2 \left(4k^2 + q_D^2\right)} \tag{3.5}$$

where ε, Ze and N_I represent the dielectric constant, the electric charge of ion and the ionized impurity density, respectively, and $N(E_{\mathbf{k}})$ is the density-of-states (DOS) for carriers at the final state after scattering. q_D is the inverse of the Debye length, which is given by:

$$q_D = \sqrt{\frac{e^2 n}{\varepsilon k_B T}} \tag{3.6}$$

where n is the carrier density, k_B is the Boltzmann constant, and T is the temperature, which is assumed to be 300 K in this study. The II scattering is known to be anisotropic, and the scattering angle θ is determined by using a random number r in [0, 1], as expressed in [3.11]:

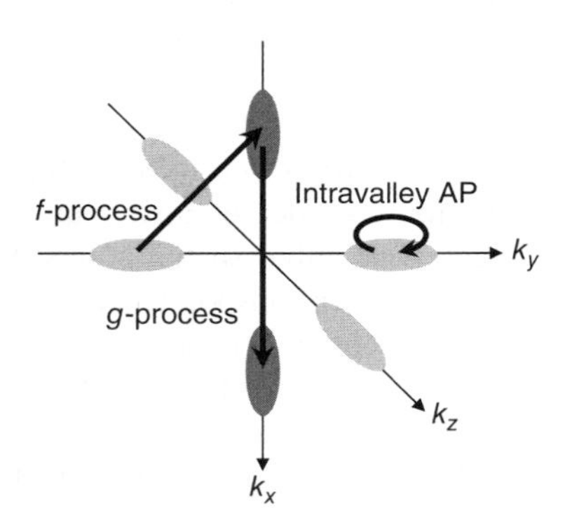

Figure 3.3 Intravalley AP and intervalley *f*- and *g*-phonon scattering considered in the simulation.

$$\cos\theta = 1 - \frac{2r}{1 + (1-r)\left(\frac{2k}{q_D}\right)^2} \tag{3.7}$$

The intravalley AP scattering is treated as an elastic process, with a scattering rate given by:

$$W(\mathbf{k}) = \frac{2\pi D_{ac}^2 k_B T}{\hbar \rho v^2} N(E_\mathbf{k}) \tag{3.8}$$

where D_{ac} is the deformation potential, ρ is the crystal density, v is the sound velocity. In Equation (3.8), the AP energy is assumed to be much smaller than $k_B T$ at room temperature and, hence, the AP scattering is treated as an elastic process. Since the AP scattering is isotropic, the scattering angles are determined as $\phi = 0 \sim 2\pi$ and $\cos\phi = -1 \sim 1$ by using a random number in [0, 1]. The IP scattering, also known as non-polar-optical phonon scattering, is treated by considering three *f*- and three *g*-phonons, with a scattering rate given by:

$$W(\mathbf{k}) = \frac{\pi N_v D_{op}^2}{\rho \omega_{op}} \left(N_q + \frac{1}{2} \mp \frac{1}{2} \right) N\left(E_\mathbf{k} \pm \hbar\omega_{op} - \Delta E_{vv'}\right) \tag{3.9}$$

where N_v is the number of possible final valleys, D_{op} the deformation field, $\hbar\omega_{op}$ the IP energy, $\Delta E_{vv'}$ the energy difference between valleys v and v', and N_q the average number of phonons at a given temperature. The complex symbols ($\mp, \pm$) represent phonon absorption and emission processes, respectively. As shown in Table 3.1, the IP energies are comparable or larger than $k_B T$ at room temperature, so the IP scattering is treated as an inelastic process. For II, AP and IP scattering, we used the values shown in Table 3.1, which are taken from [3.12].

The SR scattering at the gate oxide interfaces is modeled as a mixture of specular reflections and elastic diffusions, each occurring with a probability of 0.5 at every hitting interface [3.13]. The electron-electron scatterings are incorporated by splitting the Coulomb interactions into short-range and long-range potentials, where the short-range part is included through an addition of a molecular dynamics loop, and the long-range part is treated as an additional scattering mechanism – that is, inelastic electron-plasmon scattering [3.5]. The scattering rate of the electron-plasmon scattering is given by [3.11]:

$$W(\mathbf{k}) = \frac{e^2 \omega_p}{8\pi\varepsilon} \frac{k}{E_\mathbf{k}} \left(N_p + \frac{1}{2} \mp \frac{1}{2} \right) \ln\left(\frac{q_c}{q_{\min}} \right) \tag{3.10}$$

where

$$\omega_p = \sqrt{\frac{e^2 n}{\varepsilon m}} \tag{3.11}$$

$$q_{\min} = k\left| 1 - \sqrt{1 \pm \frac{\hbar\omega_p}{E_\mathbf{k}}} \right|, q_c = \sqrt{\frac{e^2 n}{\varepsilon k_B T}} \tag{3.12}$$

Table 3.1 Values of material parameters used in the MC simulation [3.12].

Material constants		Symbol	Si	
Lattice constant		a	5.431 (Å)	
Crystal density		ρ	2329.0 (kg/m^3)	
Dielectric constant		ε	11.7 ε_0 (F/m)	
Sound velocity		v_s	9040 (m/s)	
Effective mass	Longitudinal	m_l	$0.98 m_0$	
	Transverse	m_t	$0.19 m_0$	
	Nonparabolicity	α	0.5 (eV^{-1})	
Deformation	Acoustic phonon	D_{ac}	9.5 (eV)	
potential & fields	Intervalley phonons	D_{op}	Values (eV/cm)	Energy (meV)
		$f-$TA	3×10^7	19.0
		$f-$LA	2×10^8	47.4
		$f-$TO	2×10^8	59.0
		g − TA	5×10^7	12.0
		g − LA	8×10^7	18.5
		g − LO	1.1×10^9	61.2

q_c is a cut-off wavenumber of plasmon, which means that plasmon with wavenumber larger than q_c cannot be excited because of instantaneous decay caused by carrier diffusion. The electron-plasmon scattering is anisotropic, and the scattering angle θ is determined by using a random number r in [0, 1], as follows [3.11]:

$$\cos\theta = \frac{1+f-(1+2f)^r}{f} \tag{3.13}$$

where f is a function represented using the electron energies before and after scattering, $E_{\mathbf{k}}$ and $E_{\mathbf{k}'}$, as given by:

$$f = \frac{2\sqrt{E_{\mathbf{k}}E_{\mathbf{k}'}}}{\left(\sqrt{E_{\mathbf{k}}}-\sqrt{E_{\mathbf{k}'}}\right)^2} \tag{3.14}$$

The degeneracy effects are also incorporated in such a way that a distribution function $f_v(\mathbf{k}, \mathbf{r})$ is tabulated at each location throughout the device, and any scattering processes are rejected if the final electron state and band index v are such that $1-f_v(\mathbf{k}, \mathbf{r}) \le \eta$, where η is a random number in [0, 1].

3.1.3 Ballistic Transport Limit

To investigate a physical mechanism determining v_{inj}, we first perform a ballistic MC simulation, which means that scattering is neglected in the channel region, whereas scattering in the source and drain is considered. Figure 3.4 shows the potential and sheet electron density and average electron velocity profiles computed for the fictitious ballistic MOSFETs with various

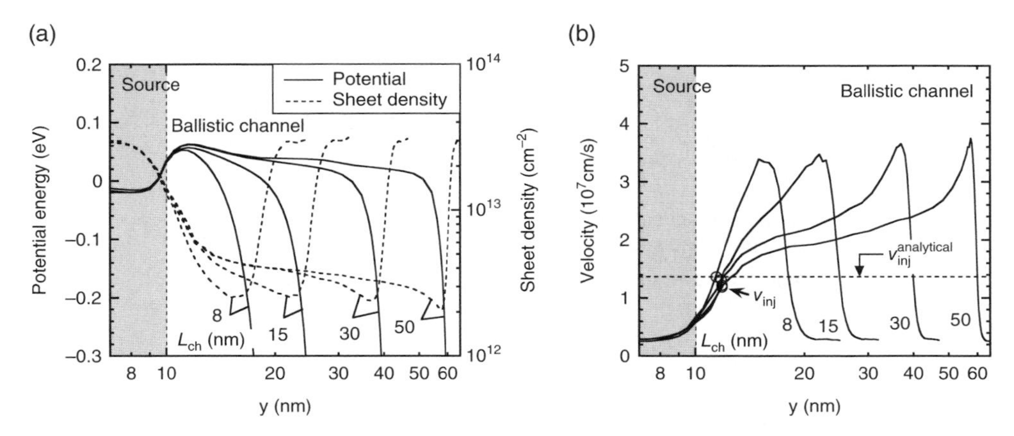

Figure 3.4 (a) Potential and sheet electron density and (b) average electron velocity profiles along the channel computed for ballistic DG-MOSFETs with various channel lengths. $V_{DS}=0.6$V and V_G is adjusted to keep the sheet electron density at the bottleneck point equal to $N_s=7\times10^{12}\,\text{cm}^{-2}$. Scattering in the source and drain regions is considered. The horizontal dashed line represents the injection velocity, calculated using the analytical model.

channel lengths, where the source-drain voltage $V_{DS}=0.6$V (current saturation region) and the gate voltage V_G are adjusted to keep the sheet electron density at the bottleneck point equal to $N_s=7\times10^{12}\,\text{cm}^{-2}$.

It should be noted that the average electron velocity increases along the channel as approaching the bottleneck point, and the injection velocity is almost independent of L_{ch}. This is due to the fact that the electrons with negative velocities, which represent electrons reflected by the potential barrier, vanish away upon approaching the bottleneck point, as schematically shown in Figure 3.5.

The position of the bottleneck point in the ballistic MOSFETs is almost unchanged, with L_{ch} as shown in Figure 3.4, and v_{inj} is found to be independent of L_{ch}. Figure 3.6 shows the computed momentum distribution function for the ballistic MOSFET with $L_{ch}=30$ nm, where the channel region extends from $y=10$–40 nm. The thick solid line indicates the distribution function at the bottleneck point. It is found that the distribution function, having negative velocity, rapidly decays in a few nanometers at the source-end of the channel. As a result, an asymmetric distribution function is formed at the bottleneck point and, thus, the average electron velocity increases as mentioned above.

To examine the shape of Figure 3.6 more closely, the distribution functions at the inside of the channel and at the bottleneck point are shown in Figures 3.7 (a) and (b), respectively.

The open circles in Figures 3.7(a) and (b) represent the Fermi-Dirac function and hemi-Fermi-Dirac function, integrated with respect to k_z, respectively, which are calculated by using the following equation:

$$\begin{aligned} f_{FD}\left(k_y\right) &= \sum_{v=1}^{6}\sum_{n_x=1}^{20}\frac{1}{\pi}\int dk_z\frac{1}{1+\exp\left\{\left[E_x^{n_x,v}+E_y\left(k_y\right)+E_z\left(k_z\right)-e\phi_{FS}\right]/k_BT\right\}} \\ &= \sum_{v=1}^{6}\sum_{n_x=1}^{20}\frac{\sqrt{2m_z^v k_BT}}{\pi\hbar}F_{-1/2}\left(\left[e\phi_{FS}-E_x^{n_x,v}-E_y\left(k_y\right)\right]/k_BT\right) \end{aligned} \quad (3.15)$$

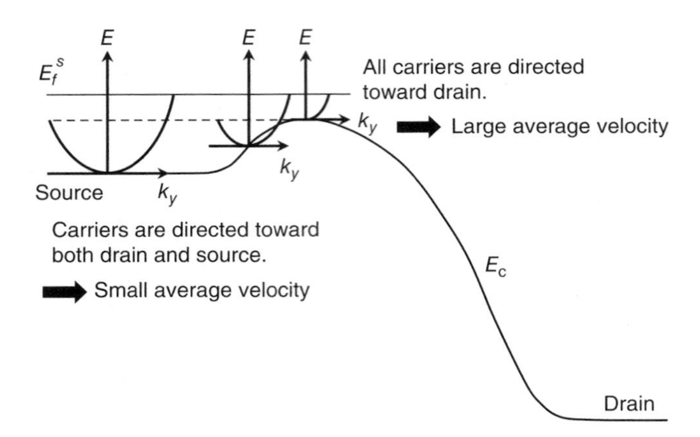

Figure 3.5 Schematic diagram explaining the increase in average electron velocity at the source-end of the channel.

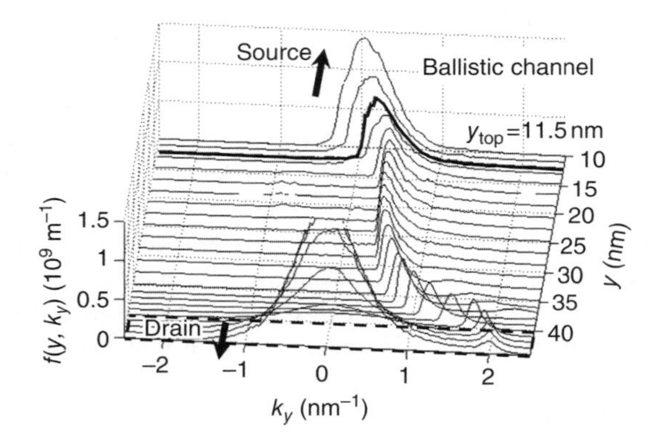

Figure 3.6 Distribution function in momentum space computed for ballistic MOSFET with $L_{ch} = 30$ nm. The channel region extends from $y = 10$–40 nm. The bottleneck point is located at $y = 11.5$ nm.

where $F_{-1/2}$ denotes the –1/2-order Fermi-Dirac integral represented by the Fermi energy $e\phi_{FS}$, the quantized energy levels $E_x^{n_x,v}$, and the free kinetic energy in the y-direction $E_y(k_y)$. $e\phi_{FS}$ and $E_x^{n_x,v}$ were obtained by solving the 1-D Schrödinger-Poisson equations at the corresponding sheet electron density. An analytical model for perfectly ballistic MOSFETs discussed below usually assumes a hemi-Fermi-Dirac function at the bottleneck point [3.1, 3.14].

It is found from Figure 3.7(a) that the distribution function is quasi-symmetric inside the source, because of the thermalization due to the scattering. On the other hand, the significantly asymmetric distribution function without negative momentum component is formed at the bottleneck point, as shown in Figure 3.7(b). Note that the asymmetric distribution function obtained from the MC simulation agrees well with the hemi-Fermi-Dirac function denoted by the open circles. The discrepancy from a discontinuous shape at zero momentum is considered to be due to a non-negligible driving force at the bottleneck point shown in the inset of Figure 3.7(c), which is unavoidable in the finite difference scheme used in the MC simulation.

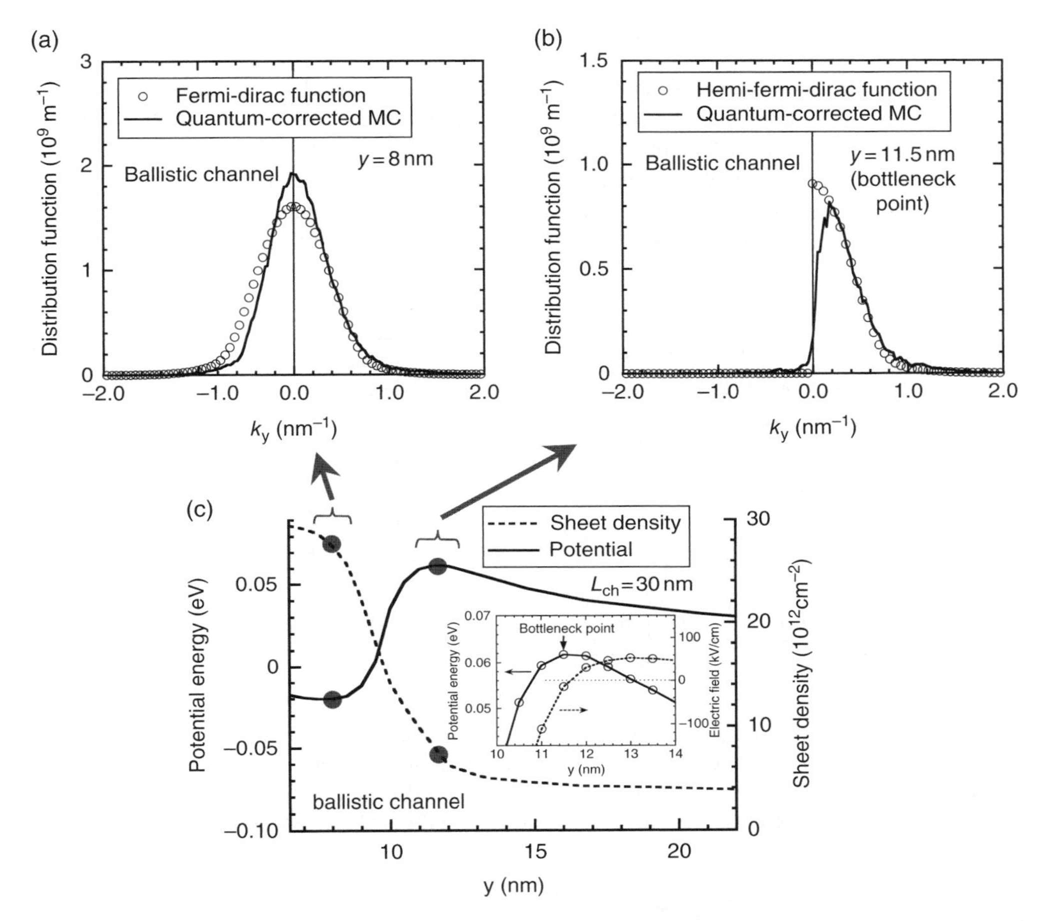

Figure 3.7 Distribution functions at (a) $y=8$; and (b) 11.5 nm of Figure 3.6. The open circles in (a) and (b) denote the Fermi-Dirac function and hemi-Fermi-Dirac function, respectively. (c) represents the magnified profiles of the corresponding potential and sheet electron density, where the inset indicates the electric field distribution with spatial mesh points; denoted by the open circles.

We also compared with an injection velocity estimated from the analytical model for perfectly ballistic MOSFETs proposed by Natori [3.1, 3.14]. The horizontal dashed line in Figure 3.4(b) denotes the injection velocity calculated using the Natori model, based on the hemi-Fermi-Dirac function, where we considered the 20 quantized energy levels from the lowest one for both the two-fold and four-fold valleys. It is found that the injection velocity from the analytical model $v_{inj}^{analytical}$ corresponds well to the ones obtained by using the quantum-corrected MC simulation. The above results mean that the analytical model based on the hermi-Fermi-Dirac function is valid for the estimation of injection velocity and saturation current in perfectly ballistic MOSFETs. Incidentally, if the quantum corrections are not included, the calculated injection velocity will be underestimated for the same electron density, because the dominant electron transport in the two-fold valleys with a higher mobility along the channel is not taken into account.

3.1.4 Quasi-ballistic Transport

Next, we discuss the influence of scattering. Figure 3.8 shows the electron transport properties computed for real MOSFETs with various channel lengths, where "real" means that the scattering in the channel is considered. The bias conditions are the same as in the previous section.

First, it is found that the average electron velocity at the bottleneck point v_s depends on the channel length for the real MOSFETs. In other words, as shown Figure 3.9, v_s increases as L_{ch} reduces, where the solid and the open circles represent the results for the real and the ballistic

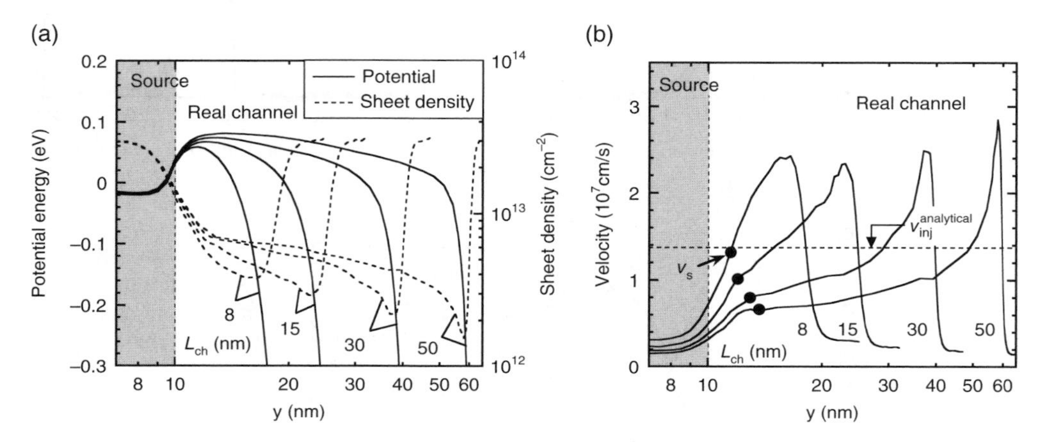

Figure 3.8 (a) Potential and sheet electron density; and (b) average electron velocity profiles along the channel, computed for real DG-MOSFETs with various channel lengths, where "real" means that scattering in the channel is considered. The bias conditions are the same as in Figure 3.4. The horizontal dashed line represents the injection velocity calculated using the analytical model.

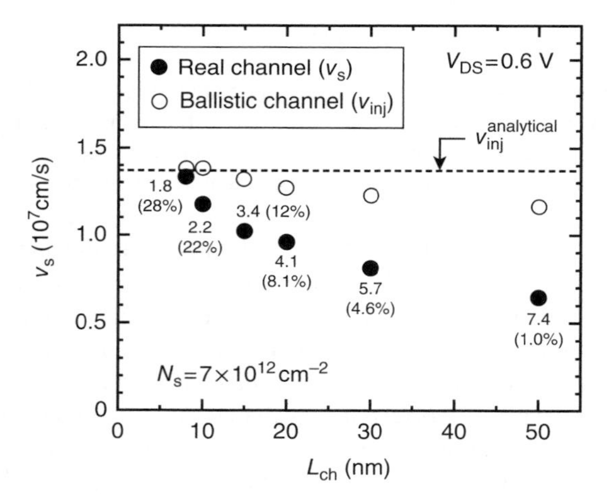

Figure 3.9 Channel length dependences of average electron velocity at the bottleneck point, where the solid and open circles represent results for real and ballistic MOSFETs, respectively. The numbers denote the average frequencies of scattering events in the channel (percentage of ballistic electrons). Note that scattering due to the short-range Coulomb interaction using MD routine is not included in these statistics.

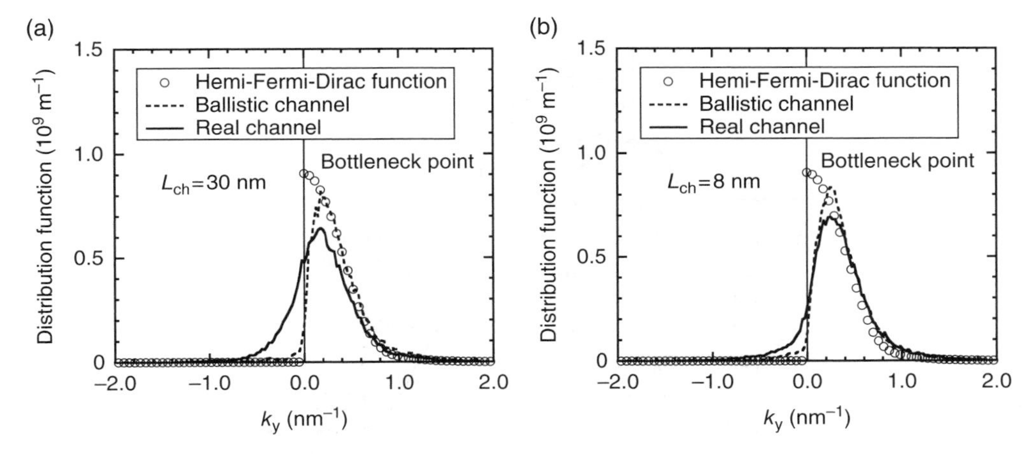

Figure 3.10 Role of scattering in the form of distribution function at the bottleneck point, where the channel length is (a) 30, and (b) 8 nm. The solid and dashed lines correspond to the results for real and ballistic MOSFETs, respectively. The open circles denote hemi-Fermi-Dirac functions.

MOSFETs, respectively. The numbers denote the average frequencies of scattering events that electrons encounter during their flight through the channel region (percentage of ballistic electrons). It is clearly found that, when the channel length becomes shorter than 30 nm, the average electron velocity at the source end of the channel enhances according to the increase in ballistic electrons, then it approaches the ballistic limit in the sub-10 nm regime.

The role of backscattering in the form of distribution function at the bottleneck point is shown in Figure 3.10.

It should be emphasized that the increase in the negative momentum component represents the presence of backward channel current due to the scattering, whereas the decrease in the positive component is to keep the electron density induced by the gate bias voltage, which explains the qualitative roles of the numerator and the denominator in Equation (3.1), respectively. This may be simply called the thermalization effects of electrons by scattering at the source-end of the channel. As a result, the distribution function broadens, and the average electron velocity decreases. We notice that the thermalization by scattering becomes less significant for $L_{ch} = 8$ nm.

3.1.5 *Role of Elastic and Inelastic Phonon Scattering*

Figure 3.11 shows the characteristics of elastic and inelastic phonon scattering related to carrier transport. Carriers are injected into the channel with kinetic energy on the order of the thermal energy kT, and they suffer elastic and inelastic scattering events in the channel. The total kinetic energy of a carrier is conserved before and after elastic scattering, whereas it is significantly changed by inelastic scattering via phonon absorption and emission processes.

In Figure 3.11, the inelastic phonon energies $\hbar\omega$s and their deformation potentials Ds are indicated [3.12]. As is well known, the g-longitudinal optical (LO) phonon with $\hbar\omega = 61$ meV is dominant. As a consequence, the inelastic phonon emission is suppressed at the beginning of the channel, because the carrier energy is initially too small. In addition, phonon absorption is rare at ordinary temperatures, so that elastic scattering (such as that of acoustic phonon)

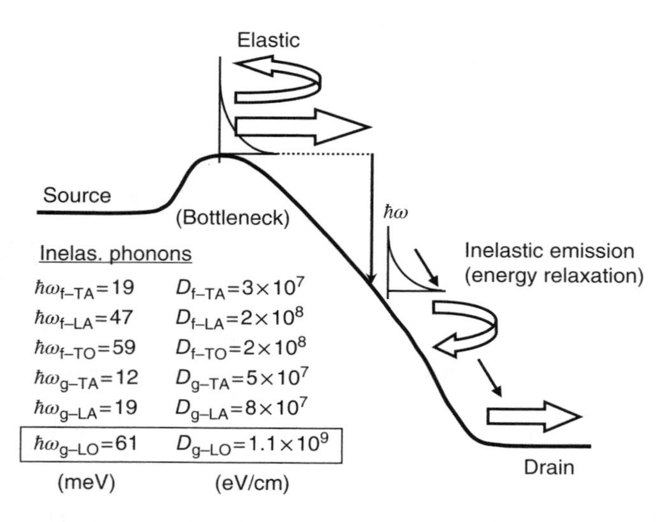

Figure 3.11 Characteristics of elastic and inelastic phonon scattering on carrier transport. The inelastic phonon energies $\hbar\omega$s and deformation potentials Ds are also indicated.

dominates at the beginning of the channel and occasionally returns carriers back to the source. If a carrier survives this region, it is subsequently exposed to frequent inelastic phonon emission, and immediately loses energy in multiples of kT (mostly $\hbar\omega_{g\text{-}LO}=61$ meV). Because the carrier then has little chance of returning to the source, and is eventually absorbed into the drain, the current is reduced from its ballistic limit value only due to the backscattering at the beginning of the channel.

Based on these ideas, Natori has suggested that the inelastic emission process suppresses backscattering and improves the ballisticity [3.14–3.17]. This interesting phenomenon is important for the practical design of ballistic MOSFETs, so we quantitatively examine Natori's prediction using the quantum-corrected MC device simulation [3.18].

In this study, we consider only phonon scattering, in order to confirm Natori's prediction. Surface roughness scattering is not included. We performed MC simulations under the following different scattering conditions in the channel region:

1. no scattering processes
2. only elastic processes
3. elastic + inelastic emission processes
4. elastic + inelastic emission and absorption processes

Elastic processes represent intravalley AP scattering, whereas inelastic processes are IP scattering, due to the inelastic phonons shown in Figure 3.11. Cases **1** and **4** correspond to a ballistic channel and a real channel, respectively. In the source and drain electrodes, all scattering processes (including electron-plasmon scattering [3.5]) are considered to activate the rapid decay of hot electrons. In particular, electron-plasmon scattering ensures that hot electrons entered into the drain relax to an equilibrium state in a very short distance [3.19]. The redistribution of electric charges, and of the electrostatic potential due to each scattering mechanism, was examined by solving the 2-D Poisson equation self-consistently. The device

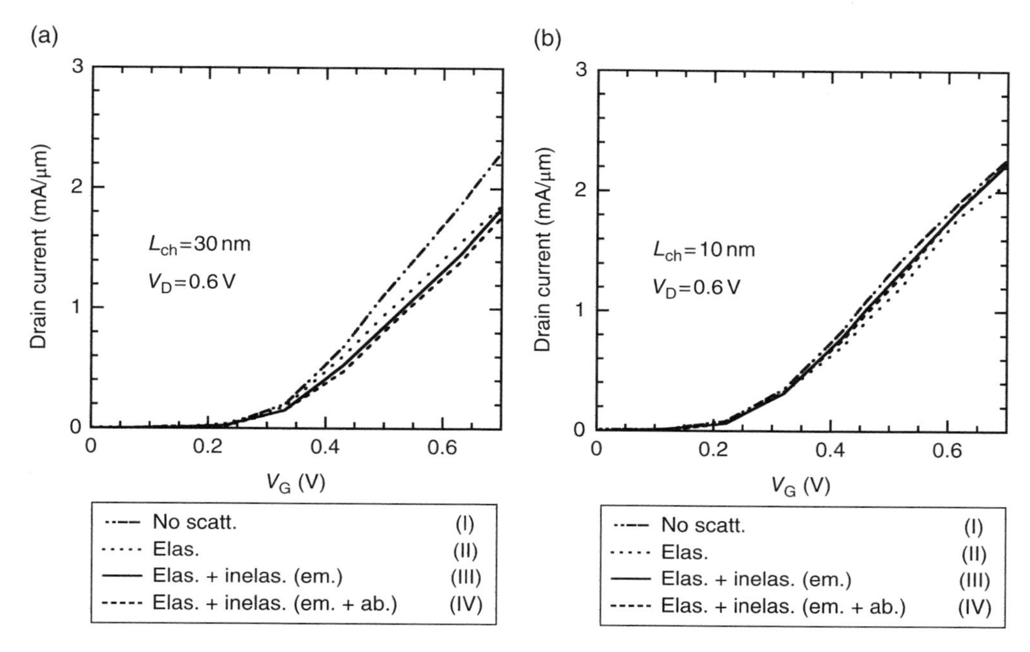

Figure 3.12 $I_D - V_G$ characteristics computed for (a) L_{ch} = 30 and (b) 10 nm, where the threshold voltage is 0.3 V for no-scattering cases. V_D = 0.6 V.

structure shown in Figure 3.2, which employs DG and UTB structures, is used in this study. The channel length is taken as 30 and 10 nm, and the temperature is 300 K.

Figure 3.12 shows $I_D - V_G$ characteristics computed for L_{ch} = 30 and 10 nm, where the threshold voltage is fixed at 0.3 V for the no-scattering cases. The drain voltage V_D is 0.6 V. The drain current for L_{ch} = 30 nm decreases monotonically as each scattering mechanism is added, as shown in Figure 3.12(a). On the other hand, as seen in Figure 3.12(b), the drain current for L_{ch} = 10 nm slightly increases when the inelastic emission processes are included.

A similar drain current increase, due to inelastic emission process, has also been reported for carbon nanotube MOSFETs under low-V_G conditions [3.20]. However, in the latter case, the inelastic process significantly decreases the drain current when the gate voltage exceeds a well-defined threshold. Such a drain current decrease for large V_G is not observed in Figure 3.12(b) but, instead, the drain current approaches the ballistic limit. This may be because the 3-D carrier transport is simulated here with a quantum correction.

We computed the average electron velocity profiles, as shown in Figure 3.13. The lower portion of each figure summarizes the source-end velocities v_s, defined as the electron velocity at the bottleneck point. It is found that the elastic and inelastic absorption processes decrease v_s for both channel lengths.

On the other hand, *the inelastic emission processes* have the opposite effect on v_s for the longer channel length. That is, the inelastic emission process increases v_s for L_{ch} = 10 nm, which is in agreement with Natori's prediction [3.16, 3.17], whereas it decreases v_s for L_{ch} = 30 nm. To further probe the dependence on inelastic emission process, we examined variations in the potential profiles, as shown in Figure 3.14 for L_{ch} = 30 and 10 nm. The plotted profiles are averaged over the electron density in the vertical direction. The effects on local

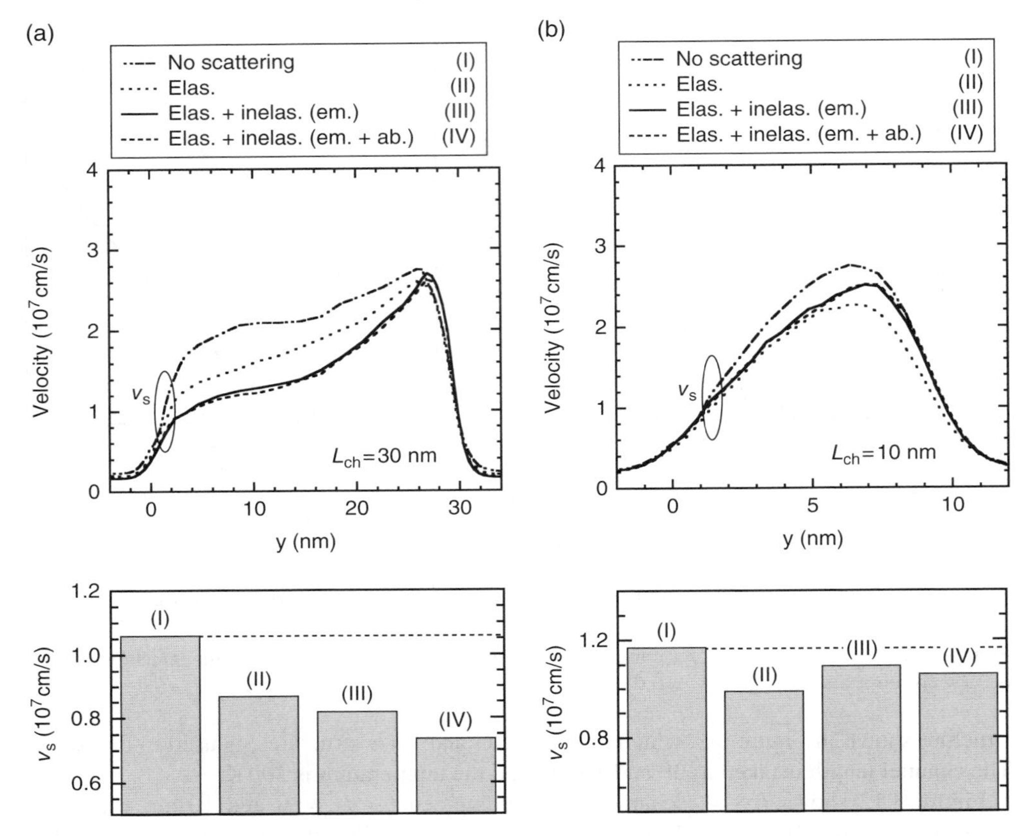

Figure 3.13 Average electron velocity profiles computed for (a) $L_{ch}=30$ and (b) 10 nm. $V_G=0.5$ V and $V_D=0.6$ V. The lower figures summarize the source-end velocities v_s.

phonon emission in the drain side of the channel are also plotted, to clarify the influence of charge accumulation during the inelastic emission processes.

For $L_{ch}=30$ nm, the bottleneck barrier is found to broaden, due to the accumulated charges, even due to the local phonon emission in the drain side of the channel ($y=20$–30 nm). Therefore, the inelastic phonon emission influences the carrier transport via the modulation of the source-end potential profile in a long-channel device and, consequently, the drain current decreases.

On the other hand, the source-end potential profile hardly changes, due to scattering for the shorter channel device, as shown in Figure 3.14(b), implying that the charge accumulation has a negligible influence on the potential in ultrashort-channel MOSFETs. Consequently, the suppression of backscattering due to inelastic phonon emission can give rise to a ballistic current in the short channel device. It is likely that the different dependence of inelastic phonon emission on L_{ch} results from the different lengths of the bottleneck barrier. Therefore, it may be possible to enhance the ballisticity by artificially reducing the bottleneck barrier length, as demonstrated in [3.18].

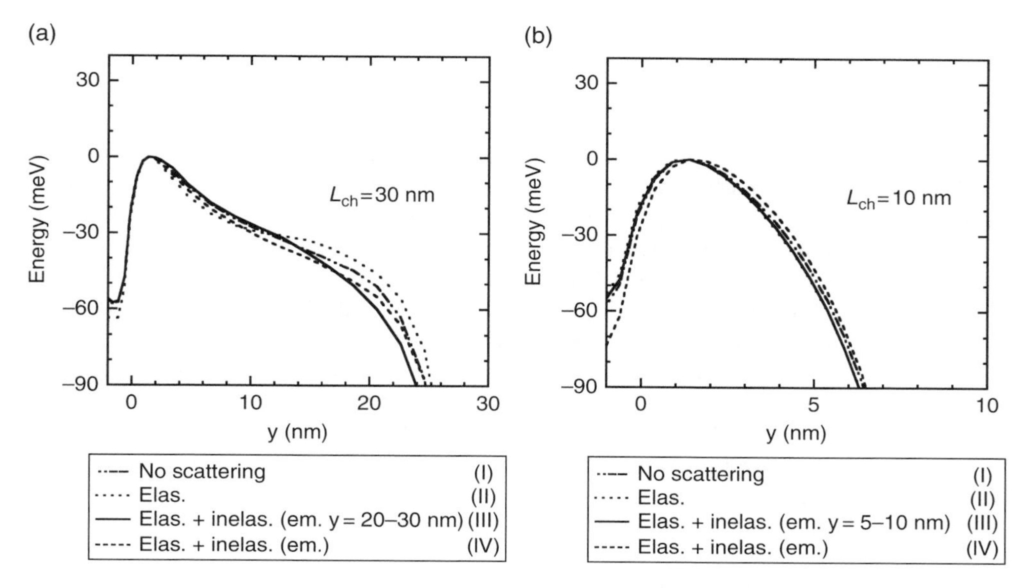

Figure 3.14 Potential profile variations due to elastic and inelastic scattering processes, where L_{ch} = (a) 30 nm; and (b) 10 nm. The effects of local phonon emission in the drain side of the channel are also plotted, to clarify the influence of charge accumulation during the inelastic emission process.

3.2 Multi-sub-band Monte Carlo Simulation Considering Quantum Confinement in Inversion Layers

As discussed in Section 3.1, ballistic transport has been expected to boost the device performance of MOSFETs. In the quasi-ballistic transport regime, the drain saturation current is expressed using the injection velocity v_{inj}, the backward channel velocity v_{back} and the backscattering coefficient R, instead of conventional mobility and saturation velocity, as follows [3.1]:

$$I_{sat} = Qv_{inj} \times \frac{1-R}{1+R\left(v_{inj} / v_{back}\right)} \tag{3.16}$$

On the other hand, UTB-SOI, DG, Fin, and GAA nanowire MOSFETs with an intrinsic channel shown in Figure 1.3 are considered to have an advantage for achieving ballistic transport, since impurity scattering is absent in the channel region [3.14]. However, such ultrathin channel devices suffer from increased AP scattering [3.21, 3.22] and a new type of SR scattering, caused by spatial fluctuation of quantized sub-bands, as shown in Figure 1.4 [3.23, 3.24]. Because realistic and atomistic treatment of scattering processes is required, an accurate calculation of R is usually nontrivial. In this section, we present an extraction of quasi-ballistic transport parameters, using a multi-sub-band MC (MSB-MC) simulation, considering AP, IP, II and SR scatterings for two-dimensional electron gas (2-DEG) in inversion layers, where the channel-thickness-dependent deformation potential D_{ac} in AP scattering and the contribution from spatial fluctuation of quantized sub-bands in SR scattering were considered [3.25, 3.26]. Figure 3.15 shows an algorithm of MSB-MC simulation and its application to a DG-MOSFET.

In the MSB-MC simulation, the Schrödinger-Poisson solver and MC simulator are self-consistently solved. Specifically, the device is meshed into vertical slices, as shown in Figure 3.15(b), and the 1D Schrödinger equations are solved to compute sub-band levels and wavefunctions in each slice. Then, by connecting the obtained sub-band levels along the channel direction, so that a potential energy distribution $U(x)$ can be formed, carrier transport is simulated by the MSB-MC techniques separately for different sub-bands in each of three pairs of the conduction valleys, $v = 1$–2, 3–4, and 5–6, as shown in Figure 3.2(b), along the channel direction. This so-called mode-space approximation largely reduces the computational resources, and its validity for the present uniform channel configuration has been confirmed previously [3.27].

The developed MC simulator is validated by comparisons with electron mobilities calculated for bulk MOS and ultrathin SOI MOSFETs. In our approach, quasi-ballistic transport parameters are derived using particle-based MC techniques – that is, we carefully monitored particle trajectories crossing over the bottleneck point from the source to the channel and vice versa, as shown in Figure 3.1, and collected necessary information for evaluating the quasi-ballistic transport parameters. We further discuss the gate and drain bias voltage dependencies of the backscattering coefficient [3.26].

3.2.1 Scattering Rates for 2-D Electron Gas

Since quasi-ballistic transport is related to carrier mobility in the on-state, an accurate modeling of scattering mechanisms is indispensable. In this section, we consider bulk, SOI, and DG MOSFETs. In all cases, a triangular potential well or a quantum well is formed within the channel, and electrons are strongly confined to behave as a 2-DEG. To describe the electron transport of 2-DEG in inversion layers, we employed the self-consistent multi-sub-band MC approach, in which the Schrödinger-Poisson solver and MC simulator are self-consistently solved. Scattering processes considered are AP, OP, II, and SR scatterings, whose scattering rates are formulated for 2-DEG, as described below. Note that nonparabolicity of the conduction band is also taken into account.

When the confinement direction is set as z, the elastic intravalley AP scattering rate we used is expressed as:

$$\Gamma_{\mathrm{ac}} = \frac{D_{\mathrm{ac}}^2 k_{\mathrm{B}} T \sqrt{m_x m_y}}{2\hbar^3 \rho v^2} \left(1 + 2\alpha E'\right) \int_0^{T_{\mathrm{Si}}} \varphi_m^2(z) \varphi_n^2(z) dz \tag{3.17}$$

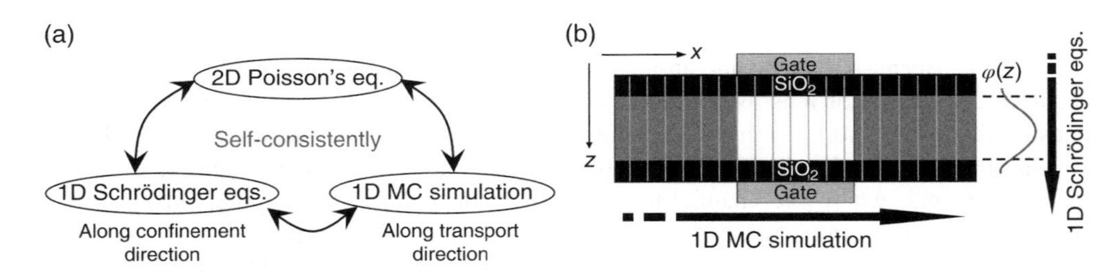

Figure 3.15 MSB-MC method. (a) Algorithm of MSB-MC simulation; (b) its application to DG MOSFET.

where D_{ac} is the deformation potential of AP, whose treatment will be discussed in Section 3.2.2. α is a nonparabolicity parameter and set as $0.5\,\mathrm{eV}^{-1}$. E' is the kinetic energy of the final state, and $\varphi_m(z)$ and $\varphi_n(z)$ are electron wave functions of the initial and final states along the confinement direction, respectively. T_{Si} is the channel thickness, and the Si/SiO_2 interfaces are set at $z=0$ and T_{Si}. The other parameters are the same as those defined in Sections 3.1.1 and 3.1.2.

The inelastic IP scattering rate is expressed as:

$$\Gamma_{\text{inelastic}} = \frac{N_v D_{\text{op}}^2 \sqrt{m_x m_y}}{2\hbar^2 \rho \omega_{\text{op}}} \left(1+2\alpha E'\right) \left[N_q + \frac{1}{2} + \sigma \right] \int_0^{T_{\text{Si}}} \varphi_m^2(z) \varphi_n^2(z)\, dz \tag{3.18}$$

where σ is given as 1/2 for phonon emission and –1/2 for phonon absorption. For the inelastic IP scattering, we adopted the values shown in Table 3.1 [3.12].

The expression we employed for the II scattering rate is shown below. This was reported to express experimental mobility well over a wide range of electron densities [3.28, 3.29]:

$$\Gamma_{\text{imp}} = \frac{e^4 \sqrt{m_x m_y} \left(1+2\alpha E'\right)}{8\pi \hbar^3 \varepsilon^2} \int_0^{2\pi} d\theta \left\{ \int_0^{T_{\text{Si}}} dz_0 N_{\text{imp}}(z_0) \left| \int_0^{T_{\text{Si}}} dz \frac{\varphi_m(z) e^{-\Delta k(\theta)|z-z_0|} \varphi_n(z)}{\Delta k(\theta) + Q_{\text{scr}}(\Delta k)} \right|^2 \right\} \tag{3.19}$$

where θ is the scattering angle between the initial and final wave vectors $\mathbf{k}$ and $\mathbf{k}'$, respectively, $\Delta k = |\mathbf{k} - \mathbf{k}'|$, ε is the dielectric permittivity of Si and assumed to be $11.7\,\varepsilon_0$, $N_{imp}(z_0)$ is the impurity density (donor or acceptor concentration) at position z_0, and Q_{scr} is the temperature-dependent screening function given as: [3.29, 3.30]

$$Q_{\text{scr}}(\Delta k) = \frac{e^2 n_{\text{scr}}}{2\varepsilon k_B T} g_1\left(\frac{\Delta k \lambda_{\text{th}}}{4\sqrt{\pi}}\right) \int_0^{T_{\text{Si}}} \int_0^{T_{\text{Si}}} dz' dz\, \varphi_1^2(z) \varphi_1^2(z') e^{-\Delta k|z-z'|} \tag{3.20}$$

where n_{scr} is the sheet electron density of the lowest sub-band, which contributes to the screening, λ_{th} stands for the thermal wavelength, and the function $g_1(x)$ is defined as:

$$g_1(x) = \frac{1}{x} e^{-x} \int_0^x e^{t^2} dt \tag{3.21}$$

which is derived from the real part of the plasma dispersion function [3.30].

As for SR scattering, we introduced the so-called Prange-Nee terms [3.31]. Although this approach is simple and further advanced approaches are available [3.32–3.34], we considered the approach accurate enough to express SR scattering in SOI devices, since it can include both scatterings, due to the confining electric field at the Si/SiO_2 interfaces and spatial fluctuation of quantized sub-bands. We used an exponential spectrum with root-mean-square $\Delta = 0.5\,\text{nm}$ and correlation length $\Lambda = 1.0\,\text{nm}$ for the roughness spectrum. The expression of the SR scattering rate is given as:

$$\Gamma_{\text{sr}} = \frac{\left(1+2\alpha E'\right)\sqrt{m_x m_y}}{\pi \hbar^3} \int_0^{2\pi} d\beta \frac{\pi \Delta_m^2 \Lambda^2}{\left(1+|\Delta k|^2 \Lambda^2/2\right)^{3/2}} \left| \frac{\hbar^2}{2m_z} \frac{d\varphi_n(0)}{dz} \frac{d\varphi_m(0)}{dz} \right|^2 \tag{3.22}$$

where $(\hbar^2/2m_z)(d\varphi_n(0)/dz)(d\varphi_m(0)/dz)$ is the Prange-Nee term [3.31]. When an SOI or DG structure is considered, SR scattering at the back interface must also be included. In this case, the Prange-Nee term for the back interface, i.e., $(\hbar^2/2m_z)(d\varphi_n(T_{\text{Si}})/dz)(d\varphi_m(T_{\text{Si}})/dz)$, is added to Equation (3.22).

3.2.2 *Increase in* D_{ac} *for SOI MOSFETs*

D_{ac} is an important parameter for characterizing AP scattering. However, D_{ac} has a fundamentally anisotropic nature and, in addition, D_{ac} for 2-DEG exhibits kinetic energy dependence and is, strictly speaking, influenced by the screening effect [3.35–3.37]. Therefore, an exact treatment of D_{ac} for 2-DEG becomes very complicated; in many, cases D_{ac} is assumed to be isotropic and independent of energy for simplicity [3.12, 3.38, 3.39]. The *constant* D_{ac}, which agrees with the experimental bulk Si mobility, is in the 9.0–9.5 eV range [3.12, 3.38, 3.39], but these values cannot replicate the universal curve for bulk MOS inversion layer mobility for the aforementioned reasons, and thus $D_{\text{ac}} = 12$–13 eV [3.39, 3.40] is widely used in MOS simulations (cD_{ac} model).

However, in SOI MOSFETs, the use of $D_{\text{ac}} = 12$–13 eV results in the overestimation of electron mobility, suggesting that the effective isotropic D_{ac} is even higher than 13 eV in SOI channels, and may exhibit SOI thickness (T_{Si}) dependence. In this study, we employed a spatially *variable* D_{ac} model (vD_{ac} model), which has been proposed by Ohashi and coworkers [3.21, 3.22], to consider T_{Si}-dependent D_{ac}. In the vD_{ac} model, D_{ac} is treated as spatially variable – that is, it sharply increases at MOS interfaces. This is shown in Figure 3.16, which has been demonstrated to well reproduce the experimental mobility in SOI MOSFETs [3.21]. Note that we must include the effects from both the front and back interfaces for SOI and DG structures; hence, the average D_{ac} inside the Si channel increases compared with that in the bulk MOS structure.

For the bulk MOS structure depicted in Figure 3.16(a), the following expression has been proposed [3.21, 3.22]:

$$D_{\text{ac}}(z) = \Delta D_{\text{ac}} \exp(-z / L_{\text{ac}}) + D_{\text{ac}}^{\min} \tag{3.23}$$

$D_{\text{ac}}^{\min}$ represents the bulk deformation potential (9.0 eV), and ΔD_{ac} is defined as 7.0 eV. $L_{\text{ac}} = 2.5$ nm is a damping factor and z is a position measured from the Si/SiO_2 interface along the confinement direction. For the SOI or DG structure simulation, D_{ac} is calculated using the following equation:

$$D_{\text{ac}}(z) = \Delta D_{\text{ac}} \exp(-z / L_{\text{ac}}) + \Delta D_{\text{ac}} \exp\{(z - T_{\text{Si}}) / L_{\text{ac}}\} + D_{\text{ac}}^{\min} \tag{3.24}$$

In the actual calculation of AP scattering rate (3.17), we introduced an expectation value of vD_{ac} using electron wave function $\varphi_m(z)$ for each sub-band as follows:

$$vD_{\text{ac}} = \int_0^{T_{\text{Si}}} \frac{\varphi_m^*(z) D_{\text{ac}} \varphi_m(z)}{\varphi_m^*(z)\varphi_m(z)} dz \tag{3.25}$$

We have implemented both cD_{ac} and vD_{ac} models in the calculation of AP scattering rate, and then calculated the electron mobilities in bulk Si and SOI MOSFETs.

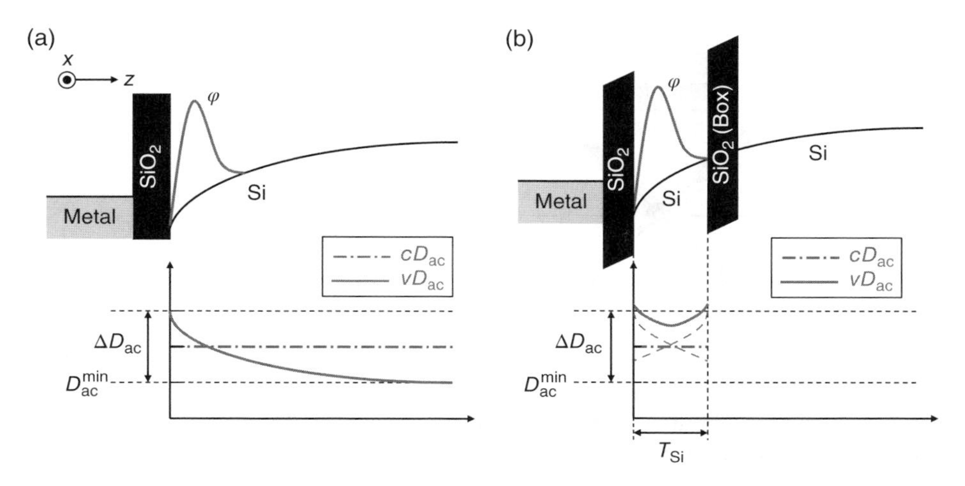

Figure 3.16 Constant D_{ac} (cD_{ac}) and variable D_{ac} (vD_{ac}) models for (a) bulk MOS; and (b) SOI MOS structures. Drastic increase in D_{ac} near the interfaces is depicted. We used expressions for $D_{ac}(z)$ proposed in References 3.21 and 3.22. Note that we must include effects from both the front and back interfaces for SOI and DG structures, so the averaged D_{ac} inside the Si channel increases, compared with that in the bulk MOS structure.

3.2.3 Simulated Electron Mobilities in Bulk Si and SOI MOSFETs

To confirm the validity of our MC simulator, we have calculated the electron mobility in bulk Si MOS and SOI MOS structures, as shown in Figure 3.16, and made comparisons with experimental results. Note that the scattering processes considered in this section are AP, IP, and SR scatterings, while II scattering is not considered, because the effect of impurity scattering is insignificant unless the effective normal electric field (E_{eff}) is sufficiently low. A (001) surface is assumed and the transport direction is set as <110>. Accordingly, the conduction band structure is split into two-fold valleys with a high electron mobility and four-fold valleys with a low one, as shown in Figure 3.17.

First, Figure 3.18 shows the calculated electron mobility in the bulk MOS inversion layer versus E_{eff}.

As pointed out in Ref. 3.41, the mobility in the medium-E_{eff} region is mainly determined by AP scattering and, thus, the approach using cD_{ac} (9.0 eV) obviously overestimates electron mobility, except for the high-E_{eff} region. On the other hand, it can be confirmed that both vD_{ac} and cD_{ac} with 13 eV well reproduce the experimental electron mobility in a bulk Si MOSFET, which indicates that vD_{ac} approach is as accurate as the conventional cD_{ac} with 13 eV for the bulk MOS structure [3.22].

Next, we have calculated electron mobility of SOI MOS structure with various T_{Si}, where the SOI region is assumed to be undoped, while the Si substrate is doped as $N_A = 2\times10^{16}\,cm^{-3}$. Figure 3.19 shows the T_{Si} dependence of SOI electron mobility computed for a medium electric field (E_{eff}=0.3 MV/cm) and also a low electric field (E_{eff}=0.05 MV/cm).

First, in Figure 3.19 (a), we note a significant mobility enhancement at approximately T_{Si}=3 – 4 nm in cD_{ac} (13 eV) model, which is explained by the modulation in electron

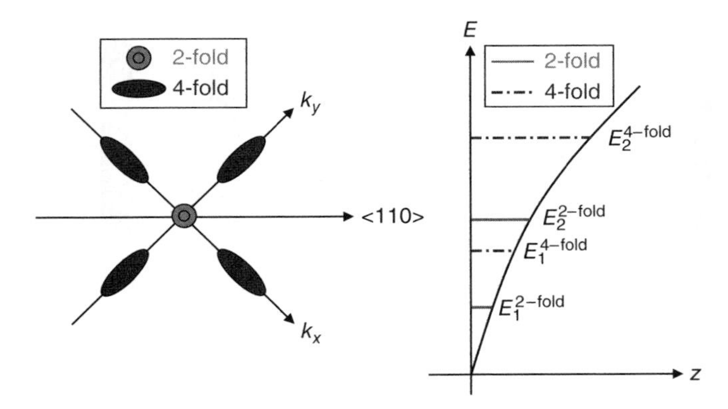

Figure 3.17 Conduction band structure of Si used in the simulation. The ellipsoidal multivalleys and its band nonparabolicity are taken into account. Typical sub-band splitting is also depicted.

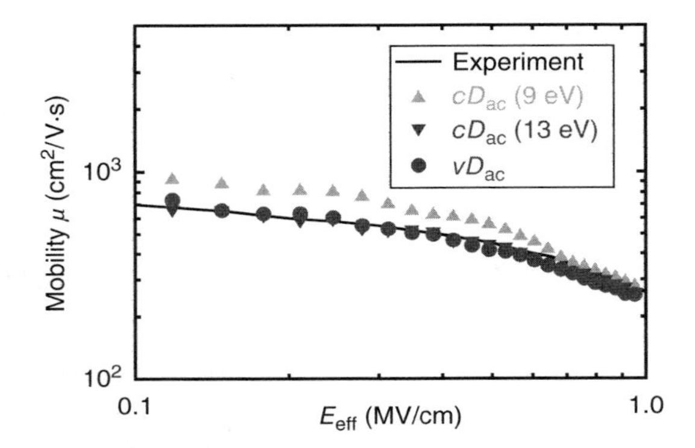

Figure 3.18 Electron mobility in bulk MOS inversion layer versus effective normal field E_{eff}. Results obtained using cD_{ac} (9.0 eV), cD_{ac} (13 eV), and vD_{ac} are plotted as triangles, reversed triangles, and circles, respectively. The experimental curve [3.41] for substrate acceptor concentration $N_A = 2 \times 10^{16}\,\text{cm}^{-3}$ is also plotted as a solid line.

occupancy of each conduction band valley – that is, when $T_{Si} < 4$ nm, most electrons occupy the two-fold valleys with a higher electron mobility, owing to the formation of a sufficiently lower quantized energy sub-band in the two-fold valleys [3.43, 3.44]. However, the vD_{ac} model indicates a slighter mobility enhancement, and it also indicates a decrease in mobility from $T_{Si} = 8$ to 6 nm, both of which tendencies are similar to those of the experiment.

The above results mean that the cD_{ac} (13 eV) model overestimates SOI electron mobility, because the increase in D_{ac} stemming from the back interface cannot be incorporated in the cD_{ac} model. Consequently, the vD_{ac} model is more accurate than the cD_{ac} model with respect to the reproducibility of experimental mobilities, especially for ultrathin body SOI MOSFETs with $T_{Si} < 8$ nm. Here, the discrepancy in SOI thickness exhibiting the mobility peak may be due to the assumption taken in the simulation – that is, an infinitely high potential barrier at the Si/SiO_2 interfaces for solving Schrödinger equations. The assumption

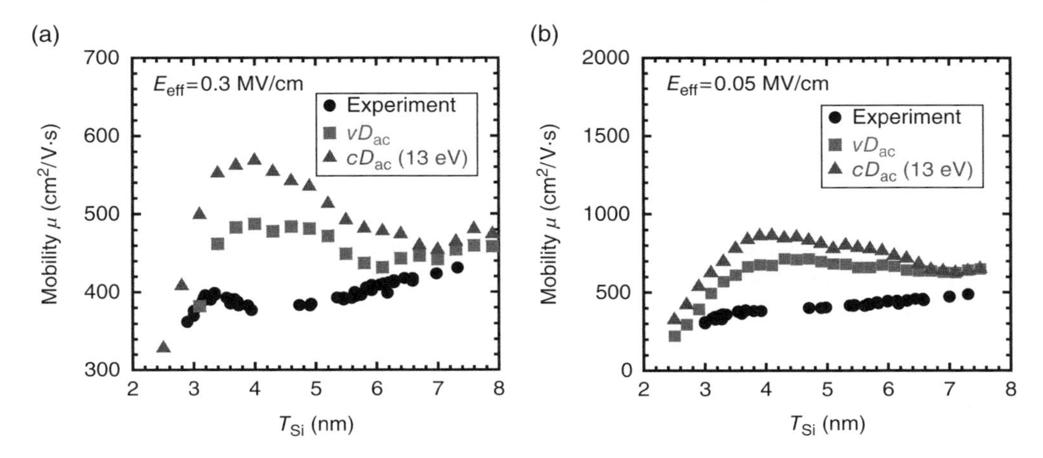

Figure 3.19 T_{Si} dependence of SOI electron mobility computed for (a) $E_{eff}=0.3$ and (b) 0.05 MV/cm. Circles, squares, and triangles represent the experimental [3.42], vD_{ac}, and cD_{ac} (13 eV) results, respectively.

should increase the eigen-energies of confined electrons, leading to an increased energy gap between split sub-bands.

Also, in a low E_{eff} region, the vD_{ac} model provides a more reasonable result than the cD_{ac} model, as shown in Figure 3.19 (b). In this case, mobility enhancement is not observed in the experiment and vD_{ac} model. However, the cD_{ac} model still predicts a clear mobility enhancement in this low E_{eff} condition, since AP scattering rate is significantly underestimated. The present results suggest that the vD_{ac} model is more suitable for precise device design with ultrathin SOI channel such as $T_{Si} < 5$ nm. On the contrary, it is also worth noting that cD_{ac} and vD_{ac} models are coincident for $T_{Si} \approx 8$ nm, indicating that the cD_{ac} model is applicable for thicker channel devices, and we will further discuss this point in the next section. We have to add that the higher electron mobility obtained in the simulation, compared with the experimental result, may be due to the neglect of Coulomb scatterings, which are known to be important scattering processes in low E_{eff} region [3.41].

3.2.4 Electrical Characteristics of Si DG-MOSFETs

In this section, the electrical characteristics of Si DG-MOSFETs are simulated using the MSB-MC method. The structure of the simulated DG-MOSFETs is shown in Figure 3.20.

As in the previous section, a (001) surface is assumed, and the transport direction is set as <110>. Note that the channel is undoped; thus, carrier backscattering in the channel is caused by phonon and SR scatterings. The gate oxide is assumed to be SiO_2, and its thickness is given as 0.5 nm. The doping concentrations of the source and drain are both set as $N_D = 1.0 \times 10^{20}$ cm^{-3}, and we also consider II scattering in the source and drain regions. In this study, to suppress the short-channel effects, the channel thickness T_{Si} was chosen by following the empirical rule $T_{Si} = L_{ch}/3$, with the channel length L_{ch} varied from 30 to 6 nm. Indeed, we confirmed that the threshold voltage (V_{th}) lowering due to L_{ch} scaling can be successfully suppressed by introducing this T_{Si} scaling, as shown in Figure 3.21, where the results obtained using $T_{Si}=L_{ch}/2$ scaling and fixed T_{Si} (=3 nm) are also plotted for comparison.

As clearly shown in this figure, the $T_{Si}=L_{ch}/3$ scaling almost completely suppresses the V_{th} lowering until $L_{ch}=6$ nm. In addition, the number of sub-bands considered in the simulation was chosen to be large enough (i.e., sub-bands with quantized energies less than 0.5 eV were all considered, because $V_D=0.5$ V was used in this study).

Next, we simulated two different channel thicknesses as $T_{Si}=6$ and 3 nm, because the effect of vD_{ac} strongly depends on T_{Si}. The simulated I_D-V_G characteristics are shown in Figure 3.22 for (a) $T_{Si}=6$ nm and $L_G=18$ nm, and (b) $T_{Si}=3$ nm and $L_G=9$ nm, where $V_D=0.5$ V. Hereafter, for simplicity, we refer to $V_G=0.5$ V as "on-state". Solid lines with ● and ■ stand for cD_{ac} (13 eV) and vD_{ac} approaches, respectively.

It is found that the drain current decrease due to vD_{ac} becomes clearer in Figure 3.22 (b), because the influence of vD_{ac} stemming from the back interface is enhanced in $T_{Si}=3$ nm. This

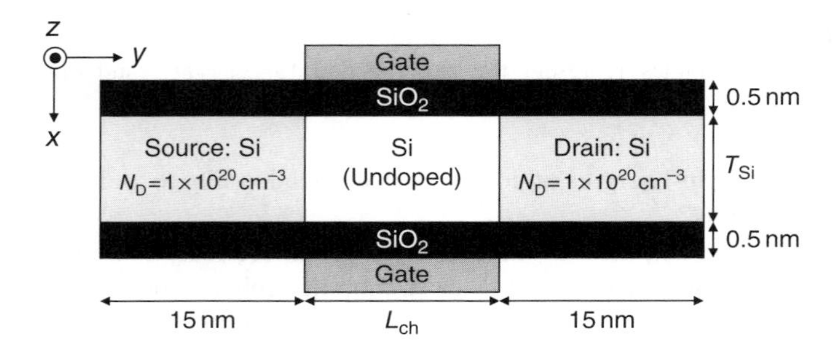

Figure 3.20 Structure of DG-MOSFET used in the simulation. Note that the channel length L_{ch} was varied from 30 to 6 nm, whereas the channel thickness T_{Si} was chosen by following the empirical rule $T_{Si}=L_{ch}/3$, which is often used as a guideline to suppress short-channel effects. The gate oxide is assumed to be SiO_2, and its thickness is given as 0.5 nm. The doping concentrations of the source and drain are both set as $N_D=1.0\times10^{20}$ cm^{-3}. A (001) surface is assumed, and the transport direction is set as <110>.

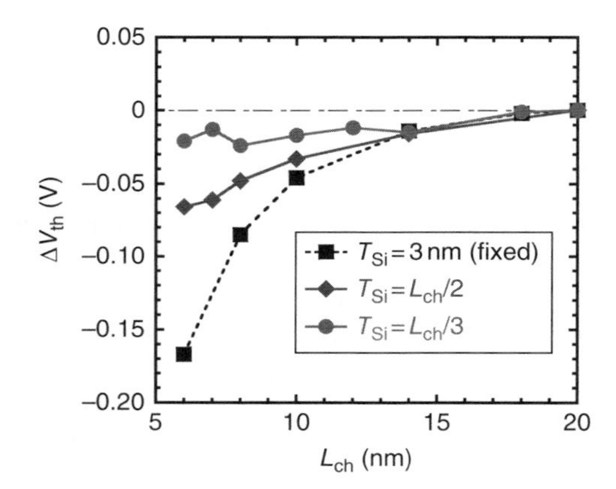

Figure 3.21 Threshold voltage lowering ΔV_{th} computed as a function of L_{ch}, where the vertical axis represents variations in V_{th}, measured from the values at $L_{ch}=20$ nm, i.e., $\Delta V_{th}=V_{th}(L_{ch})-V_{th}(20\text{ nm})$. A dashed line with ■ represents fixed T_{Si} (=3 nm), a solid line with ◆ $T_{Si}=L_{ch}/2$, and a solid line with with ● $T_{Si}=L_{ch}/3$.

can be understood by calculating the spatially averaged vD_{ac} using Equation (3.25) as a function of T_{Si}, as shown in Figure 3.23 – namely, due to increased influence of D_{ac} coming from the back interface, the averaged vD_{ac} becomes larger than that of cD_{ac} (13 eV), and increases to about 17 eV for $T_{Si}=3$ nm. However, the drain current decrease rate is only about 7% in Figure 3.22(b), which is because the IP, SR, and II scatterings in the source and drain also play important roles in practical devices. Therefore, the cD_{ac} model is considered to be applicable to the drive current analysis of ultrasmall DG-MOSFETs.

Incidentally, the black dashed line in Figure 3.23 represents an approximation formula reproducing the averaged vD_{ac} curve for the lowest sub-band, which is given in [3.25]:

$$vD_{ac} = 11.0\exp\left(-T_{Si} / L_{var}\right) + 12.0 \ (\text{eV}) \tag{3.26}$$

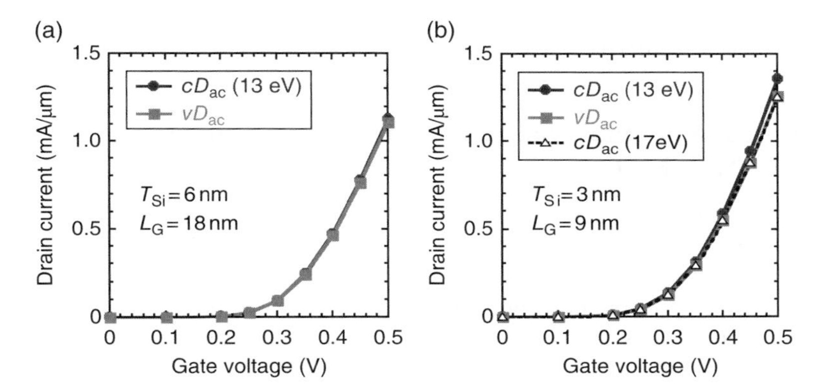

Figure 3.22 $I_D - V_G$ characteristics computed at $V_D=0.5$ V for DG-MOSFETs with (a) $T_{Si}=6$ nm and $L_{ch}=18$ nm, and (b) $T_{Si}=3$ nm and $L_{ch}=9$ nm. Solid lines with ● and ■ represent results of cD_{ac} (13 eV) and vD_{ac} models, respectively. The dashed line with Δ in (b) represents the result of cD_{ac} (17 eV) model.

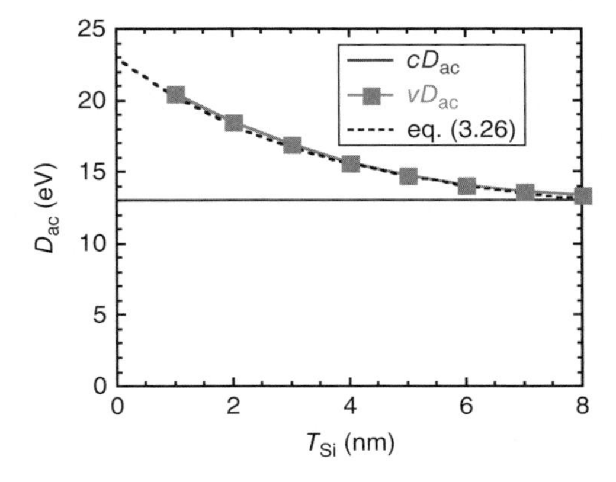

Figure 3.23 T_{Si} dependence of vD_{ac} value for $V_G=0.5$ V. Horizontal solid line and solid line with ■ represent cD_{ac} (13 eV), and the expectation value of vD_{ac} for the lowest sub-band, respectively. The black dashed line is the approximation formula for vD_{ac}, as defined in Equation (3.26).

where a decay factor L_{var} is determined to be 3.5 nm by fitting to the vD_{ac} curve, as plotted using the dashed line in Figure 3.23. Then, we can deduce an equivalent value of D_{ac} for any T_{Si} less than 8 nm, using Equation (3.26) with $L_{var}=3.5$ nm. For instance, we obtained $D_{ac}=17$ eV for $T_{Si}=3$ nm from Equation (3.26), and we actually simulated the I_D-V_G characteristics for the DG MOSFET with $T_{Si}=3$ nm by using the D_{ac} value of 17 eV, which is shown in Figure 3.22(b) with the black dashed line. It can be seen that the I_D-V_G characteristics of vD_{ac} model is completely reproduced with the equivalent value of D_{ac}, namely, 17 eV. Therefore, Equation (3.26) may be used as an analytical formula expressing T_{Si} – dependent D_{ac}.

By the way, it may be doubted if the D_{ac} value deduced from Equation (3.26) is also applicable to simulations in subthreshold region, because Equation (3.26) has been obtained by fitting to on-state property. However, scattering is shown to have negligible effect on subthreshold properties [3.45, 3.46] and, therefore, the use of equivalent D_{ac} value of Equation (3.26) would be acceptable for an efficient device simulation of ultra-scaled DG MOSFETs. In the next section, the quasi-ballistic transport in ultrasmall DG-MOSFETs is examined, using the MSB-MC approach described in Section 3.2.

3.3 Extraction of Quasi-ballistic Transport Parameters in Si DG-MOSFETs

Next, we extract quasi-ballistic transport parameters in the DG-MOSFET [3.25, 3.26]. As indicated in Figure 3.1, the backscattering coefficient R is defined as the ratio between the backward and forward channel currents. We evaluated them by directly monitoring particle trajectories crossing over the bottleneck point from the source to the channel and from the channel to the source; thus, we believe that the intrinsic values for R, v_{inj}, and v_{back} have been extracted by the present simulation.

3.3.1 *Backscattering Coefficient*

Figure 3.24 shows the simulated R as a function of L_{ch} and T_{Si}, where the results simulated using the cD_{ac} (13 eV) and vD_{ac} models are plotted.

Comparing the cD_{ac} (13 eV) results with the vD_{ac} results, the vD_{ac} model is found to predict a larger R, as expected. However, the difference between these two D_{ac} results is negligibly small in terms of the L_{ch} dependence of R, which is consistent with the results shown in Figure 3.22. Hence, we employed the cD_{ac} (13 eV) model for subsequent simulations in this section.

As for the L_{ch} dependence of R, R decreases with decreasing L_{ch} until $L_{ch}=10$ nm, which means that ballistic transport is enhanced, owing to the channel length scaling down to 10 nm. On the other hand, R drastically increases when L_{ch} becomes shorter than 10 nm. Since it is observed in ultrathin channel devices with $T_{Si}<3$ nm, the reason for this is considered to be due to the influence of SR scattering. To confirm it, the rates of AP and SR scatterings (because these elastic processes mainly cause backscattering to the source) are plotted in Figure 3.25 for $L_{ch}=10$ and 6 nm; note that they are for the lowest sub-band electrons in the two-fold valleys, which most electrons occupy.

As can be seen in Figure 3.25(b), SR scattering (or, more precisely, SR scattering caused by the spatial fluctuation of quantized sub-bands) becomes dominant in the sub-10 nm channel length with T_{Si} < 3 nm. The AP scattering rate also increases with decreasing L_{ch}, because a form factor included in the AP scattering rate becomes larger for a smaller T_{Si} [3.42]. However, since the increase in SR scattering rate is much more significant, the increase in R observed for L_{ch} < 10 nm is certainly caused by the intensified SR scattering. Furthermore, by closely examining the present results shown in Figure 3.24, we notice that R starts to decrease more steeply when L_{ch} < 14 nm (T_{Si} < 4.67 nm), which is attributed to the mobility increase, as shown in Figure 3.19 (a). This means that R has a close relationship with electron mobility defined in a diffusive transport regime; thus, a higher mobility channel would lead to a higher ballistic efficiency of nanoscale MOSFETs.

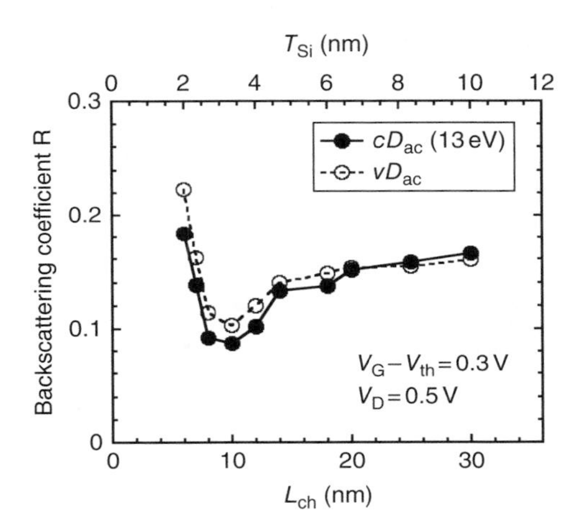

Figure 3.24 Simulated R as a function of L_{ch} and T_{Si}, where results simulated using cD_{ac} (13 eV) and vD_{ac} models are plotted. The lower and upper horizontal axes represent L_{ch} and T_{Si}, respectively.

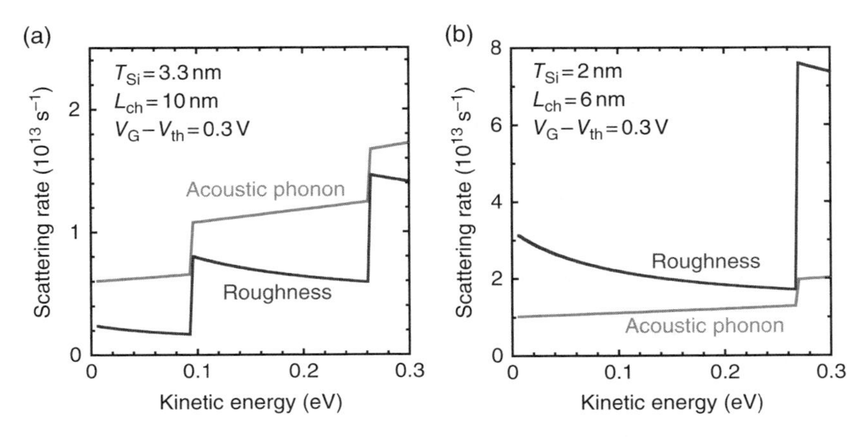

Figure 3.25 Rates of AP (cD_{ac} (13 eV) model) and SR scatterings calculated for L_{ch} = (a) 10; and (b) 6 nm. Note that these elastic processes mainly cause backscattering of carriers to the source.

3.3.2 Current Drive

Figures 3.26 and 3.27 show L_{ch} dependences of the on-current I_{ON} and the $I_D - V_G$ characteristics simulated using the MSB-MC simulator, respectively, where I_{ON} in Figure 3.26 was calculated at $V_G - V_{th} = 0.3$ V in Figure 3.27, and V_{th} is defined as the gate voltage which corresponds to $I_D = 0.01$ mA/μm.

You may already notice that I_{ON} in Figure 3.26 varies almost inversely with R shown in Figure 3.24. Therefore, the on-current increase or decrease in ultrascaled DG-MOSFETs is found basically to be determined by the quasi-ballistic transport parameter R. Accordingly, the on-current degradation in the sub-10 nm regime is confirmed, owing to the SR scattering intensified by the spatial fluctuation of quantized sub-bands. Hence, significant improvement of the quality of gate oxide interfaces is indispensable to receiving the benefits of ballistic transport in sub-10 nm DG/SOI-MOSFETs.

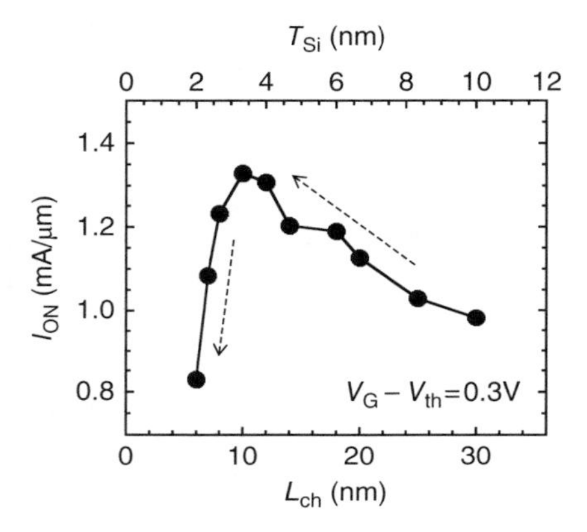

Figure 3.26 L_{ch} dependence of I_{ON}, where I_{ON} was calculated at $V_G - V_{th} = 0.3$ V of $I_D - V_G$ characteristics in Figure 3.27. I_{ON} increases with reducing L_{ch} until $L_{ch} = 10$ nm, and then, it sharply decreases in the sub-10 nm regime. This tendency is coincident with that in an inverse of R shown in Figure 3.24.

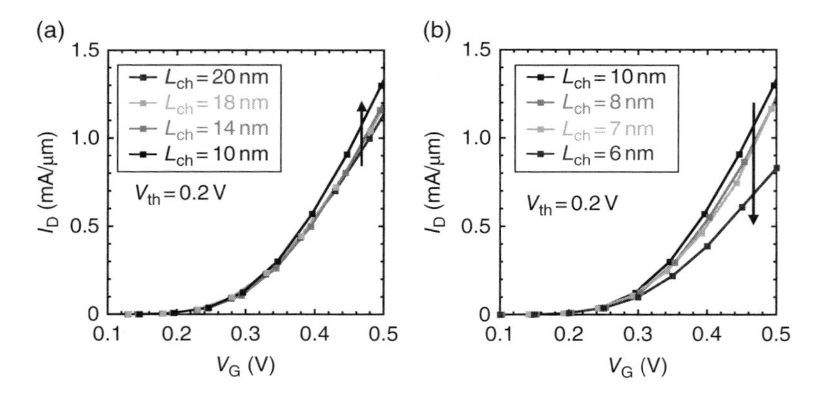

Figure 3.27 L_{ch} dependence of $I_D - V_G$ characteristics simulated using the MSB-MC simulator. V_{th} was set as 0.2 V, at which $I_D = 0.01$ mA/μm. For all gate voltages considered in the simulation, I_D increases with reducing L_{ch} until 10 nm in (a), and it turns to decrease in the sub-10 nm regime in (b).

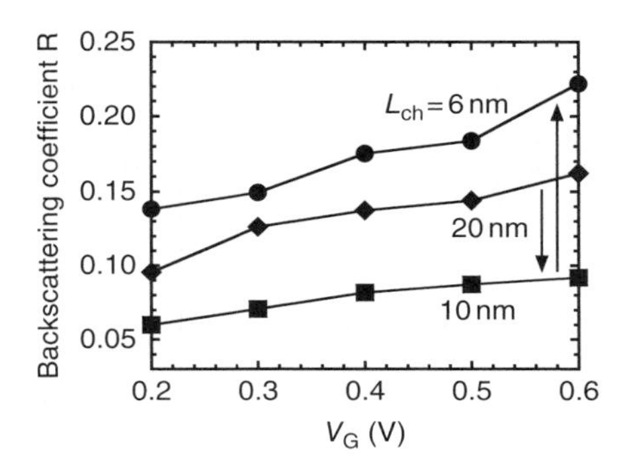

Figure 3.28 Gate voltage dependencies of R calculated for $L_{ch}=6$, 10, and 20 nm. $V_D=0.5$ V.

3.3.3 *Gate and Drain Bias Dependences*

Since the scattering rate depends on the electron kinetic energy, as expressed in Equations (3.17) to (3.22), R may be a function of the gate and drain bias voltages. In order to demonstrate that the L_{ch} dependence of R found in Figure 3.24 is unaltered by the bias conditions, we further examined the gate and drain bias voltage dependencies of R, using the MC simulator. Figure 3.28 shows the gate voltage dependencies of R calculated for three channel lengths (i.e., $L_{ch}=6$, 10, and 20 nm). In accordance with the L_{ch} dependence of R shown in Figure 3.24, the DG-MOSFET with $L_{ch}=10$ nm exhibits the smallest R for the gate voltages considered in the present simulation.

It is found that R increases with gate voltage for all channel lengths. To determine the reason, we plotted the variations due to gate voltage in the lowest sub-band energy profile and in the average electron kinetic energy profile in Figures 3.29 (a) and 3.29 (b), respectively, for $L_{ch}=10$ nm, and in Figures 3.29 (c) and 3.29 (d), respectively, for $L_{ch}=6$ nm, where $V_D=0.5$ V. Note that for both channel lengths, the average kinetic energy increases on the left side of the channel as the channel potential decreases with increasing gate voltage.

As a result of the increased average kinetic energy, phonon scattering is enhanced at the source end of the channel. It is also found that the increase in the average kinetic energy on the left side of the channel occurs similarly for both channel lengths, which means that phonon scattering is, similarly, also enhanced. In addition, SR scattering is also enhanced with gate voltage. Both of these lead to R increasing with gate voltage, as shown in Figure 3.28. However, for $L_{ch}=6$ nm, SR scattering becomes dominant over acoustic phonon scattering, as shown in Figure 3.25(b), and then R more rapidly increases with gate voltage in the case of $L_{ch}=6$ nm. According to the above results, the L_{ch} dependence of R in Figure 3.24 is found to be unaltered by the gate bias condition in the present DG MOSFETs.

Next, Figure 3.30 shows I_D–V_D characteristics and the drain voltage dependencies of R, calculated for $L_{ch}=10$ nm. At $V_D=0$ V, $R=1$ was obtained for all gate voltages, which verifies the validity of our R extraction method. Here, it is found that R is almost constant when the drain voltage is in the saturation region, which corresponds to the saturation of the drain current. This can also be understood using the variations due to the drain voltage in the lowest sub-band energy profile, and in the average electron kinetic energy profile, as shown in Figures 3.31(a) and (b), respectively.

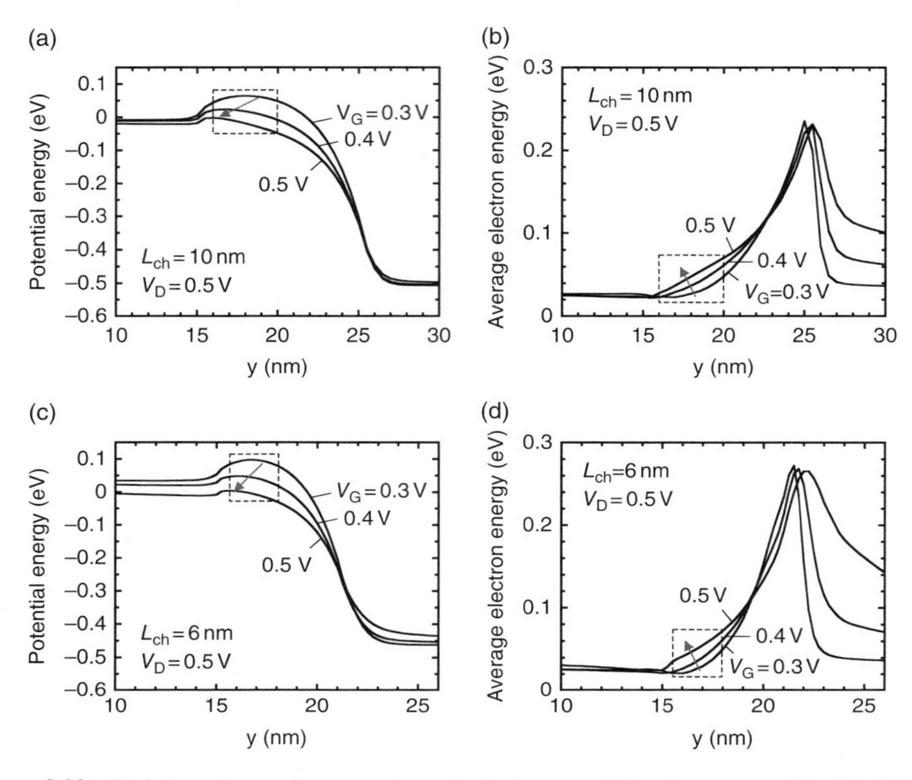

Figure 3.29 Variations due to the gate voltage in the lowest sub-band energy profile ((a), (c)) and in the average kinetic energy profile ((b), (d)), where $L_{ch} = 10$ nm for (a) and (b), and 6 nm for (c) and (d). $V_D = 0.5$ V.

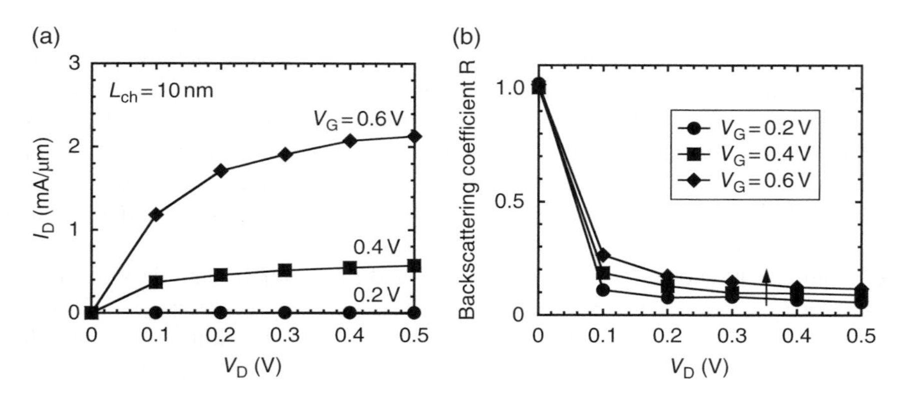

Figure 3.30 (a) I_D–V_D characteristics and (b) drain voltage dependencies of R, calculated for $L_{ch} = 10$ nm.

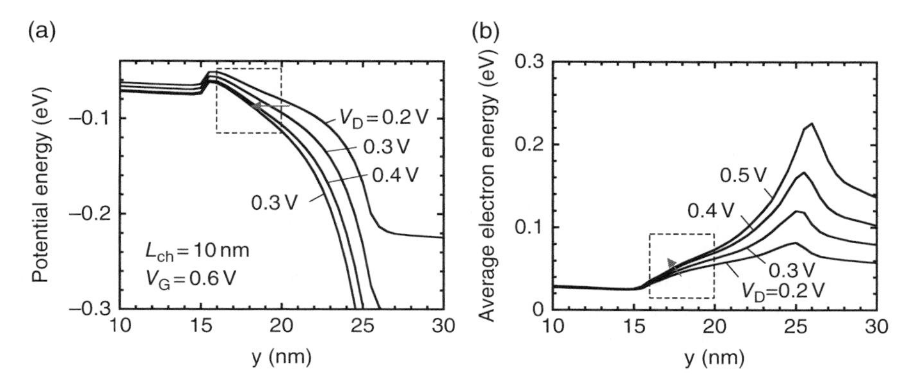

Figure 3.31 Variations due to drain voltage in: (a) the lowest sub-band energy profile; and (b) the average electron kinetic energy, where L_{ch} = 10 nm and V_G = 0.6 V.

First, the average kinetic energy slightly increases on the left side of the channel with the drain voltage, as shown in Figure 3.31(b). Next, the width of the potential bottleneck barrier (i.e., kT-layer length) reduces with drain voltage, as shown in Figure 3.31 (a). This is because, as the drain voltage increases, the potential energy around the drain side of the channel is significantly decreased, and this distinctly reduces the kT-layer length. Since the reduction in kT-layer length offsets the increase in scattering number, owing to increased electron kinetic energy, R eventually becomes almost constant with respect to the drain voltage. Consequently, the drain bias condition is also found not to alter the L_{ch} dependence of R, as shown in Figure 3.24.

3.4 Quasi-ballistic Transport in Si Junctionless Transistors

Recently, a junctionless transistor (JLT), a novel MOSFET device with no pn junctions, has attracted tremendous interest [3.47–3.52]. A JLT is a gated resistor that is heavily and uniformly doped with a single type of dopant. There is no doping gradient between the channel and the source/drain region, which can prevent the formation of atomically shallow and abrupt pn junctions, thus relaxing the drastic requirements on doping techniques and thermal budget. Besides this, owing to its lower sensitivity to the channel-gate oxide interface, a JLT can have a carrier mobility higher than the inversion layer mobility [3.48, 3.53]. In other words, operation in the accumulation mode of a JLT keeps carriers away from the interface to avoid surface-related scattering, which means that a JLT exploits bulk conduction and, thus, can minimize mobility degradation caused by surface-related scattering or high-k surface phonon scattering.

These facts make it a very promising candidate for future technology nodes. On the other hand, since the working principle of the JLT requires a much higher doping concentration ($\approx 10^{19}$–$10^{20}\,cm^{-3}$), there are concerns about degradation in performance due to strong ionized impurity scattering. However, recent experimental and theoretical investigations have demonstrated that the mobility and performance degradation caused by ionized impurity scattering in JLTs can be largely mitigated by the screening effect due to free carriers [3.54, 3.55] and the forward-scattering properties by high-speed carriers [3.56, 3.57], which means that most carriers are forward-scattered by ionized impurities as the electron kinetic energy increases with the applied bias voltage.

Elaborate analytical compact current-voltage models needed for circuit simulations, and to understand the fundamentals of device characteristics, have already been developed [3.49, 3.58, 3.59]. The carrier mobility for JLTs has also been calculated by solving a 1D Schrödinger-Poisson equation perpendicular to the gate [3.60]. However, there have been few studies to date on device characteristics of ultrashort-channel JLTs under nonequilibrium transport conditions, such as on-state drain current properties and quasi-ballistic transport.

In this study, we applied the MSB-MC technique described in Section 3.2 to investigate the actual influence of the surface roughness and ionized impurity scattering on the drain current properties in JLTs [3.61]. In particular, we considered a surface roughness-related scattering caused by a channel thickness fluctuation in UTB DG-channels [3.24], which can prevent carriers from being entirely ballistic, even in ultrashort channel devices, as presented in Section 3.3. Furthermore, we calculate a backscattering coefficient for JLTs that characterizes the quasi-ballistic transport of carriers in ultrasmall MOSFETs, in order to examine the gate bias dependence of ionized impurity scattering.

3.4.1 Device Structure and Simulation Conditions

The device model used in the simulation is shown in Figure 3.32. We employed a DG structure and simulated both a JLT and conventional inversion-mode MOSFET, using the MSB-MC simulator. Since the electrical characteristics of such an UTB MOSFET are seriously degraded by a parasitic resistance in the source and drain, we took care to minimize both the parasitic resistance and the short-channel effects at the same time. The resultant device structures are as follows.

For JLTs, we used a p^+ poly Si gate electrode with work function of 5.20 eV, and assumed a constant doping concentration of $1 \times 10^{20}\,cm^{-3}$ in the source, channel, and drain. For conventional MOSFETs, we used an n^+ poly Si gate electrode with work function of 3.95 eV, and the doping concentration was set at zero in the channel. The temperature is 300 K, to focus on the device properties at normal operating condition. The gate oxide was SiO_2, with a thickness of 0.5 nm. A (001) surface was assumed and the channel direction was set as <110> throughout the present study. The channel length L_{ch} was taken to be 20 and 10 nm, and the channel thickness T_{ch} was 3 nm. Scattering processes considered are AP, OP, II, and SR scatterings, as described in Section 3.2.1. We employed the cD_{ac} (13 eV) model for AP scattering in this study.

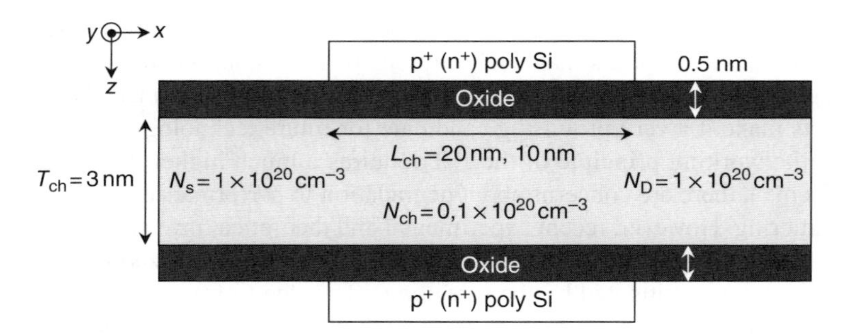

Figure 3.32 Device model used in the simulation. We employed a DG structure and simulated both a JLT and conventional inversion-mode MOSFET, using the MSB-MC simulator.

In the present study, to focus on the influences of SR and II scatterings in JLTs, we performed simulations under the following two conditions:

- Case 1: without II scattering
- Case 2: with II scattering

In both cases 1 and 2, the AP, IP, and SR scattering processes are considered. Hence, the only difference between case 1 and case 2 is whether II scattering is considered or ignored. As will be discussed in subsequent sections, these model calculations help to quantify the influences of not only II scattering, but also SR scattering.

3.4.2 *Influence of SR Scattering*

We present the electron transport properties in the absence of II scattering to elucidate the influence of SR scattering. Figure 3.33 shows the drain current versus gate voltage ($I_D - V_G$) characteristics of JLTs and conventional MOSFETs in the absence of II scattering, computed for $L_{ch}=20$ nm and 10 nm. $V_D=0.5$ V, and the horizontal axis represents the gate overdrive, V_G-V_{th}. The threshold voltage V_{th} was defined as the gate voltage, which corresponds to $I_D=0.01$ mA/μm.

First of all, it is worth noting that JLTs exhibit a drain current larger than that of conventional MOSFETs for both channel lengths. To find out why, the lowest sub-band energy, sheet electron density, and average electron velocity distributions of a JLT and conventional MOSFET are compared in Figure 3.34, where $L_{ch}=20$ nm and 10 nm, both at $V_G-V_{th}=0.4$ V.

The plots on the left-hand side of Figures 3.34(a) and (b) confirm that almost the same electron densities are induced in the channels of JLTs and conventional MOSFETs by the gate voltage. On the other hand, the average velocity in the channels is found to be larger for JLTs,

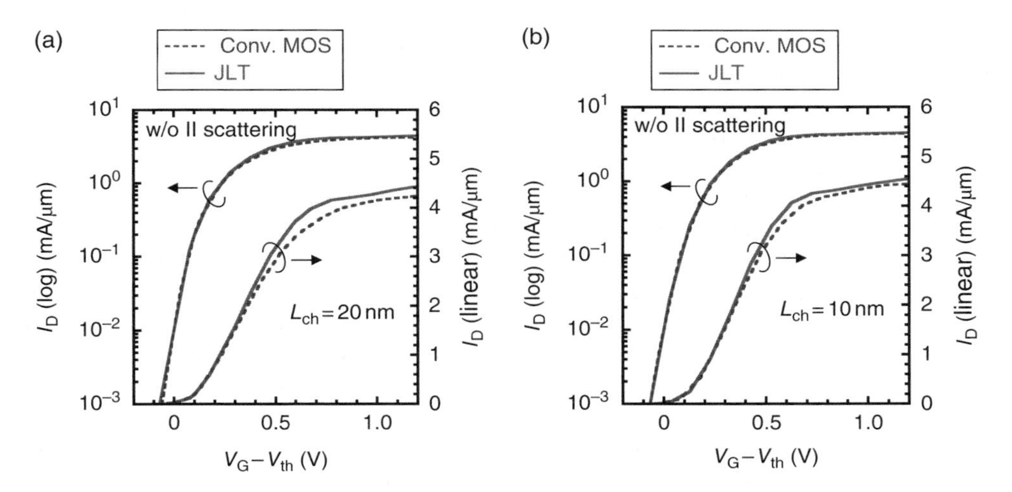

Figure 3.33 $I_D - V_G$ characteristics of JLTs and conventional MOSFETs in the absence of II scattering, computed for: (a) $L_{ch}=20$ nm; and (b) 10 nm. $V_D=0.5$ V, and the horizontal axis represents the gate overdrive, V_G-V_{th}.

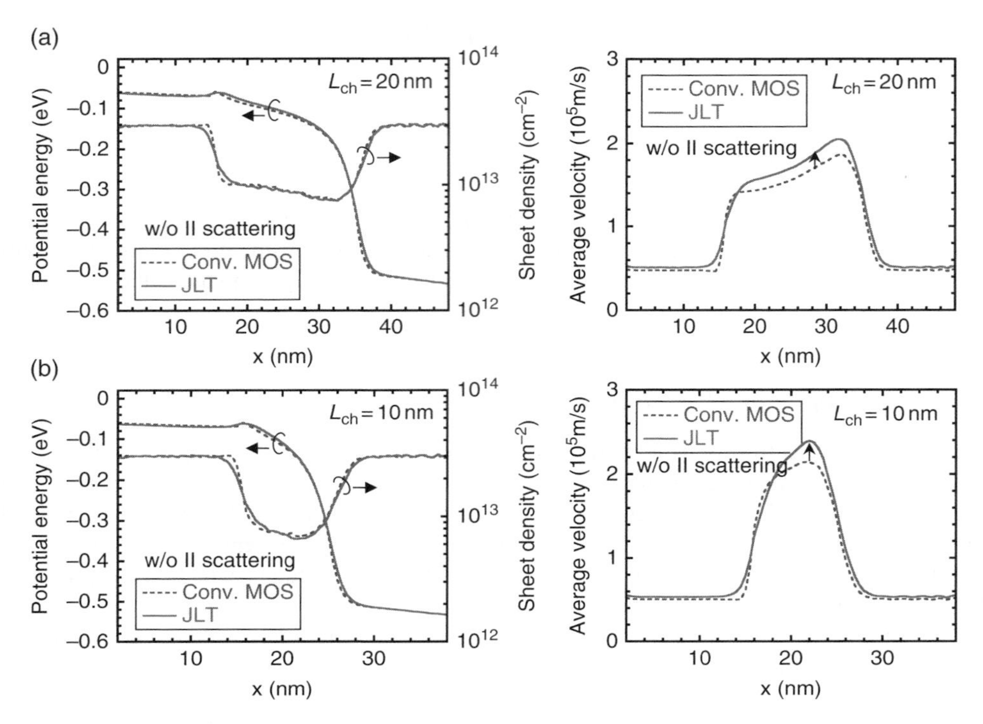

Figure 3.34 The lowest sub-band energy, the sheet electron density, and the average electron velocity distributions for JLTs and conventional MOSFETs, where: (a) $L_{ch}=20$ nm; and (b) 10 nm, both at $V_G-V_{th}=0.4$ V. II scattering is ignored.

as seen in the plots on the right-hand side of Figures 3.34(a) and (b). This difference in average velocity suggests that phonon scattering and/or SR scattering has a different effect on electron transport in JLTs vs. conventional MOSFETs, because II scattering is ignored in Figure 3.34.

First, we check the influence of phonon scattering. Figure 3.35 shows AP and IP scattering rates for both JLTs and conventional MOSFETs, calculated at the bottleneck point and at $V_G-V_{th}=0.4$ V for the lowest sub-band.

Note that the zero in the horizontal axis represents the lowest quantized sub-band energy. $1\rightarrow 2$ denotes the intersub-band transition from the lowest sub-band to a first higher one. Since the strength of phonon scattering for 2-DEG is principally determined by the carrier confinement, the scattering rates crucially depend on the channel thickness in both JLTs and conventional MOSFETs. Accordingly, the AP and IP scattering rates are hardly affected by the difference in device structure, as shown in Figure 3.35. We should add that the AP and IP emission rates are very close in the high-energy region, because a large effective deformation potential, D_{ac}, of 13 eV is used in the present calculation.

Next, we discuss the influence of SR scattering. Figure 3.36 shows SR scattering rates and the lowest sub-band wave function and conduction band edge profiles in the confinement direction, calculated for the same conditions as in Figure 3.35.

As described in Section 3.2.1, the SR scattering rate depends on the Prange-Nee term at the interfaces – that is, $(\hbar^2/2m_z)\,(d\varphi_n(0)/dz)\,(d\varphi_m(0)/dz)$ and $(\hbar^2/2m_z)\,(d\varphi_n(T_{Si})/dz)\,(d\varphi_m(T_{Si})/dz)$.

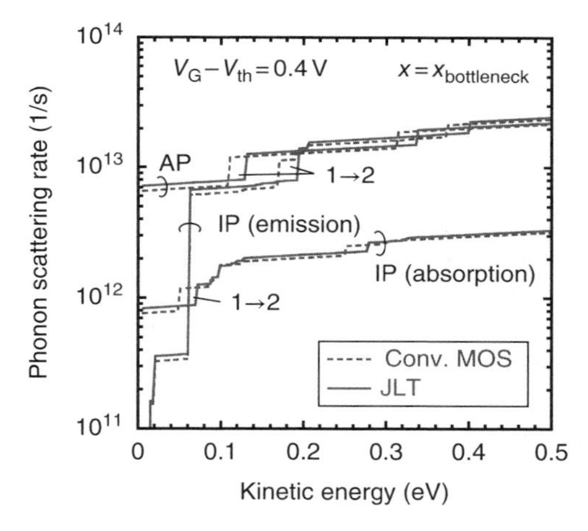

Figure 3.35 AP and IP scattering rates for both JLTs and conventional MOSFETs, calculated at the bottleneck point and at $V_G - V_{th} = 0.4$ V for the lowest sub-band. The zero in the horizontal axis represents the lowest quantized sub-band energy. $1 \rightarrow 2$ denotes the intersub-band transition from the lowest sub-band to a first higher one.

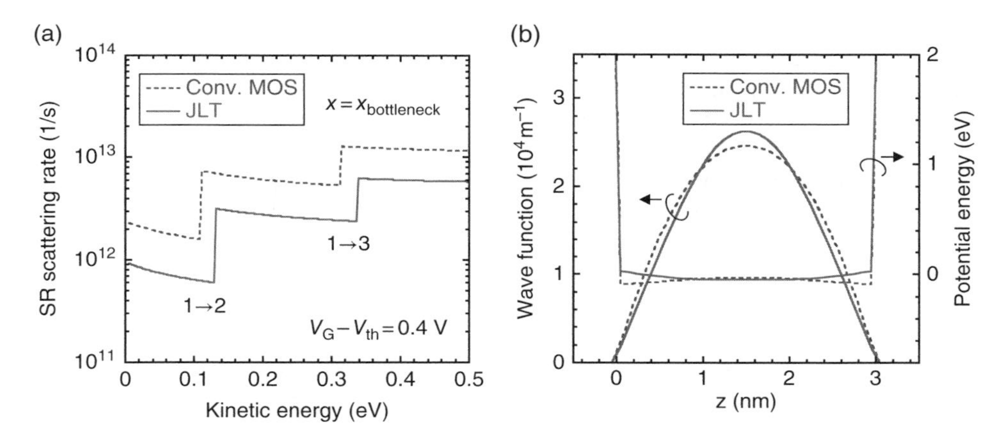

Figure 3.36 (a) SR scattering rates; and (b) the lowest sub-band wave function and conduction band edge profiles in the confinement direction, calculated under the same conditions as in Figure 3.35. In (a), the zero in the horizontal axis represents the lowest quantized sub-band energy. $1 \rightarrow 2$ and $1 \rightarrow 3$ denote the inter-sub-band transitions from the lowest sub-band to the higher ones.

Therefore, the lower slope of the wave function at the interfaces (see Figure 3.36(b)) results in smaller scattering rates in JLTs, as shown in Figure 3.36(a). In general, owing to the accumulation-mode operation, carriers in JLTs get distributed farther away from the interfaces, compared with those in conventional MOSFETs. Consequently, the present results demonstrate that JLTs can actually minimize the mobility degradation caused by SR scattering and improve current drivability.

3.4.3 Influence of II Scattering

In this section, we present the electron transport properties in the presence of II scattering. Figure 3.37 shows the II scattering rates calculated using Equations (3.19) to (3.21) for a wide range of electron densities, n, of 10^{17}–$10^{20}\,cm^{-3}$, where the donor concentration, N_D, is fixed at $10^{20}\,cm^{-3}$.

As is well known, the II scattering rate decreases with electron kinetic energy, since higher-speed electrons pass by impurities more swiftly, without changing direction. In addition, the II scattering rate also decreases with n, as a result of the increased screening effect. In the following simulations, we consider the abovementioned electron kinetic energy and electron density dependences of the II scattering rate.

Figure 3.38 shows the $I_D - V_G$ characteristics of JLTs and conventional MOSFETs in the presence of II scattering, computed for $L_{ch} = 20\,nm$ and $10\,nm$. The calculation conditions are the same as those in Figure 3.33.

It will be found that the drain currents of both devices decrease when II scattering is considered and, as expected, JLTs exhibit a smaller drain current than that of conventional MOSFETs. This represents that a conventional DG MOSFET is superior to a DG JLT in terms of the current drive. However, as reported in recent experimental [3.54, 3.55] and theoretical [3.56, 3.57] investigations, the performance degradation in JLTs due to II scattering is insignificant, owing to the screening effect of free carriers and the forward-scattering properties of high-speed carriers.

Here, we should point out that a similar drain current behavior – that is, the drain current of the JLT asymptotically approaching the conventional one with increasing gate voltage, was observed in the experimental measurement [3.48]. Therefore, the linearity in the transconductance $(g_m) - V_G$ curve of JLTs is comparable to, and even better than, that of conventional MOSFETs, although the maximum g_m is slightly higher in conventional MOSFETs. Incidentally, the decrease in drain current in JLTs due to II scattering is caused by a reduction in the average electron velocity, rather than in the induced electron density [3.61].

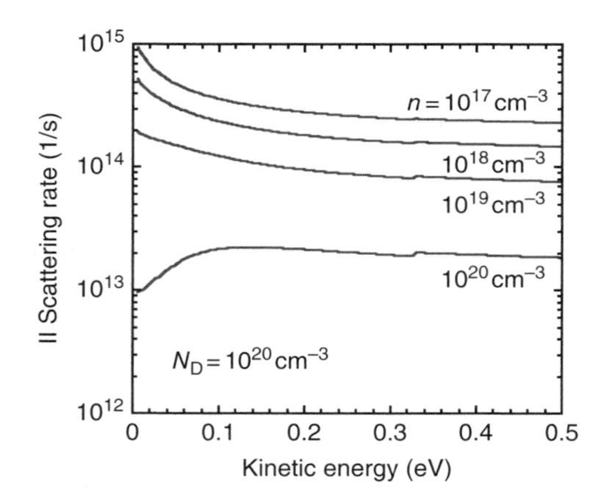

Figure 3.37 II scattering rates calculated for a wide range of electron densities, n, of 10^{17}–$10^{20}\,cm^{-3}$, where the donor concentration, N_D, is fixed at $10^{20}\,cm^{-3}$. The zero in the horizontal axis represents the lowest quantized sub-band energy.

3.4.4 Backscattering Coefficient

We now discuss gate-bias-dependent II scattering using a quasi-ballistic transport parameter, the backscattering coefficient R. The same techniques mentioned in Section 3.3 are applied to JLTs in this study. Figure 3.39 shows the backscattering coefficients in JLTs and conventional MOSFETs with $L_{ch}=20\,\text{nm}$ and $10\,\text{nm}$, computed as a function of the gate voltage, where the results with and without II scattering are indicated.

Without II scattering, R increases with the gate voltage for both devices, which corresponds to the results for Si DG-MOSFETs with intrinsic channel presented in Section 3.3.3.

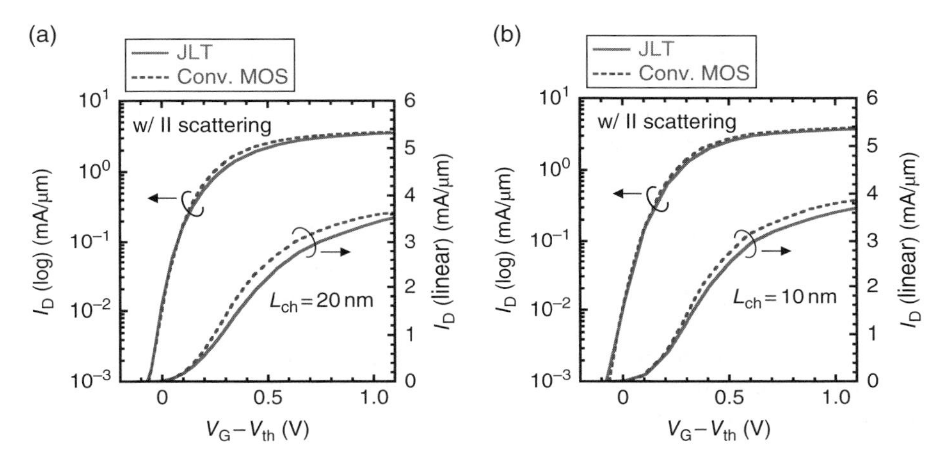

Figure 3.38 I_D-V_G characteristics of JLTs and conventional MOSFETs in the presence of II scattering, computed for: (a) $L_{ch}=20\,\text{nm}$; and (b) $10\,\text{nm}$. $V_D=0.5\,\text{V}$, and the horizontal axis represents the gate overdrive, V_G-V_{th}.

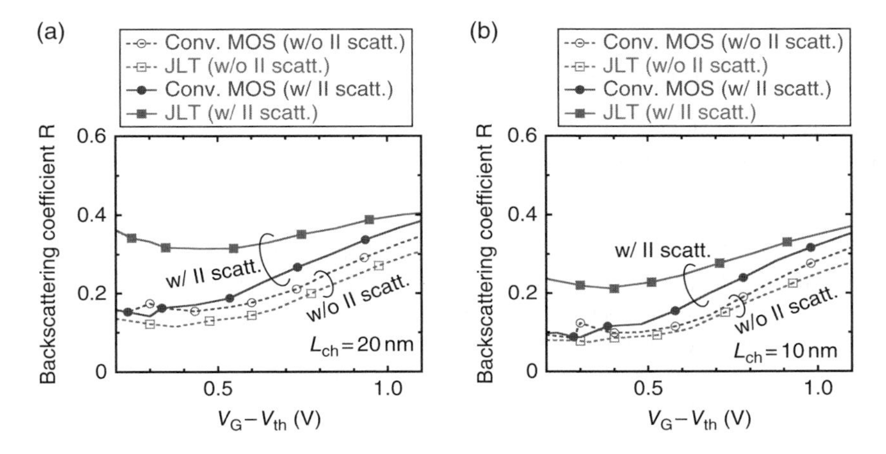

Figure 3.39 Backscattering coefficients in JLTs and conventional MOSFETs with (a) $L_{ch}=20\,\text{nm}$; and (b) $10\,\text{nm}$, computed as a function of the gate voltage, where the results with and without II scattering are indicated.

The increase in R is caused by the strengthening in SR and AP scatterings with increasing gate voltage. On the other hand, for JLTs with II scattering, the slope of the curve is markedly lower than those of the other curves. As a result, the difference in R between JLTs and conventional MOSFETs with II scattering becomes larger (smaller) as the gate voltage decreases (increases). This is because the screening effect is weak under low gate bias conditions and, hence, II scattering becomes active. On the other hand, increasing the gate voltage enhances the screening effect as well as the forward scattering, because the electric field inside the channel increases with increasing gate voltage [3.56, 3.57]. Thus, they contribute to mitigating II scattering under high gate bias conditions.

3.5 Quasi-ballistic Transport in GAA-Si Nanowire MOSFETs

In the sub-10 nm regime, Si nanowire (SNW)-MOSFETs are strong candidates for ultra-scaled CMOS technologies because of their superior electrostatic integrity provided by a gate-all-around (GAA) architecture, as described in Section 1.2. So far, a number of one-dimensional (1-D) transport studies in SNW-MOSFETs have been performed by taking into consideration electron-phonon interactions [3.62–3.65], random dopant fluctuation [3.66], and interface roughness [3.67, 3.68]. The backscattering of mobile carries in 1-D nanotransistors has been also examined, by performing simple numerical simulations with model potential profiles [3.69].

In this section, we investigate the quasi-ballistic transport in ultra-scaled SNW-MOSFETs by the three-dimensional multi-sub-band ensemble Monte Carlo (3DMSB-MC) method, which includes all relevant scattering mechanisms, such as scattering by AP and IP, as well as II and SR scatterings. We have performed the transport simulations in SNW-MOSFETs with channel lengths ranging from 10–20 nm, and calculated the backscattering coefficient as a function of the gate voltage. Our results show that characteristic backscattering properties of SNW-MOSFETs are essentially different from those of Si planar DG MOSFETs, discussed in Sections 3.3 and 3.4, due to a dimensionality effect on the carrier scattering.

3.5.1 Device Structure and 3DMSB-MC Method

In this study we consider a SiNW GAA MOSFETs, shown in Figure 3.40(a) with cross-section 3 nm × 3 nm and gate oxide thickness 1 nm. The donor concentration in the source and drain regions are taken as $N_D = 10^{20}\,\mathrm{cm}^{-3}$, and the channel region is undoped. The conduction band valleys in Si are shown in Figure 3.40(b) with <100> taken as the current direction. For simplicity, the coordinates (y, z) are chosen along two other directions <010> and <001>.

Usage of the conduction band structure of bulk Si is warranted by the fact that the effective mass in a Si-NW with cross-section of 3 nm remains relatively close to that in bulk Si [3.64, 3.70]. In the present work we have developed a 3DMSB-MC simulator which enables us to obtain fin aspect-ratio and study the role of crystallographic orientation and various scattering mechanisms in NW-MOSFETs with strong quantum confinement. In the simulator, the 3-D problem has been decoupled into the 1-D transport equation, along the S-D (x) direction and the 2-D Schrödinger equation in the transverse (y, z) cross-section, as shown in Figure 3.41. Then, the carrier transport is computed by the MSB-MC technique separately for different sub-bands in each of three pairs of the conduction valleys, v = 1-1', 2-2', and 3-3', as shown in Figure 3.40(b).

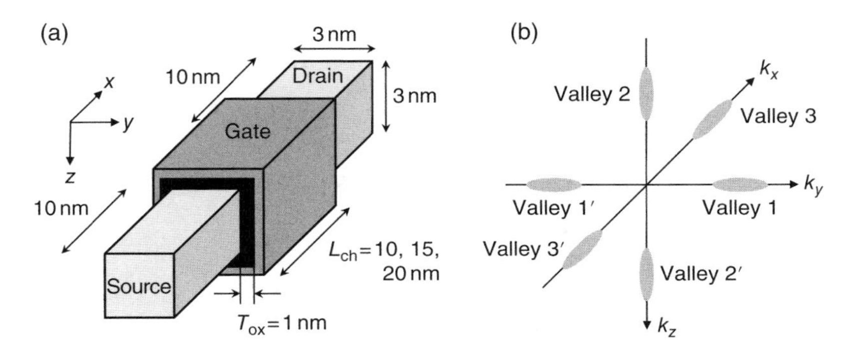

Figure 3.40 (a) SNW-MOSFET model; and (b) conduction band valleys of Si.

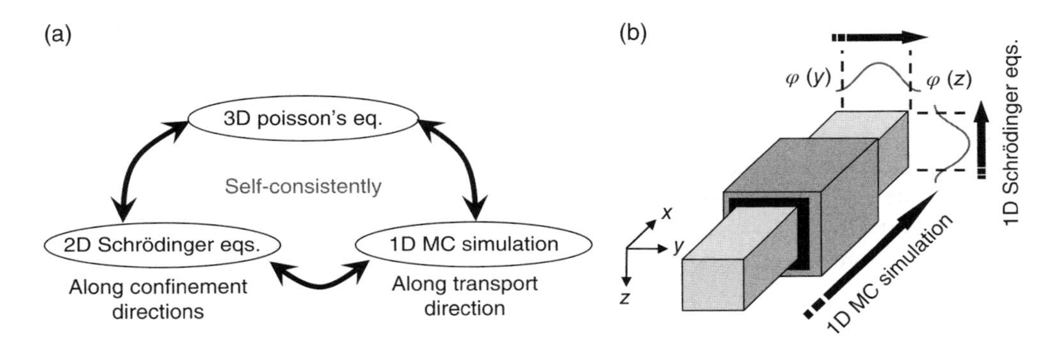

Figure 3.41 3DMSB-MC method. (a) Algorithm; (b) its application to SNW GAA MOSFET.

3.5.2 *Scattering Rates for 1-D Electron Gas*

In this work we include AP, IP, II, and SR scattering mechanisms. The scattering rates have been taken from the corresponding model of 1-D electron gas. When the confinement directions are set as y and z, as shown in Figure 3.41, the elastic intravalley AP scattering rate we used is expressed as:

$$\Gamma_{\mathrm{ac}} = \frac{\sqrt{2}\,D_{\mathrm{ac}}^2 k_{\mathrm{B}} T \sqrt{m_x}}{8\pi^2\hbar^2 c_L}\frac{2}{\sqrt{E'}}\left(1+2\alpha E'\right)\int_0^{T_{\mathrm{Si}}^y}\int_0^{T_{\mathrm{Si}}^z}\varphi_m^2(y,z)\varphi_n^2(y,z)\,dydz \tag{3.27}$$

where T_{Si}^y and T_{Si}^z are the channel thicknesses along the y- and z-directions, respectively. Similarly, the inelastic IP scattering rate is expressed as:

$$\Gamma_{\mathrm{inelastic}} = \frac{\sqrt{2}\,D_{\mathrm{op}}^2 \sqrt{m_x}}{8\pi^2\hbar^2 \rho\omega_{\mathrm{op}}}\left[N_q+\frac{1}{2}+\sigma\right]\frac{2}{\sqrt{E'}}\left(1+2\alpha E'\right)\int_0^{T_{\mathrm{Si}}^y}\int_0^{T_{\mathrm{Si}}^z}\varphi_m^2(y,z)\varphi_n^2(y,z)\,dydz \tag{3.28}$$

A simple continuum model [3.73] has been used to incorporate II scattering, as shown in Equations (3.29) to (3.32), which only occurs in the S/D regions, and thus does not play an essential part in quasi-ballistic transport studied here.

$$\Gamma_{\text{imp}} = \frac{Z^2 e^4 N_a \sqrt{m_x}}{16\sqrt{2}\pi^2 \hbar^2 \varepsilon} \frac{(1+2\alpha E')}{\sqrt{E'(1+\alpha E')}} \int d\mathbf{R} I_{nm}^2 (q_x, \mathbf{R}) \tag{3.29}$$

$$I_{nm}(q_x, \mathbf{R}) = \int_0^{T_{\text{Si}}^y} \int_0^{T_{\text{Si}}^z} \varphi_n(y,z) K_0(q_x, \mathbf{R}) \varphi_m(y,z) dydz \tag{3.30}$$

$$K_0(q_x, \mathbf{R}) = \int \frac{e^{iq_x x} \cdot \exp\left[-\sqrt{(\mathbf{r}-\mathbf{R})^2 + x^2}/L_D\right]}{\sqrt{(\mathbf{r}-\mathbf{R})^2 + x^2}} dx \tag{3.31}$$

where L_D is the Debye length, which is given by:

$$L_D = \sqrt{\frac{\varepsilon k_B T}{e^2 n}} \tag{3.32}$$

It is assumed that SR scattering induces only intravalley transitions and that the four interfaces are uncorrelated. Therefore, the scattering rate at each interface is calculated separately. We describe the scattering rate at the top interface below. Those associated with SR at other interfaces can be obtained in a similar way. The SR is characterized by two parameters – the root-mean-square deviation Δ and the correlation length Λ, and the SR scattering rate is given by

$$\Gamma_{\text{sr}} = \sum_n \frac{\sqrt{2m_x}}{\hbar^2} \frac{\sqrt{2}\Delta^2 \Lambda}{2 + q_x^2 \Lambda^2} |M_{mn}|^2 \frac{\Theta[\varepsilon_n(k_x) - E_m]}{\sqrt{\varepsilon_n(k_x) - E_m}} \tag{3.33}$$

with $q_x = k_x \pm k'_x$, the difference between the initial (k_x) and the final (k'_x) electron wavevectors, and the top (bottom) sign is for backward (forward) scattering. M_{mn} is the generalized Prange-Nee formula [3.71] with model auto-correlation function [3.72], given by:

$$\begin{aligned} M_{mn} = & -\frac{\hbar^2}{T_{\text{Si}}^y} \int_0^{T_{\text{Si}}^y} dy \int_0^{T_{\text{Si}}^z} dz \varphi_n(y,z) \left[\frac{\partial}{\partial z} \frac{1}{m_x} \frac{\partial \varphi_m(y,z)}{\partial z} \right] \\ & + \int_0^{T_{\text{Si}}^y} dy \int_0^{T_{\text{Si}}^z} dz \varphi_n(y,z) \left(1 - \frac{z}{T_{\text{Si}}^z}\right) \frac{\partial V}{\partial z} \varphi_m(y,z) \\ & + (E_n - E_m) \int_0^{T_{\text{Si}}^y} dy \int_0^{T_{\text{Si}}^z} dz \varphi_n(y,z) \left(1 - \frac{z}{T_{\text{Si}}^z}\right) \frac{\partial \varphi_m(y,z)}{\partial z} \end{aligned} \tag{3.34}$$

M_{mn} is composed of three terms which correspond to the fluctuations of wave functions, electrostatic potential, and energy levels induced by the surface roughness, respectively. The scattering rates for the other interfaces are obtained by modifying the term M_{mn}. The scattering rates are calculated along the transport direction for each slice corresponding to the mode-space decomposition, and they are updated after the solution of 2D Schrödinger equations, as shown in Figure 3.41(a).

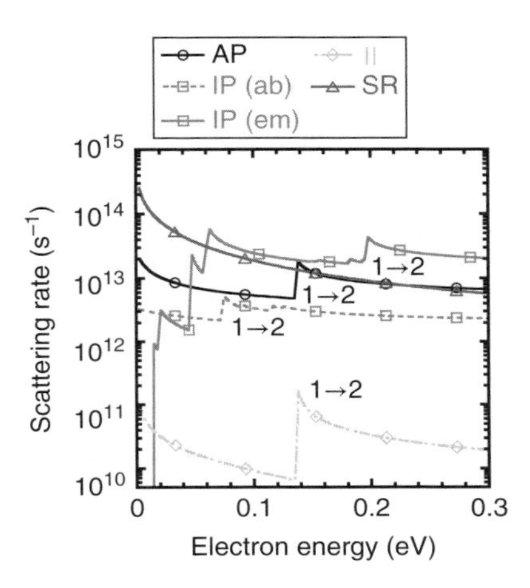

Figure 3.42 Partial scattering rates in the lowest sub-band for all scattering mechanisms in the source region of a SNW-MOSFET, with $L_{ch}=10\,nm$, $V_G=0.3\,V$ and $V_D=0.01\,V$. Zero energy in the horizontal axis represents the bottom of the lowest sub-band.

Figure 3.42 shows all the partial scattering rates for the lowest sub-band in the source region of SNW-MOSFET, with $L_{ch}=10\,nm$ as a function of the electron kinetic energy. The gate and drain bias conditions are taken as $V_G=0.3\,V$ and $V_D=0.01\,V$ [3.74]. The II scattering rate corresponds to the electron density of $n=N_D=10^{20}\,cm^{-3}$. An acoustic deformation potential of 13 eV has been taken for silicon [3.26]. For the SR scattering, we have used $\Delta=0.25\,nm$ and $\Lambda=1.3\,nm$ from Ref. 3.74.

It is found that intersub-band transitions are represented. Thus, the scattering rates exhibit a characteristic saw-edged shape, in agreement with quasi-one-dimensional nature of the electron gas in SNW channel. We also add that partial scattering rates hardly change along the S-D direction and for different channel lengths.

3.5.3 $I_D - V_G$ *Characteristics and Backscattering Coefficient*

The $I_D - V_G$ characteristics of SNW-MOSFETs, with $L_{ch}=10$, 15, and 20 nm at $V_D=0.2\,V$, are shown in Figure 3.43. Zero on the horizontal axis corresponds to the threshold voltage V_{th} taken as the value of gate voltage for $I_D=0.01\,\mu A$. In the present 3DMSB-EMC simulations, we have found nearly ideal subthreshold behavior for all channel lengths. Thus, we have estimated the subthreshold slopes, which are 73, 70 and 69 mV/decade, for $L_{ch}=10$, 15 and 20 nm, respectively. This means that the larger value of on-current in devices with shorter channels correlates with lesser number scattering events at smaller L_{ch}.

Next, we have calculated R by using the same technique for Si DG-MOSFETs as in Sections 3.3 and 3.4. Figure 3.44 shows the backscattering coefficients in the SNW-MOSFETs with $L_{ch}=10$, 15, and 20 nm, as a function of the gate voltage. The backscattering coefficient R decreases with L_{ch}, which indicates better ballistic efficiency in shorter channels.

We also observe a characteristic cusp in the backscattering coefficient around $V_G=0.1\sim0.15\,V$. This is caused by stronger AP and SR scattering, due to larger 1-D density-of-states at the lowest sub-band edge, which represent a threshold energy value for the incoming source

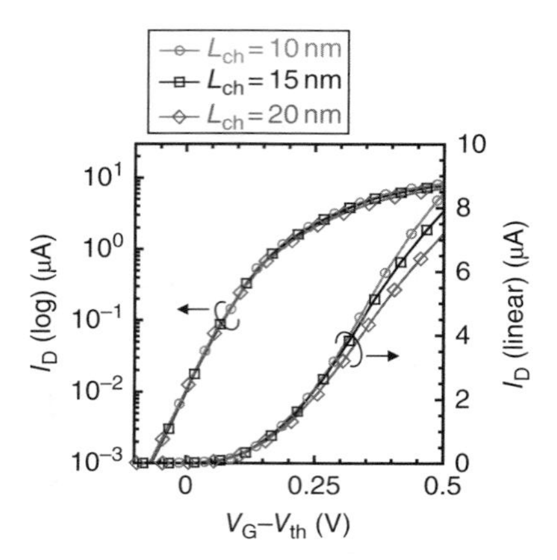

Figure 3.43 $I_D - V_G$ characteristics of SNW-MOSFETs, with L_{ch} = 10, 15, and 20 nm, calculated at V_D = 0.2 V. The estimated subthreshold slopes are: 73, 70 and 69 mV/decade, for L_{ch} = 10, 15 and 20 nm, respectively.

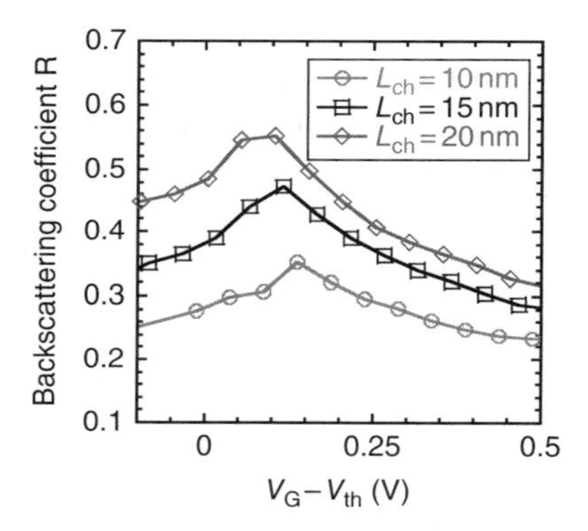

Figure 3.44 Backscattering coefficients in SNW-MOSFETs with L_{ch} = 10, 15, and 20 nm, as a function of the gate voltage.

electrons. In fact, the incoming source electrons are found to start increasing drastically around $V_G \approx 0.14$ V in the case of L_{ch} = 10 nm, as shown in Figure 3.45(a), where the lowest sub-band energy profiles along the channel are plotted for various gate voltages, with an energy distribution function of electrons inside the source.

Then, by further increasing V_G above ≈ 0.2 V, the lowest sub-band edge is moved down and the average energy of the incoming source electrons increases, as shown in Figure 3.45(b). As a result, R decreases for V_G above ≈ 0.2 V, due to the reduction of the 1-D scattering rates with increasing energy, as shown in Figure 3.42. This decreasing behavior of R is a distinctive characteristics of 1-D NW MOSFETs, in contrast to Si planar DG MOSFETs, where R grows with V_G because of stronger AP scattering at larger electron kinetic energy, as discussed in Sections 3.3 and 3.4 [3.26, 3.61]. Here, it should be noted that such a nonmonotonic behavior in quasi-ballistic SNW-MOSFETs has been reported in [3.62], which has demonstrated that 1-D scattering mechanisms produce series of cusps and dips in nanowire mobility as a function of the gate voltage.

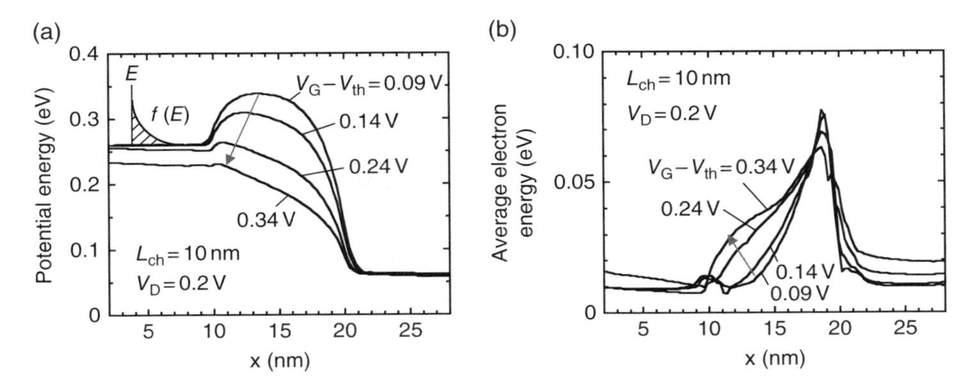

Figure 3.45 (a) The lowest sub-band energy profiles along the channel for various gate voltages, with an energy distribution function of electrons inside the source. (b) The average energy distributions of electrons for gate voltages corresponding to (a).

Finally, *R*s in SNW-MOSFETs are found to be several times larger, compared to the previous values obtained for Si planar DG MOSFETs (See Figures 3.28 and 3.39). This is caused by much stronger SR scattering in the presently considered narrow SNW-MOSFETs with 3 nm × 3 nm cross section (see Figure 3.42).

References

[3.1] K. Natori (Aug. 2001). Scaling limit of the MOS transistor – A ballistic MOSFET–. *IEICE Transactions on Electronics* **E84-C**(8), 1029–1036.

[3.2] H. Tsuchiya, K. Fujii, T. Mori and T. Miyoshi (Dec. 2006). A quantum-corrected Monte Carlo study on quasi-ballistic transport in nanoscale MOSFETs. *IEEE Transactions on Electron Devices* **53**(12), 2965–2971.

[3.3] S. Takagi, J. Koga and A. Toriumi (Mar. 1998). Mobility enhancement of SOI MOSFETs due to subband modulation in ultrathin SOI films. *Japanese Journal of Applied Physics* **37**(3B), 1289–1294.

[3.4] H. Tsuchiya, M. Horino, M. Ogawa and T. Miyoshi (Dec. 2003). Quantum transport simulation of ultrathin and ultrashort silicon-on-insulator metal-oxide-semiconductor field-effect transistors. *Japanese Journal of Applied Physics* **42**(12), 7238–7243.

[3.5] H. Tsuchiya, A. Oda, M. Ogawa and T. Miyoshi (Nov. 2005). A quantum-corrected Monte Carlo and molecular dynamics simulation on electron-density-dependent velocity saturation in silicon metal-oxide-semiconductor field-effect transistors. *Japanese Journal of Applied Physics* **44**(11), 7820–7826.

[3.6] M. Ogawa, H. Tsuchiya and T. Miyoshi (Mar. 2003). Quantum electron transport modeling in nano-scale devices. *IEICE Transactions on Electronics* **E86-C**(3), 363–371.

[3.7] H. Tsuchiya and T. Miyoshi (June 1999). Quantum transport modeling of ultrasmall semiconductor devices. *IEICE Transactions on Electronics* **E82-C**(6), 880–888.

[3.8] D. Ferry, R. Akis and D. Vasileska (2000). Quantum effects in MOSFETs: Use of an effective potential in 3D Monte Carlo simulation of ultra-short channel devices. *IEDM Technical Digest*, 287–290.

[3.9] D. Ferry (Nov. 2000). Effective potentials and the onset of quantization in ultrasmall MOSFETs. *Superlattices and Microstructures* **28**(5/6), 419–423.

[3.10] H. Tsuchiya, A. Svizhenko, M. Anantram, M. Ogawa and T. Miyoshi (2005). Comparison of non-equilibrium Green's function and quantum-corrected Monte Carlo approaches in nano MOS simulation. *Journal of Computational Electronics* **4**(1/2), 35–38.

[3.11] K. Tomizawa (1993). *Numerical Simulation of Submicron Semiconductor Devices*. Artech House.

[3.12] C. Jacoboni and L. Reggiani (July 1983). The Monte Carlo method for the solution of charge transport in semiconductors with applications to covalentmaterials. *Reviews of Modern Physics* **55**(3), 645–705.

[3.13] M. V. Fischetti and S. E. Laux (Nov. 1988). Monte Carlo analysis of electron transport in small semiconductor devices including band-structure and space-charge effects. *Physical Review B* **38**(14), 9721–9745.

[3.14] K. Natori (Oct. 1994). Ballistic metal-oxide-semiconductor field effect transistor. *Journal of Applied Physics* **76**(8), 4879–4890.

[3.15] K. Natori (Nov. 2002). Ballistic MOSFET reproduces current-voltage characteristics of an experimental device. *IEEE Electron Device Letters* **23**(11), 655–657.

[3.16] K. Natori and T. Kurusu (Sep. 2004). Nanoscale quasi-ballistic MOSFETs in reflection-transmission model. *Extended Abstracts of International Conference on Solid State Devices and Materials (SSDM)*, Tokyo, 728–729.

[3.17] K. Natori (July 2008). Ballistic/quasi-ballistic transport in nanoscale transistor. *Applied Surface Science* **254**(19), 6194–6198.

[3.18] H. Tsuchiya and S. Takagi (Sep. 2008). Influence of elastic and inelastic phonon scattering on the drive current of quasi-ballistic MOSFETs. *IEEE Transactions on Electron Devices* **55**(9), 2397–2402.

[3.19] P. Palestri, D. Esseni, S. Eminente, C. Fiegna, E. Sangiorgi and L. Selmi (Dec. 2005). Understanding quasi-ballistic transport in nano-MOSFETs: Part I – Scattering in the channel and in the drain. *IEEE Transactions on Electron Devices* **52**(12), 2727–2735.

[3.20] S. Koswatta, S. Hasan, M. Lundstrom, M. Anantram and D. Nikonov (2006). Ballisticity of nanotube field-effect transistors: Role of phonon energy and gate bias. *Applied Physics Letters* **89**(2), p. 023125.

[3.21] T. Ohashi, T. Takahashi, N. Beppu, S. Oda and K. Uchida (2011). Experimental evidence of increased deformation potential at MOS interface and its impact on characteristics of ETSOI FETs. *IEDM Technical Digest*, 390–393.

[3.22] T. Ohashi, T. Takahashi, T. Kodera, S. Oda and K. Uchida (Sep. 2012). Experimental observation of record-high electron mobility of greater than 1100 $cm^2V^{-1}s^{-1}$ in unstressed Si MOSFETs and its physical mechanisms. *Extended Abstracts of International Conference on Solid State Devices and Materials (SSDM)*, Kyoto, Japan, 807–808.

[3.23] H. Sakaki, T. Noda, K. Hirakawa, M. Tanaka and T. Matsusue (Dec. 1987). Interface roughness scattering in GaAs/AlAs quantum wells. *Applied Physics Letters* **51**(23), 1934–1936.

[3.24] K. Uchida, H. Watanabe, A. Kinoshita, J. Koga, T. Numata and S. Takagi (2002). Experimental study on carrier transport mechanism in ultrathin-body SOI n- and p-MOSFETs with SOI thickness less than 5 nm. *IEDM Technical Digest*, 47–50.

[3.25] S. Koba, R. Ishida, Y. Kubota, H. Tsuchiya, Y. Kamakura, N. Mori and M. Ogawa (2013). The impact of increased deformation potential at MOS interface on quasi-ballistic transport in ultrathin channel MOSFETs scaled down to sub-10 nm channel length. *IEDM Technical Digest*, 312–315.

[3.26] S. Koba, R. Ishida, Y. Kubota, H. Tsuchiya, Y. Kamakura, N. Mori and M. Ogawa (Oct. 2014). Effects of increased acoustic phonon deformation potential and surface roughness scattering on quasi-ballistic transport in ultrascaled Si-MOSFETs. *Japanese Journal of Applied Physics* **53**, p. 114301.

[3.27] R. Venugopal, Z. Ren, S. Datta and M. S. Lundstrom (Oct. 2002). Simulating quantum transport in nanoscale transistors: Real versus mode-space approaches. *Journal of Applied Physics* **92**(7), 3730–3739.

[3.28] D. Querlioz and P. Dollfus (2010). *The Wigner Monte Carlo Method for Nanoelectronic Devices*, Chapter 3. Wiley, New York.

[3.29] F. Monsef, P. Dollfus and S. G. Retailleau (Apr. 2004). Electron transport in Si/SiGe modulation-doped heterostructures using Monte Carlo simulation. *Journal of Applied Physics* **95**(7), 3587–3593.

[3.30] A. L. Fetter (Nov. 1974). Electrodynamics and thermodynamics of a classical electron surface layer. *Physical Review B* **10**(9), 3739–3745.

[3.31] R. E. Prange and T.-W. Nee (Apr. 1968). Quantum spectroscopy of te low-field oscillations in the surface impedance. *Physical Review B* **168**(3), 779–786.

[3.32] D. Esseni (Mar. 2004). On the modeling of surface roughness limited mobility in SOI MOSFETs and its correlation to the transistor effective field. *IEEE Transactions on Electron Devices* **51**(3), 394–401.

[3.33] S. Jin, M. V. Fischetti and T.-W. Tang (Sep. 2007). Modeling of surface-roughness scattering in ultrathin-body SOI MOSFETs. *IEEE Transactions on Electron Devices* **54**(9), 2191–2203.

[3.34] P. Toniutti, D. Esseni and P. Palestri (Nov. 2010). Failure of the scalar dielectric function approach for the screening modeling in double-gate SOI MOSFETs and in FinFETs. *IEEE Transactions on Electron Devices* **57**(11), 3074–3083.

[3.35] M. V. Fischetti and S. E. Laux (Jul. 1993). Monte Carlo study of electron transport in silicon inversion layers. *Physical Review B* **48**(4), 2244–2274.

[3.36] M. V. Fischetti and S. E. Laux (Aug. 1996). Band structure, deformation potentials and carrier mobility in strained Si, Ge and SiGe alloys. *Journal of Applied Physics* **80**(4), 2234–2252.

[3.37] M. V. Fischett (Jan. 2001). Long-range Coulomb interactions in small Si devices. Part II. Effective electron mobility in thin-oxide structures. *Journal of Applied Physics* **89**(2), 1232–1250.

[3.38] K. Hess (2000). *Advanced Theory of Semiconductor Devices*, Chapter 7. Wiley-IEEE Press, New York.

[3.39] S. Takagi, J. L. Hoyt, J. J. Welser and J. F. Gibbons (Aug. 1996). Comparative study of phonon-limited mobility of two-dimensional electrons in strained and unstrained Si metal-oxide-semiconductor field-effect transistors. *Journal of Applied Physics* **80**(3), 1567–1577.

[3.40] D. Esseni, P. Palestri and L. Selmi (2011). *Nanoscale MOS Transistors*, p. 112. Cambridge University Press, New York.

[3.41] S. Takagi, A. Toriumi, M. Iwase and H. Tango (Dec. 1994). On the universality of inversion layer mobility in Si MOSFETs: Part I – Effects of substrate impurity concentration. *IEEE Transactions on Electron Devices* **41**(12), 2357–2362.

[3.42] K. Uchida, J. Koga and S. Takagi (Oct. 2007). Experimental study on electron mobility in ultrathin-body silicon-on-insulator metal-oxide-semiconductor field-effect transistors. *Journal of Applied Physics* **102**(7), p. 074510.

[3.43] S. Takagi, J. Koga and A. Toriumi (Mar. 1998). Mobility enhancement of SOI MOSFETs due to subband modulation in ultrathin SOI films. *Japanese Journal of Applied Physics* **37**(3B), 1289–1294.

[3.44] S. Takagi, J. Koga and A. Toriumi (1997). Subband structure engineering for performance enhancement of Si MOSFETs. *IEDM Technical Digest*, 219–222.

[3.45] Y. Yamada, H. Tsuchiya and M. Ogawa (Jul. 2009). Quantum transport simulation of silicon-nanowire transistors based on direct solution approach of the Wigner transport equation. *IEEE Transactions on Electron Devices* **56**(7), 1396–1401.

[3.46] D. Querlioz, J. S.-Marti, K. Huet, A. Bournel, V. A.-Fortuna, C. Chassat, S. G.-Retailleau and P. Dollfus (Sep. 2007). On the ability of the particle Monte Carlo technique to include quantum effects in nano-MOSFET simulation. *IEEE Transactions on Electron Devices* **54**(9), 2232–2242.

[3.47] C.-W. Lee, A. Afzalian, N. D. Akhavan, R. Yan, I. Ferain and J.-P. Colinge (Feb. 2009). Junctionless multigate field-effect transistor. *Applied Physics Letters* **94**(5), p. 053511.

[3.48] J.-P. Colinge, C.-W. Lee, A. Afzalian, N. D. Akhavan, R. Yan, J. Ferain, P. Razavi, B. O'Neill, A. Blake, M. White, A.-M. Kelleher, B. McCarthy and R. Murphy (Mar. 2010). Nanowire transistors without junctions. *Nature Nanotech* **5**, 225–229.

[3.49] E. Gnani, A. Gnudi, S. Reggiani and G. Baccarani (Sep. 2011). Theory of the junctionless nanowire FET. *IEEE Transactions on Electron Devices* **58**(9), 2903–2910.

[3.50] B. Sorée, W. Magnus and W. Vandenberghe (Dec. 2011). Low-field mobility in ultrathin silicon nanowire junctionless transistors. *Applied Physics Letters* **99**(23), p. 233509.

[3.51] R. Rios, A. Cappellani, M. Armstrong, A. Budrevich, H. Gomez, R. Pai, N. Rahhal-orabi and K. Kuhn (Sep. 2011). Comparison of junctionless and conventional trigate transistors with *Lg* down to 26 nm. *IEEE Electron Device Letters* **32**(9), 1170–1172.

[3.52] S. Barraud, M. Berthomé, R. Coquand, M. Cassé, T. Ernst, M.-P. Samson, P. Perreau, K. K. Bourdelle, O. Faynot and T. Poiroux, (Sep. 2012). Scaling of trigate junctionless nanowire MOSFET with gate length down to 13 nm. *IEEE Electron Device Letters* **33**(9), 1225–1227.

[3.53] N. Kadotani, T. Ohashi, T. Takahashi, S. Oda and K. Uchida (Sep. 2011). Experimental study on electron mobility in accumulation-mode silicon-on-insulator metal-oxide-semiconductor field-effect transistors. *Japanese Journal of Applied Physics* **50**, p. 094101.

[3.54] K. Goto, T.-H. Yu, J. Wu, C. H. Diaz and J. P. Colinge (Aug. 2012). Mobility and screening effect in heavily doped accumulation-mode metal-oxide-semiconductor field-effect transistors. *Applied Physics Letters* **101**(7), p. 073503.

[3.55] T. Rudenko, A. Nazarov, I. Ferain, S. Das, R. Yu, S. Barraud and P. Razavi (Nov. 2012). Mobility enhancement effect in heavily doped junctionless nanowire silicon-on-insulator metal-oxide-semiconductor field-effect transistors. *Applied Physics Letters* **101**(21), p. 213502.

[3.56] J. Choi, K. Nagai, S. Koba, H. Tsuchiya and M. Ogawa (Apr. 2012). Performance analysis of junctionless transistors based on Monte Carlo simulation. *Applied Physics Express* **5**, p. 054301.

[3.57] K. Nagai, H. Tsuchiya and M. Ogawa (Mar. 2013). Channel length scaling effects on device performance of junctionless field-effect transistor. *Japanese Journal of Applied Physics* **52**, p. 044302.

[3.58] J. P. Duarte, S.-J. Choi and Y.-K. Choi (Dec. 2011). A full-range drain current model for double-gate junctionless transistors. *IEEE Transactions on Electron Devices* **58**(12), 4219–4225.

[3.59] Z. Chen, Y. Xiao, M. Tang, Y. Xiong, J. Huang, J. Li, X. Gu and Y. Zhou (Dec. 2012). Surface-potential-based drain current model for long-channel junctionless double-gate MOSFETs. *IEEE Transactions on Electron Devices* **59**(12), 3292–3298.

[3.60] K. Wei, L. Zeng, J. Wang, G. Du and X. Liu (Aug. 2014). Physically based evaluation of electron mobility in ultrathin-body double-gate junctionless transistors. *IEEE Electron Device Letters* **35**(8), 817–819.

[3.61] M. Ichii, R. Ishida, H. Tsuchiya, Y. Kamakura, N. Mori and M. Ogawa (Apr. 2015). Computational study of effects of surface roughness and impurity scattering in Si double-gate junctionless transistors. *IEEE Transactions on Electron Devices* **62**(4), 1255–1261.

[3.62] R. Kotlyar, B. Obradovic, P. Matagne, M. Stettler and M. D. Giles (June 2004). Assessment of room-temperature phonon-limited mobility in gated silicon nanowires. *Applied Physics Letters* **84**(25) 5270–5272.

[3.63] N. D. Akhavan, A. Afzalian, A. Kranti, I. Ferain, C.-W. Lee, R. Yan, P. Razavi, R. Yu and J.-P. Colinge (Apr. 2011). Influence of elastic and inelastic electron-phonon interaction on quantum transport in multigate silicon nanowire MOSFETs. *IEEE Transactions on Electron Devices* **58**(4), 1029–1037.

[3.64] Y. Yamada, H. Tsuchiya and M. Ogawa (Mar. 2012). Atomistic modeling of electron-phonon interaction and electron mobility in Si nanowires. *Journal of Applied Physics* **111**, p. 063720.

[3.65] Y.-M. Niquet, C. Delerue, D. Rideau and B. Videau (May 2012). Fully atomistic simulation of phonon-limited mobility of electrons and holes in <001>-, <110>- and <111>-oriented Si nanowires. *IEEE Transactions on Electron Devices* **59**(5), 1480–1487.

[3.66] N. Mori, G. Mil'nikov, H. Minari, Y. Kamakura, T. Zushi, T. Watanabe, M. Uematsu, K. Itoh, S. Uno and H. Tsuchiya (Dec. 2013). Nano-device simulation from an atomistic view. *IEDM Technical Digest*, 116–119.

[3.67] M. Luisier, A. Schenk and W. Fichtner (Mar. 2007). Atomistic treatment of interface roughness in Si nanowire transistors with different channel orientations. *Applied Physics Letters* **90**, p. 102103.

[3.68] S.G. Kim, M. Luisier, A. Paul, T. B. Boykin and G. Klimeck (May 2011). Full three-dimensional quantum transport simulation of atomistic interface roughness in silicon nanowire FETs. *IEEE Transactions on Electron Devices* **58**(5), 1371–1380.

[3.69] R. Kim and M. S. Lundstrom (Jan. 2009). Physics of carrier backscattering in one- and two-dimensional nanotransistors. *IEEE Transactions on Electron Devices* **56**(1), 132–139.

[3.70] N. Neophytou, A. Paul, M. S. Lundstrom and G. Klimeck (June 2008). Bandstructure effects in silicon nanowire electron transport. *IEEE Transactions on Electron Devices* **55**(6), 1286–1297.

[3.71] S. Barraud, E. Sarrazin and A. Bournel (2011). Temperature and size dependencies of electrostatics and mobility in gate-all-around MOSFET devices. *Semiconductor Science and Technology* **26**, p. 025001.

[3.72] S. M. Goodnick, D. K. Ferry and C. W. Wilmsen (Dec. 1985). Surface roughness at the Si(100)-SiO_2 interface. *Physical Review B* **32**(12), 8171–8186.

[3.73] E. B. Ramayya (2010). *Thermoelectric properties of ultrascaled silicon nanowires*. PhD dissertation at the University of Wisconsin-Madison.

[3.74] S. Barraud (Nov. 2011). Dissipative quantum transport in silicon nanowires based on Wigner transport equation, *Journal of Applied Physics* **110**, p. 093710.

4

Phonon Transport in Si Nanostructures

Growing heat dissipation has become one of the main issues limiting the reliability and performance of today's IC chips [4.1]. Considering the trend of nanoscale CMOS technology, new structures such as Fin, ultra-thin body, and nanowire structures all aim to maintain transistor performance even on the smallest scales [4.2]. In these structures, however, more significant self-heating effects occur, compared with traditional planar MOSFETs, because they are surrounded by materials with low thermal conductivity, such as SiO_2, which weakens the thermal coupling between the channel and the substrate [4.3–4.6]. Furthermore, it is expected that the thermal conductivity of nanostructure materials is strongly reduced, due to frequent phonon scattering at boundaries [4.7, 4.8]. This property is preferable for (for example) thermoelectric applications, and now attracts not only academic but also industrial interest [4.9–4.11]. Therefore, accurate simulations are required for analyzing the heat transfer problems in nanoscale semiconductors.

When modeling heating behavior on such a small scale, the conventional approach, based on Fourier's law, is no longer appropriate. For example, if the characteristic size of the device structure is significantly smaller than the phonon mean free path (i.e., ballistic transport limit), scattering events are so rare that a thermodynamic equilibrium of phonons may not exist in such structures [4.12]. Molecular dynamics (MD) is known as a rigorous simulation method for analyzing heat transfer properties in atomic-scale structures; however, its application to realistic device structures would be impractical, because of the huge computational cost [4.13, 4.14]. In addition, the nonequilibrium Green's function (NEGF) method has been recently adapted to the phonon transport simulation, which is suitable to assess the heat transfer in the ballistic limit [4.15–4.18]. However, the computational burden is so tremendous that some of the physical accuracy has to be sacrificed to speed up simulation.

The Monte Carlo (MC) method for directly solving the phonon Boltzmann transport equation (BTE) is considered to be a useful approach for simulating the heat transfer in realistic nanoscale devices structures with an adequate balance of accuracy and computational

Carrier Transport in Nanoscale MOS Transistors, First Edition. Hideaki Tsuchiya and Yoshinari Kamakura.

efficiency and expansion for realistic complex device structures. It has the potential to include various physical mechanisms relating to phonon scattering and kinetics, similar to the MC method for electron transport, which has been extensively studied [3.12, 3.13]. Furthermore, realistic treatment of the boundary scattering process is possible, not only for the uniform system (e.g., films or wires), but also for more complex geometries of realistic devices.

There have been many reports on the calculation of phonon transport, either by solving the BTE analytically [4.8, 4.19–4.23] or by using an MC method [4.24–4.32]. The phonon MC simulation was first reported by Peterson *et al.*, assuming the Debye model without considering the phonon dispersion and various polarizations [4.24]. Mazumder *et al.* [4.25] then presented the comprehensive algorithm for the phonon MC method with the inclusion of phonon dispersion and polarization, and Lacroix *et al.* [4.26] developed the scattering algorithm by using the Kirchhoff's law during the scattering phase. By using the MC method, Mittal *et al.* [4.29] investigated the influence of the optical phonons on the heat conduction, and Bera *et al.* [4.31] reported that phonon confinement effect becomes important in Si nanowires less than 37 nm. McGaughey *et al.* [4.32] used the MC scheme to sample the free path of phonon-phonon scattering with a Poisson distribution, and added boundary scattering by launching phonons in a random position in the structure modeled in the simulation.

On the other hand, in the recent analytical approaches based on the relaxation time approximation, rigorous physical models, such as full phonon dispersion curves, were included [4.21–4.23]. However, as for the MC analysis, the physical models are not necessarily so sophisticated; for example, the approximated phonon dispersion relationships using the fitted curves for the [100] direction only were commonly assumed in the previous works. Furthermore, there is still ambiguity in the treatment of the boundary scattering process, which may significantly affect the heat conduction analysis in nanostructured devices. For example, MC analysis has been recently done by considering the surface roughness in ultrathin nanowires (<8 nm) [4.33].

In this study, a MC method for solving the phonon BTE was developed to perform more accurate heat transfer simulations in nanoscale Si devices [4.34–4.37]. To this end, information about realistic phonon dispersion relationships for bulk Si was included in the MC simulation, and the results were compared with an approach using an approximated dispersion model [4.34, 4.35]. Furthermore, phonon boundary scattering was treated by tracing particle trajectories, and the validity of the reflection algorithms was verified through comparison with experimental data for the thermal conductivity of Si films and nanowires [4.36, 4.37]. For very thin films and wires (< ≈ 20 nm), the phonon density of states (DOS) is different from the bulk DOS [4.38], and this effect is not discussed here. Our focus is on the boundary scattering effect in nanostructures with bulk-like DOS. In order to evaluate the thermal conductivity, we adopted one particle MC method, which enables obtaining the statistical convergence efficiently.

Furthermore, we used the MC simulator to analyze the quasi-ballistic phonon transport effect on the heat conduction in Si. Therefore, in this work, the realistic accurate phonon dispersion model is compared with the approximated one commonly used in previous MC works, and the importance of the phonon dispersion relation on the thermal properties in the ballistic transport regime is discussed. If the phonons can travel ballistically, the transport nature is considered to be similar to the properties observed with photons exchanged between two black plates at different temperatures. In this case, the steady state temperature in the heat flow region becomes a constant value, which is equal to the Stefan-Boltzmann law. The ballistic transport features obeying this law were well calculated in the previous MC works [4.26, 4.30] and, therefore, we consider that the MC method is a good tool for studying the mixture regime between the ballistic and diffusive nature for the phonon transport.

4.1 Monte Carlo Simulation Method

4.1.1 Phonon Dispersion Model

Figure 4.1 shows the realistic dispersion relation of phonons in bulk Si calculated from the adiabatic bond charge model [4.39], whose accuracy was validated through comparison with experimental neutron scattering data [4.40, 4.41]. In many previous studies, approximate dispersion models were employed, an example of which is shown by the solid curves in Figure 4.1. In this model [4.20], the dispersion relations for the transverse acoustic (TA) and longitudinal acoustic (LA) modes along the [100] direction were fitted using linear curves, and isotropic symmetry was assumed for other directions. In addition, no dispersion was assumed for TA modes with wave vectors $q/q_{max} > 0.4$, where q_{max} is the largest wave vector of the first Brillouin zone (BZ) along the [100] direction. Although the curves along the Γ–X path are well fitted using this model, the dispersion of TA modes, particularly around the U, K, and W-points cannot be appropriately described. Considering that our aim is to simulate phonon transport at temperatures larger than 300 K, where phonons are distributed over a wide frequency range, the use of a more accurate dispersion relation is preferable.

Hence, in this work, the phonon DOS and group velocity were calculated from the realistic dispersion relation, and the data as a function of the phonon frequency ω and the phonon polarization p were included in the MC simulation as lookup tables. The phonon DOS can be expressed as an integral in the first BZ:

$$D_p(\omega) = \frac{1}{8\pi^3} \int_{BZ} d^3q\, \delta\left(\omega_p(\mathbf{q}) - \omega\right) \tag{4.1}$$

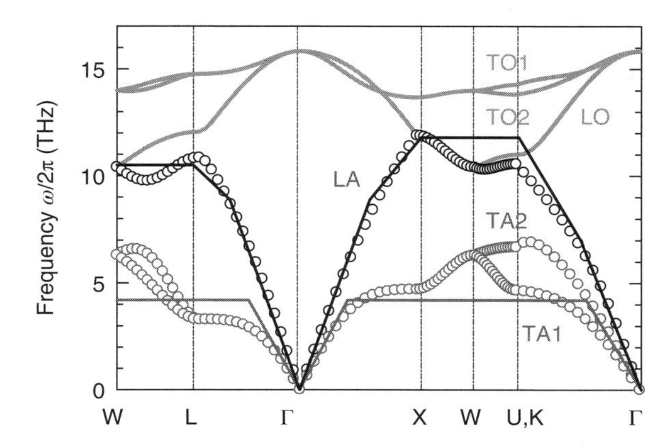

Figure 4.1 Phonon dispersion curves for bulk Si obtained using the adiabatic bond charge model [4.39] (symbols). The lines are the approximated isotropic dispersion curves fitted to the acoustic modes along the [100] (Γ–X) direction [4.20].

where $\mathbf{q}$ is the phonon wave vector and δ is the Dirac delta function. Although the phonon group velocity ($= \nabla_q \omega_p$) is actually dependent on $\mathbf{q}$, we assumed that the isotropic approximation and the average phonon group velocity on the constant-frequency surface in q-space was evaluated as:

$$\bar{v}_p(\omega) = \frac{1}{8\pi^3 D_p(\omega)} \int_{BZ} d^3q \left|\nabla_q \omega_p\right| \delta\left(\omega_p(\mathbf{q}) - \omega\right) \tag{4.2}$$

The three-dimensional integrals in Equations (4.1) and (4.2) were evaluated numerically by dividing the first BZ into small tetrahedra and using the linearly interpolated values in each element [4.42]. The mesh size along the Γ–X direction was $0.01(2\pi/a)$, where a is the lattice constant of Si, and so the number of the tetrahedron elements was 2.4×10^7 in the integral domain $q_x, q_y, q_z \geq 0$.

Figure 4.2 compares $D_p(\omega)$ and $\bar{v}_p(\omega)$, calculated using the different dispersion models. In the [100] model, a Van Hove singularity appears due to the flat dispersion of TA modes with $q/q_{max}>0.4$, and hence a delta function DOS actually exists at a frequency of 4.2 THz (which is omitted from the plot). Compared with the approximated [100] dispersion model, the realistic full-band model exhibits a larger DOS and smaller group velocity over a wide frequency range. This is caused by the limitations of the isotropic approximation, as well as the use of linear curves for fitting. Furthermore, in the [100] model, there is no DOS for TA modes in the region of 4–7 THz, because the dispersions around the U, K, and W points are not appropriately described as shown in Figure 4.1.

4.1.2 Particle Simulation of Phonon Transport

In the MC simulation, phonons in Si were modeled as particles and, by tracking the Brownian motion of each MC particle, heat conduction was analyzed. In this work, it was assumed that

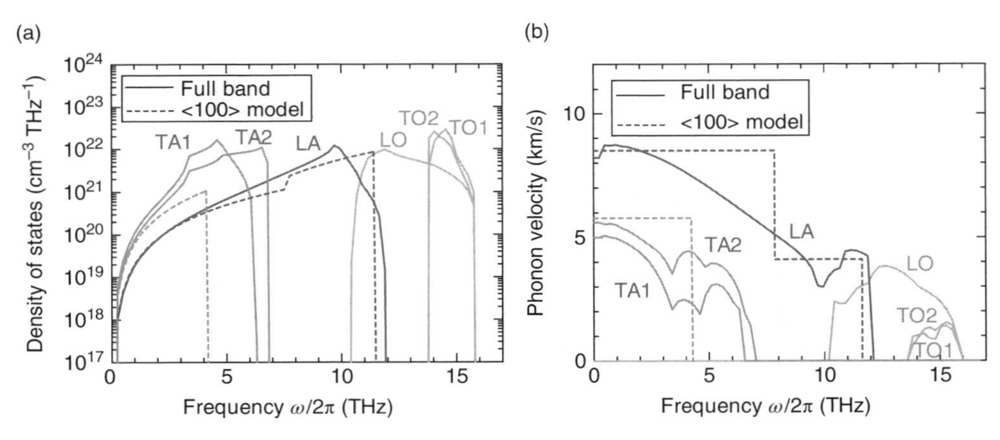

Figure 4.2 (a) Calculated phonon DOS; and (b) directionally averaged group velocity for bulk Si, plotted as a function of the frequency $\omega/2\pi$. The results calculated using the full band dispersion relation, based on the adiabatic bond charge model [4.39] (solid lines) and approximated [100] model [4.20] (dashed lines), are compared.

all the MC particles carry the same thermal energy E^*. Hence, the number of MC particles N^* can be determined by:

$$N^* = \frac{U}{E^*} \tag{4.3}$$

where U is the total phonon energy in the simulation domain, expressed as:

$$U = V\sum_p \int d\omega \hbar\omega D_p(\omega) n(\omega) \tag{4.4}$$

where V is the volume of the simulated system, and $n(\omega)$ is the Bose-Einstein distribution function:

$$n(\omega) = \frac{1}{\exp(\hbar\omega / k_B T) - 1} \tag{4.5}$$

where k_B is the Boltzmann constant and T is the lattice temperature.

For each MC particle, the phonon state (i.e., ω and p), the position $\mathbf{r}$, and the traveling direction vector $\mathbf{u}$, was assigned. It should be noted that E^* is not necessarily the same as the real phonon energy $\hbar\omega$, and thus the number of phonons carried by one MC particle is given by the following function:

$$W(\omega) = \frac{E^*}{\hbar\omega} \tag{4.6}$$

At the beginning of the simulation, MC particles were generated according to the equilibrium distribution. In order to do that, the phonon state (ω, p) was randomly selected for each particle following the probability distribution function:

$$P(\omega, p) \propto \frac{D_p(\omega) n(\omega)}{W(\omega)} \tag{4.7}$$

In this work, the anisotropy of the crystal was not taken into account, and hence the traveling direction was also randomly determined assuming the system isotropy as:

$$\mathbf{u} = (\sin\theta\cos\varphi, \sin\theta\sin\varphi, \cos\theta) \tag{4.8}$$

where $\cos\theta = 1 - 2R_1$, $\sin\theta = (1 - \cos^2\theta)^{0.5}$, $\varphi = 2\pi R_2$, and R_1 and R_2 are uniform random numbers in [0,1].

4.1.3 Free Flight and Scattering

In the MC simulation, the time evolution of the states assigned to each particle was calculated. The phonons propagate at the group velocity and undergo a series of stochastic scattering events. In this study, three types of scattering mechanisms were considered; phonon-phonon scattering due to three or more phonon Umklapp processes (τ_U), phonon-defect scattering (τ_d) and phonon-boundary scattering (τ_b). Assuming that the scattering mechanisms are not coupled with each other, the total relaxation time can be written as

$$\tau^{-1} = \tau_U^{-1} + \tau_d^{-1} + \tau_b^{-1} \tag{4.9}$$

The phonon-phonon scattering (i.e., the three or more phonon Umklapp processes) was treated as an elastic one-particle scattering event. This mechanism relaxes the phonon energy distribution in the simulation domain to the thermal equilibrium. Actually, the rigorous treatment of three-phonon process is possible in the MC method, but a huge computational cost would be required [4.25, 4.26]. Therefore, in this work, while preserving the energy conservation, the number of phonons carried by one MC particle was changed at every phonon-phonon scattering event, like the actual three or more phonon scattering processes. By using the constants A_p, χ_p, and ξ_p as determined by p, the Umklapp scattering rate was expressed as [4.20]:

$$\tau_U^{-1}(\omega, p) = A_p \omega^{\chi_p} T^{\xi_p} \exp\left(-\frac{B_p}{T}\right) \tag{4.10}$$

The scattering rate due to phonon-defect interactions can be written as [4.20]:

$$\tau_d^{-1}(\omega) = C\omega^4 \tag{4.11}$$

which is especially important at low temperatures.

Furthermore, two types of phonon-boundary scattering algorithms were considered in this study. The first was a stochastic scattering mechanism used to fit the bulk thermal conductivity at very low temperatures, but can be ignored at room temperature. The scattering rate is expressed as [4.20]:

$$\tau_b^{-1}(\omega, p) = \frac{\overline{v}_p(\omega)}{FL} \tag{4.12}$$

where L is the distance that a phonon can travel between two rough boundary surfaces, and F is a correction parameter for achieving an exact fit at low temperatures [4.19]. The second algorithm was implemented to simulate phonon scattering at boundaries in nanostructures, such as thin films and nanowires. In this study, we treated this process by actually tracing the particle trajectories. The detailed algorithm will be discussed later in Section 4.2.2.

It should be noted that in this study, the normal scattering mechanism was neglected because it does not directly contribute to heat flow. However, it is also well known that the normal scattering process is important for generating the appropriate distribution function, and indirectly affects transport [4.43]. In this study, to compare the MC simulation results to the previous analytical approach, we followed the scattering models proposed by Chantrenne *et al.* [4.20], in which the influence of the normal process would be effectively taken into account to reproduce the thermal conductivity. Of course, for more accurate simulation, the inclusion of the normal process is preferable, and it would be easily implemented by, for example, conserving the momentum, or **u**, before and after the scattering event.

In the MC simulator, the sequence of free flight and scattering was repeated. After a free flight with a duration of Δt, the position of the phonon was updated as:

$$\mathbf{r}(t+\Delta t)=\mathbf{r}(t)+\overline{v}_p(\omega)\mathbf{u}\Delta t \tag{4.13}$$

where the time step Δt should be smaller than τ. Then, the scattering mechanism i was selected according to the probability [4.26]:

$$P_i = 1-\exp\left(-\frac{\Delta t}{\tau_i(\omega,p)}\right) \tag{4.14}$$

where i is U, d, or b. When phonon-defect or boundary scattering was selected, ω was not changed, but $\mathbf{u}$ was randomly changed using Equation (4.8). Since Umklapp scattering changes the phonon energy distribution, it is necessary to appropriately select the phonon state after the scattering event. Thus, the phonon state (ω', p') after phonon-phonon scattering was randomly selected following the probability distribution function:

$$P_{\text{Kirch}}(\omega',p') \propto \left[1-\exp\left(-\frac{\Delta t}{\tau_U}\right)\right] P(\omega',p') \tag{4.15}$$

which is a modified version of Equation (4.7) considering the Kirchhoff law [4.26]. Using this, we can maintain the thermal equilibrium of the phonon system by controlling the phonon energy distribution function.

4.2 Simulation of Thermal Conductivity

4.2.1 *Thermal Conductivity of Bulk Silicon*

In this study, the thermal conductivity of bulk Si was evaluated from the product of the specific heat per unit volume (C_V) and the diffusion coefficient (D_{3D}) as [4.44]

$$\kappa = C_V D_{3D} \tag{4.16}$$

where the specific heat can be calculated from the phonon DOS by integrating over ω:

$$C_V(T)=\frac{1}{V}\frac{\partial U}{\partial T}=\sum_p \int d\omega\, c_v(\omega,p) \tag{4.17}$$

where $c_v(\omega, p) \equiv \hbar\omega D_p(\omega)(\partial n/\partial T)$ is the specific heat per mode, whose frequency dependence is plotted in Figure 4.3, and the resulting $C_V(T)$ is presented in Figure 4.4.

It was confirmed that the phonon DOS based on the adiabatic bond charge model well reproduces the measured specific heat [4.45, 4.46] if all the branches, including both acoustic and optical modes are considered. However, for simulating k, we disregarded the optical branches because of their negligible contribution; the low group velocity (see Figure 4.2(b)) and the short relaxation time [4.47] significantly impede the heat transportability of optical phonons. Figure 4.4 also compares C_V, calculated using the realistic and approximated [100] dispersion relationship for acoustic phonons. When calculating C_V using the [100] dispersion model, we also eliminated the contribution of TA modes with $q=q_{\max}>0.4$, because they have

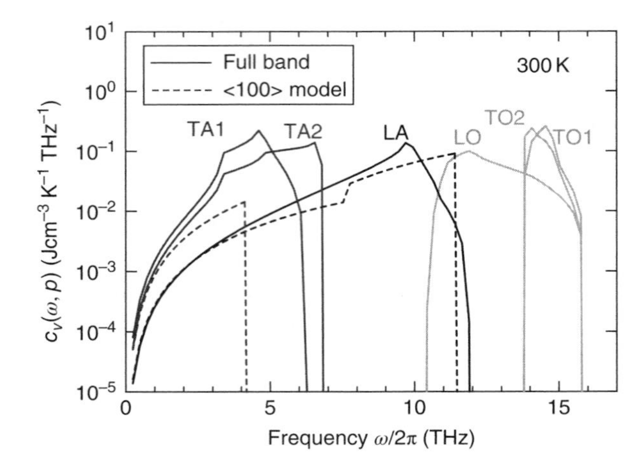

Figure 4.3 Specific heat per mode $c_v(\omega, p)$ for bulk Si at 300 K. The results calculated from the realistic full band dispersion (solid curves) and approximated [100] model (dashed curves) are compared.

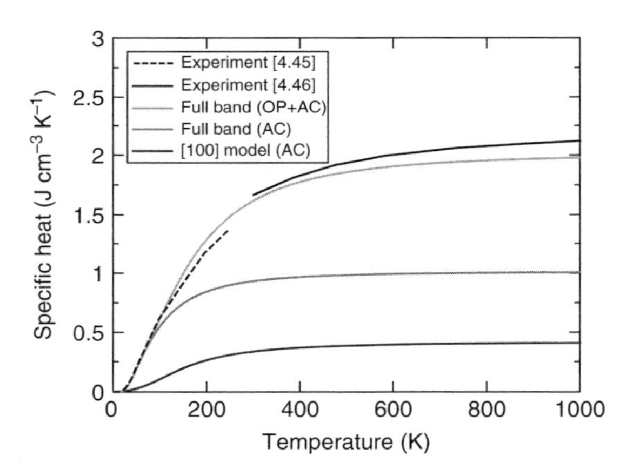

Figure 4.4 Specific heat capacity per unit volume C_V for bulk Si, plotted as a function of temperature. The results calculated from the phonon DOS presented in Figure 4.2(a) are compared to the experimental data [4.45, 4.46].

zero group velocity and do not contribute to heat conduction. Note that the two models differ by a factor of approximately two; this is mainly due to the poorer description of the TA modes in the [100] dispersion model.

Three-dimensional diffusion coefficient at a given temperature T was obtained by the MC simulation in the homogeneous system using the following formula [3.12]:

$$D_{3D} = \frac{1}{3N^* t_{sim}} \sum_{i=1}^{N^*} \frac{(x_i - \overline{x})^2 + (y_i - \overline{y})^2 + (z_i - \overline{z})^2}{2} \tag{4.18}$$

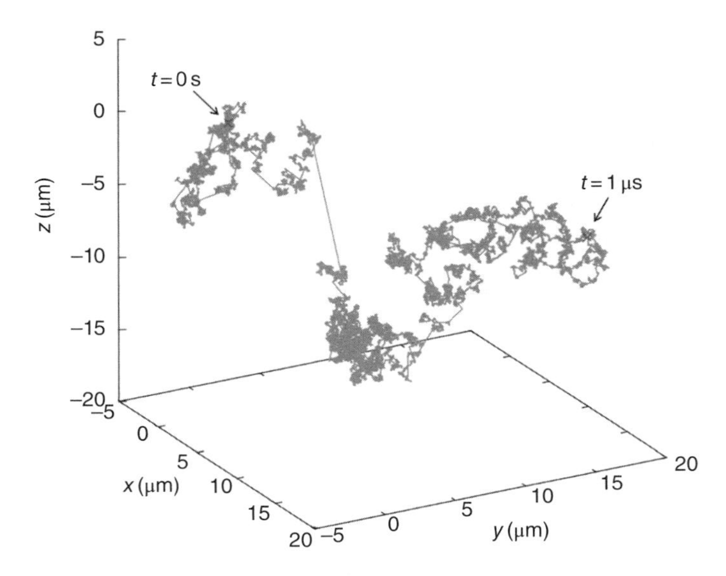

Figure 4.5 Trajectory of a particle motion simulated with the Monte Carlo method for 1 μs at 300 K.

where (x_i, y_i, z_i) express the displacement of the i-th particle during the simulation time from $t=0$ to t_{sim}, and ($\bar{x}$, $\bar{y}$, $\bar{z}$) are their ensemble average (e.g., $\bar{x} = \sum x_i / N^*$). We used $N^* = 10\ 000$ particles in this work, and t_{sim} was taken to be long enough to obtain the convergence (typically ≈ 1 μs). Figure 4.5 shows the example for the trajectory, simulated at 300 K. Note that not only the Brownian motions with frequent direction change, but also long-distance free flights are occasionally observed, as a consequence of the long-tailed nature of the distribution of phonon mean free paths λ [4.48].

In many previous MC simulations [4.24–4.28, 4.30] thermal conductivities were calculated by measuring the heat flow between the two constant-temperature baths. However, this would not be necessarily an efficient method, because it requires a trial-and-error process for determining the distance between the two baths. To eliminate the quasi-ballistic transport effect, it must be long enough compared to the phonon mean free path, which mainly depends on the temperature – while, from the viewpoint of computational cost, it is better to reduce the simulation system size. Thus, it is difficult to determine the appropriate system size to evaluate κ in the diffusive limit.

Figure 4.6 shows the temperature dependence of the thermal conductivity of bulk Si. When we used the approximated phonon dispersion, the scattering rate parameters given in the literature [4.20] well reproduced the experimental data [4.19].

When the realistic dispersion was used, a readjustment of the parameters was needed as shown in Table 4.1 and, after this procedure, good results were also obtained.

Figure 4.7 shows the phonon frequency dependence of the scattering rates at 3 K and 300 K. In Figure 4.6, we also plotted κ obtained with the analytical formula derived from BTE [4.20]:

$$\kappa = \frac{1}{3} \sum_{p \in \mathrm{AC}} d\omega \bar{v}_p(\omega)^2 \tau(\omega, p) c_v(\omega, p) \tag{4.19}$$

in which p only runs over acoustic modes (AC). Similar results were obtained regardless of the calculation method, indicating the validity of the present MC approach.

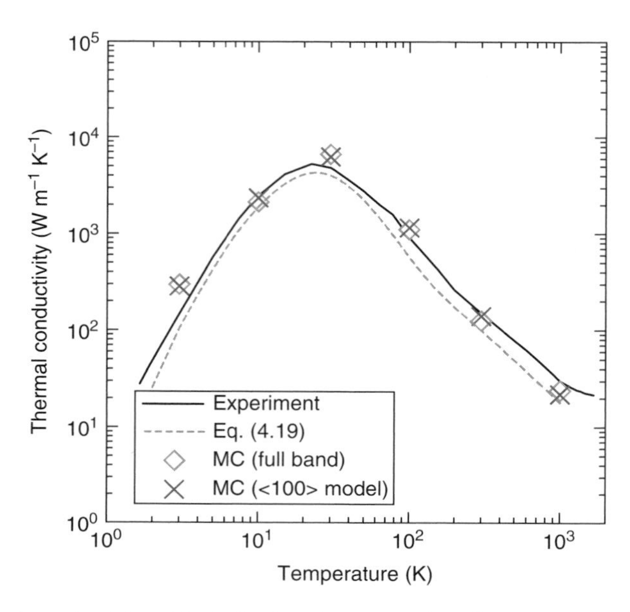

Figure 4.6 Thermal conductivity of bulk Si plotted as a function of temperature. The results obtained from the MC method (symbols) are compared with the experimental data (solid line) [4.19]. The dashed line represents the results calculated with the analytical formula given in Equation (4.19).

Table 4.1 Parameters for phonon relaxation times adjusted to yield the correct thermal conductivity of bulk Si.

Parameter	This work	[100] model[4.20]
A_{TA} ($\times 10^{-12} s^{\chi_{TA}-1}$ $K^{-\xi_{TA}}$)	1.05	0.7
A_{LA} ($\times 10^{-12} s^{\chi_{LA}-1}$ $K^{-\xi_{LA}}$)	4.5	3.0
B_{TA} (K)	0	0
B_{LA} (K)	0	0
χ_{TA}	1	1
χ_{LA}	2	2
ξ_{TA}	3.86	4
ξ_{LA}	1.36	1.5
C ($\times 10^{-45}$ s^3)	4.95	1.32
L (mm)	7.16	7.16
F	0.33	0.55

4.2.2 Thermal Conductivity of Silicon Thin Films

The thermal conductivity of Si thin films was then calculated using a MC simulation which had been calibrated to reproduce the correct bulk κ, as explained in Section 4.2.1. In this work, we did not consider such ultra-thin films that the phonon dispersion relation is modulated by the confinement effect [4.49], and the three-dimensional simulation was carried out. Along the thickness direction (the z-direction), two infinite parallel plane

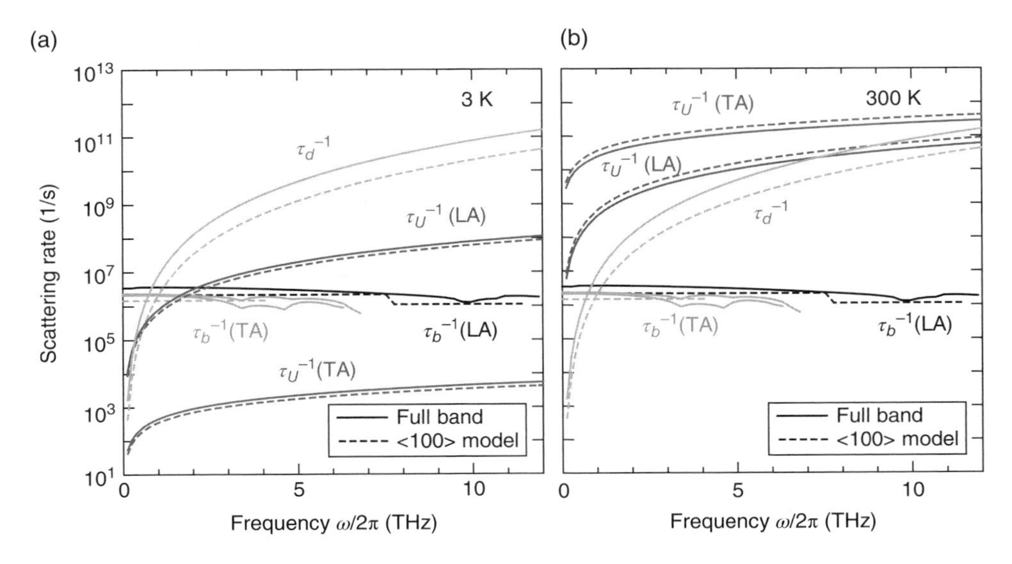

Figure 4.7 Frequency dependence of the phonon scattering rates at (a) 3 K; and (b) 300 K.

walls were placed with a separation d, and the motion of phonons confined between the two walls was simulated. Similarly to the method used for the bulk case, the thermal conductivity was obtained using the relationship $\kappa = C_v D_{2D}$, where the two-dimensional thermal diffusivity

$$D_{2D} = \frac{1}{2N^* t_{sim}} \sum_{i=1}^{N^*} \frac{(x_i - \bar{x})^2 + (y_i - \bar{y})^2}{2} \tag{4.20}$$

was obtained through the MC simulation.

In this study, the phonons were assumed to be completely reflected to random directions at the boundary walls, which resulted in the reduction of κ due to the boundary scattering. In some previous studies using the MC method [4.25, 4.27], the scattering angle θ was randomly selected using the equation:

$$\cos\theta = R \tag{4.21}$$

where R is a uniform random number in [0,1]. This method generates isotropically distributed θ values, as illustrated in Figure 4.8(a) to describe diffuse boundary scattering due to surface roughness. However, the present MC simulation results suggest that this reflection method causes unrealistic physical behavior; as shown in Figure 4.9(a), a nonuniform phonon distribution was observed along the thickness direction after the steady state had been achieved.

To avoid this difficulty, a better-suited boundary scattering algorithm should be implemented. Figure 4.8(b) illustrates another option for the reflection algorithm [4.28, 4.30, 4.50] based on Lambert's cosine law. This law is commonly used to describe the reflected light intensity at rough interfaces in the field of computer graphics [4.51], and also to simulate the

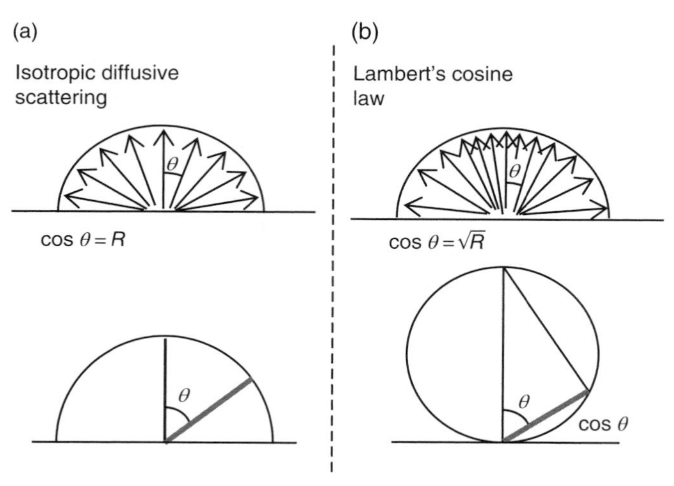

Figure 4.8 Schematic view illustrating diffuse phonon scattering at a boundary. In (a), the reflection angle θ is randomly selected based on Equation (4.21), while in (b), θ is determined based on Lambert's cosine law (Equation (4.22)). The strength of the θ distribution is proportional to the length of the thick lines.

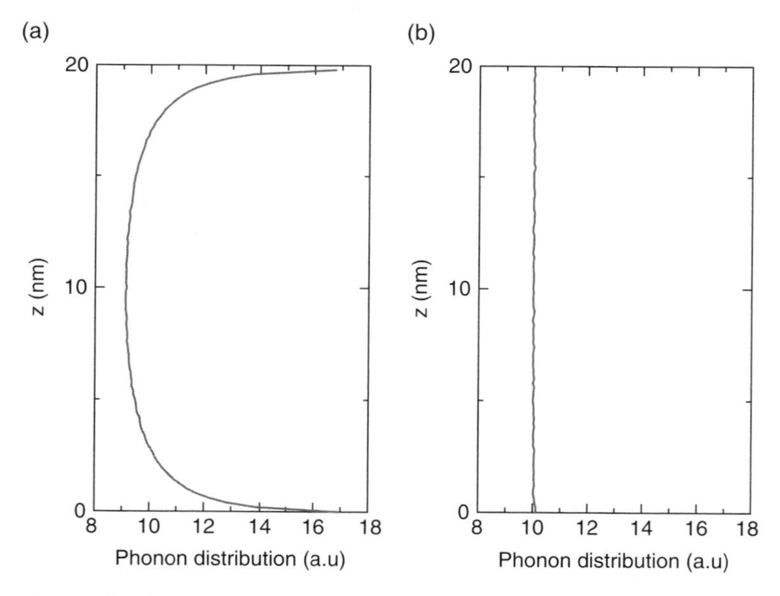

Figure 4.9 Phonon density distribution along the thickness direction z in the Si film, with $d = 20\,\text{nm}$. The steady state results obtained using the MC simulation with a boundary scattering algorithm based on (a) Equation (4.21); and (b) Equation (4.22) are compared.

flux of rarefied gas in small tubes [4.52]. It is suggested that the radiant intensity observed from an ideal diffusely reflecting surface is proportional to cos θ so that the scattering angle is given by:

$$\cos\theta = \sqrt{R} \tag{4.22}$$

instead of Equation (4.21). Note that in this case, a uniform phonon distribution was achieved, as shown in Figure 4.9(b).

Figure 4.10 shows the simulated thermal conductivities for Si films plotted as a function of the film thickness d. By assuming fully diffuse phonon scattering based on Lambert's cosine law at the top and bottom surfaces, good agreement was obtained with the experimental data measured down to 20 nm at several temperatures [4.53–4.56].

On the other hand, with a diffuse scattering algorithm based on Equation (4.21), we could not correctly reproduce the decrease in κ associated with reducing d. This could be because the scattering angles θ generated using Equation (4.21) are too large and, thus, the velocity component perpendicular to the plane tends to become smaller, which reduces the chance of the particles hitting the boundary. Phonon-boundary scattering is the dominant mechanism determining κ in thin films, and so an appropriate treatment of this process is essential for accurately simulating heat conduction.

Figure 4.11 compares κ simulated using different phonon dispersion models, plotted as a function of temperature. Here, diffuse boundary scattering based on Lambert's cosine law was assumed in both simulations. Note that κ obtained with the realistic dispersion agreed well with the measurements. On the other hand, the MC simulation using the approximated dispersion model underestimates κ, particularly in the low temperature region.

This situation could be corrected by introducing an algorithm to mix diffuse and specular reflections with some fitting parameters to describe the surface specularity [4.25], for example. However,

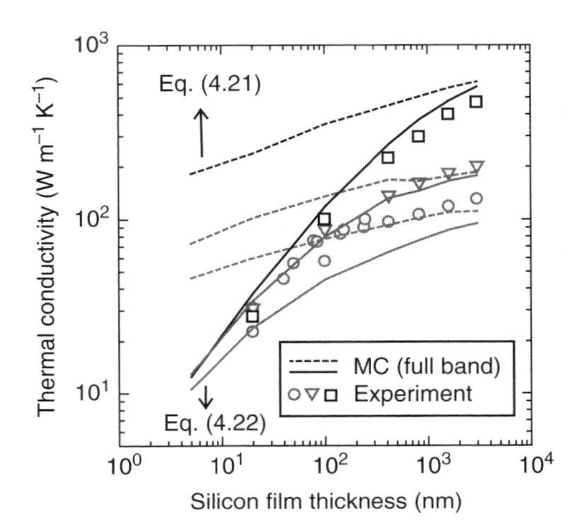

Figure 4.10 Thickness dependence of thermal conductivity of Si films at various temperatures. The results obtained using the MC simulation with a boundary scattering algorithm based on Equation (4.21) (dashed lines) and Equation (4.22) (solid lines) are compared with the experimental data [4.53–4.56].

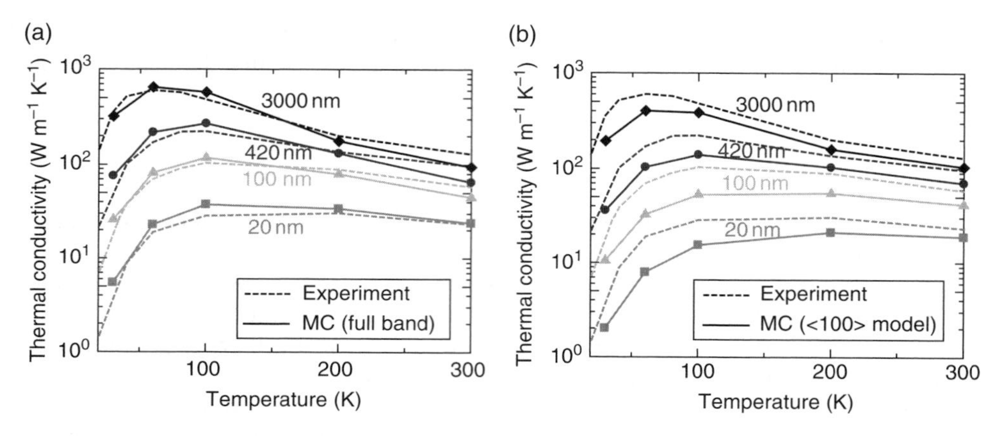

Figure 4.11 Temperature dependence of thermal conductivity of Si films with various thicknesses. The results obtained by the MC simulation with the full band dispersion relation, based on the adiabatic bond charge model [4.39] (a); and approximate [100] model [4.20] (b) are compared with the experimental data [4.53–4.56].

it has been suggested that phonon scattering at a boundary is fully diffusive at high temperatures (>≈30 K), while boundary scattering is thought to be partially specular at lower temperatures [4.57]. The present MC results with the full band model are consistent with this idea. Carefully looking at Figure 4.11(a), our simulation results underestimate the measured data, especially in the high temperature region. This would be due to the lack of consideration of the actual boundary scattering mechanism, which depends on the phonon polarization, momentum, and the features of the boundary roughness (correlation length and height) [4.23]. To obtain a more rigorous fit to the measured data, the inclusion of the partial specularity may be required; however, this would also raise κ at low temperatures. Further work is necessary to increase the accuracy of the MC simulation.

4.2.3 Thermal conductivity of silicon nanowires

The thermal conductivity of Si nanowires (SiNWs) was also calculated using the MC simulator. Here, we did not consider such thin nanowires where the phonon dispersion relationship is modulated by the confinement effect [4.58] and, hence, the three-dimensional simulation was carried out in the system illustrated in Figure 4.12. MC particles were placed inside an infinitely long cylinder with a diameter of d_{si}, and their stochastic trajectories were simulated. The particles were scattered when they hit the cylinder inner surface, and we assumed completely diffusive scattering obeying Lambert's cosine law to describe the phonon boundary scattering, as discussed in Section 4.2.2.

Figure 4.13 shows examples of the actual trajectories of the MC particles simulated for 250 ps. Looking at the case of 60 K, the particles were scattered only at the boundary, not inside the NW. This is due to the suppressed phonon-phonon scattering rate at low temperatures; thus, the boundary scattering becomes the dominant mechanism determining the thermal conductivity.

The thermal conductivity was obtained using the relationship $\kappa = C_V D_{1D}$, where the one-dimensional thermal diffusivity:

$$D_{1D} = \frac{1}{N^* t_{sim}} \sum_{i=1}^{N^*} \frac{(x_i - \bar{x})^2}{2} \tag{4.23}$$

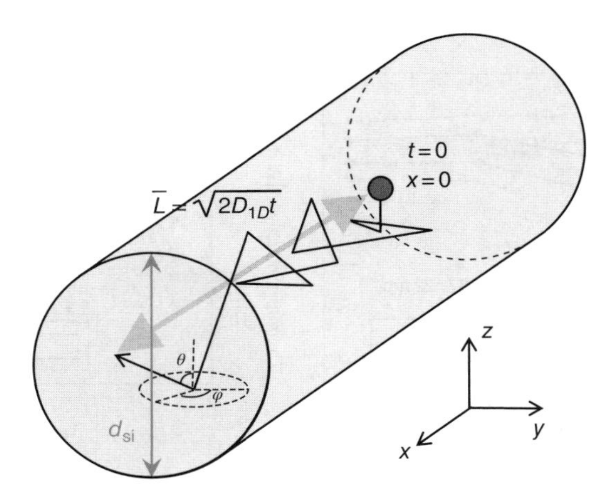

Figure 4.12 Schematic illustration of the MC method for simulating the thermal conduction in SiNWs. The phonons were modeled as particles, and their motion inside the cylindrical wall was simulated. The boundary scattering was assumed to be completely diffusive, and the scattering angle (θ, φ) was randomly selected according to Lambert's cosine law [4.29].

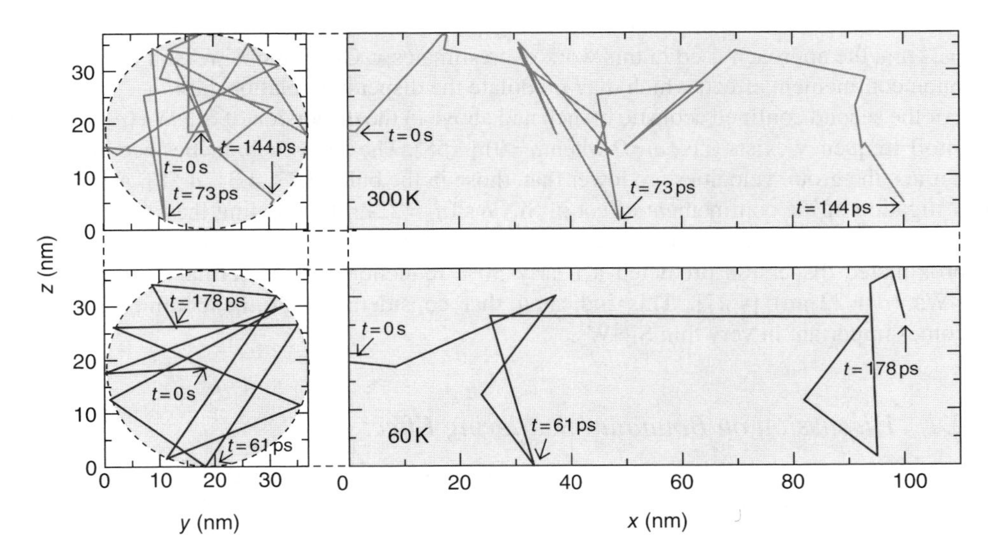

Figure 4.13 Examples of the MC particle trajectories simulated for 250 ps at 300 and 60 K in SiNW with $d_{si}=37$ nm. The initial particle position is $(x, y, z)=(0, d_{si}/2, d_{si}/2)$.

was obtained by MC simulation. The values of N^* and t_{sim} were the same as those used in the bulk simulation. Figure 4.14 shows the temperature dependence of κ in SiNWs simulated with the realistic and approximated dispersion models. Note that the experimental κ [4.59], except for the thinnest NW ($d_{si}=22$ nm), is well reproduced by the MC simulation, including the realistic phonon dispersion relationship. As mentioned above, we assumed completely diffuse scattering on the NW surface in this study, which is reasonable in accordance with the theoretical consideration [4.57]. On the other hand, for the MC simulation with the approximated

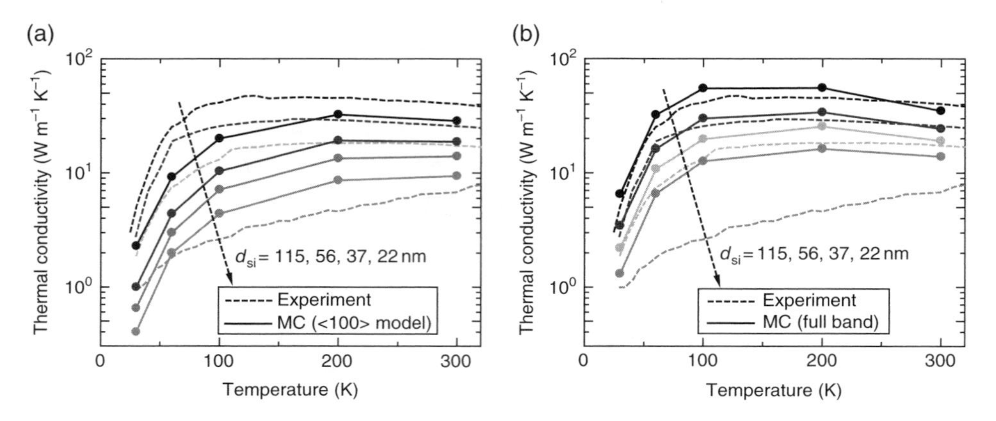

Figure 4.14 Comparison of the simulated (solid lines) and experimental [4.30] (dashed lines) thermal conductivities of SiNWs with various diameters plotted as a function of temperature. The simulated results with (a) realistic; and (b) approximated dispersion models are presented.

dispersion model, the simulated κ (d_{si}=37–115 nm) was underestimated compared with the experimental data. To obtain closer agreement, the partially specular reflection mechanism should be introduced for boundary scattering [4.27]. In the case of the thinnest SiNW of d_{si}=22 nm, the approach used in this work overestimates κ. One possible reason for this is the phonon confinement effect, which may modulate the dispersion relationship.

For the second confined acoustic branch and above in the dispersion of SiNWs (d_{si}=20 nm), a cutoff frequency exists (i.e., $\omega \neq 0$ when $q=0$) [4.58]. The acoustic branches become flatter and, thus, the group velocities are lower than those in the bulk [4.27, 4.31, 4.58]. Although an investigation of the confinement effect in SiNWs (d_{si}=22 nm), including the realistic dispersion, has not yet been carried out in detail, previous MC works taking into account the confined approximated dispersion predicted a nearly 30% reduction in the thermal conductivity of SiNWs (d_{si}=22 nm) [4.27]. This indicated that consideration of the confinement effect becomes important in very thin SiNWs.

4.2.4 Discussion on Boundary Scattering Effect

Finally, let us discuss the effect of the phonon dispersion model included in the MC simulator. Figure 4.15(a) shows the average phonon group velocity weighted by the specific heat:

$$\langle v \rangle = \frac{\sum_{p \in \mathrm{AC}} \int d\omega \, \overline{v}_p(\omega) c_v(\omega, p)}{C_V} \tag{4.24}$$

which was calculated with the full band and [100] dispersion models for bulk Si presented in Figure 4.1. Using this curve, the effective phonon mean free path can be estimated through the well-known relationship [4.21]

$$\langle \lambda \rangle = \frac{3\kappa}{C_V \langle v \rangle} \tag{4.25}$$

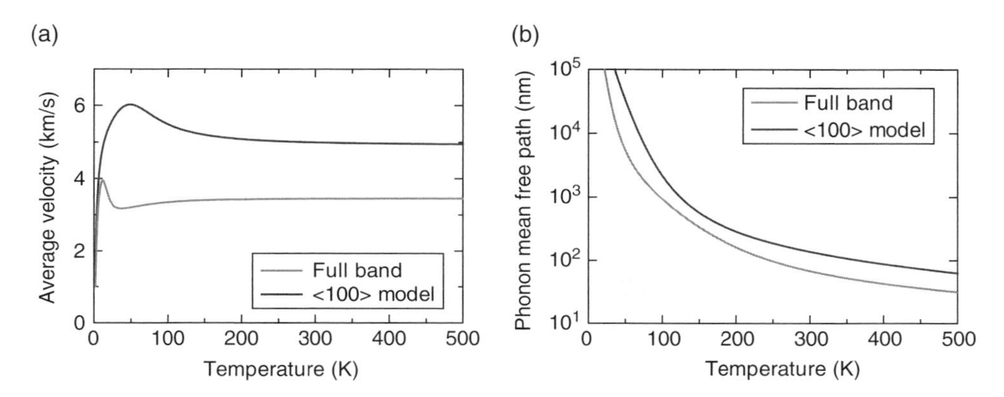

Figure 4.15 (a) Calculated averaged phonon group velocity; and (b) phonon mean free path, plotted as a function of temperature. The results obtained using the MC simulation with the full band dispersion relation, based on the adiabatic bond charge model [4.39] and approximated [100] model [4.20].

The calculated results are also plotted in Figure 4.15(b). A higher average velocity is found for the [100] model, which is expected from the comparison of $\bar{v}_p(\omega)$ shown in Figure 4.2(b). As explained in Section 4.2.1, despite the different $\langle v \rangle$ and C_V values for the two dispersion models, the MC simulator could be calibrated to yield the correct $\kappa(T)$ by adjusting the phonon relaxation times. However, the phonon mean free path would not necessarily be the same in the two models, and actually Figure 4.15 (b) exhibits a different $\langle \lambda \rangle$.

It is known that the impact of boundary scattering on κ is highly dependent on the phonon mean free path [4.60], and the discrepancy observed in Figures 4.11 and 4.14 can be attributed to the difference in $\langle \lambda \rangle$ between the two dispersion models. The longer phonon mean free path in a bulk material emphasizes the relative importance of boundary scattering, and then phonon confinement in a small space results in more significant reduction of κ.

To understand the effect of boundary scattering on thermal conductivity in more detail, the distribution of the phonon mean-free path was investigated. Figure 4.16 shows a comparison of the thermal conductivity κ per unit phonon mean free path λ between the successive collisions obtained by MC simulation in bulk Si. By integrating $\kappa(\lambda)$, the total thermal conductivity is obtained as:

$$\kappa = \int_0^{\infty} \kappa(\lambda) d\lambda, \quad \kappa(\lambda) = -\frac{1}{3} \sum_{p \in \mathrm{AC}} c_v(\omega, p) \bar{v}_p(\omega) \lambda \frac{d\omega}{d\lambda} \tag{4.26}$$

In nanostructures, phonons are frequently scattered at boundaries, which significantly impedes heat conduction. Therefore, if phonons have longer mean free paths in the bulk state, a more significant reduction in κ is expected when they are confined in a narrow space [4.48]. The simulation results shown in Figure 4.16 suggest that the MC simulation with the approximated dispersion model generates phonons with longer mean free paths. This results in a more sensitive reduction in κ when additional collisions are introduced at the boundaries.

Note that the simple analytical approach suggests that thermal conductivity is expressed as $\kappa = C_V \langle v \rangle \langle \lambda \rangle / 3$, where $\langle \lambda \rangle$ is the average mean free path. As shown in Figure 4.3, the approximated dispersion model exhibits a lower C_V, particularly in the low-temperature region, while $\langle v \rangle$ is slightly larger than that obtained by the realistic model. Considering that both models

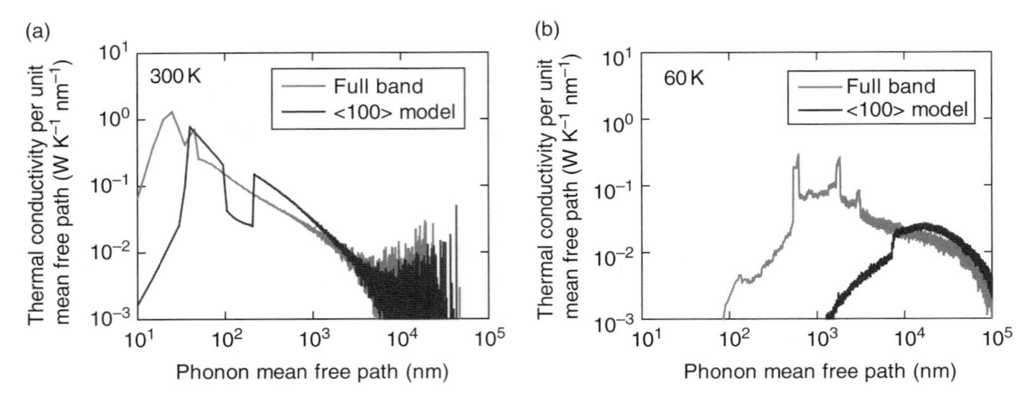

Figure 4.16 Thermal conductivity per unit phonon mean free path between the successive collisions obtained by MC simulation in bulk Si. The temperature is (a) 300 K; and (b) 60 K.

were calibrated to yield the correct κ for bulk Si, it can be understood that phonons have longer mean free paths in the MC simulation with the approximated dispersion model, as is confirmed in Figure 4.16, especially at lower temperatures. Therefore, in Figures 4.11 and 4.14, the reduction in κ in nanostructured Si is more pronounced in the approximated model at low temperatures.

For the accurate simulation of the thermal conduction in nanostructures, the distribution of the mean free path plays a significant role, and to this end, the inclusion of the realistic phonon dispersion model is essential. Recently, the dimensions of typical electronic devices have become smaller than the phonon mean free path, so it is important to model the effect of phonon boundary scattering on the heat conduction with the highest possible accuracy. Furthermore, phonon ballistic transport is also expected to have an effect [4.36, 4.61] for small dimensions, which means that correct information about not only about κ, but also $\langle v \rangle$ and $\langle \lambda \rangle$, is important. We thus believe that the MC simulation approach presented in this work can be a useful tool for accurately assessing the heat conduction properties of nanoscale devices.

4.3 Simulation of Heat Conduction in Devices

4.3.1 Simulation Method

Figure 4.17 shows an example for the one-dimensional system to simulate the heat conduction. The heat source and the constant temperature reservoir area were set in the left and right ends, respectively, and the phonons were assumed to be specularly reflected at each wall. The length L was divided into many sections with a width of Δx.

If the phonons are in thermal equilibrium at temperature T, then the thermal energy density per unit cross-sectional area stored in the section is given by:

$$\tilde{U}(T) = \Delta x \int_0^T C_V(T) dT \tag{4.27}$$

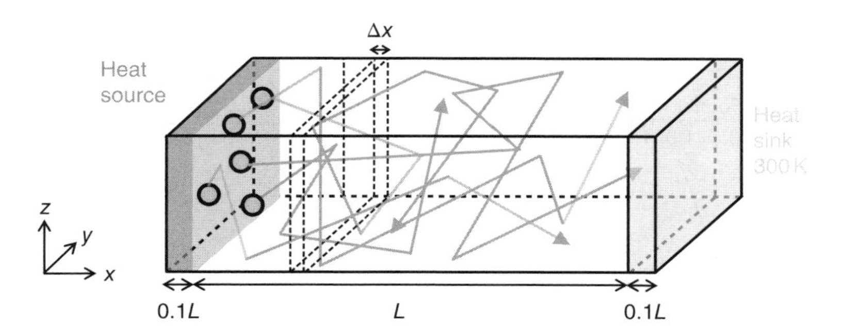

Figure 4.17 Schematic view of Si structure to investigate the heat conduction. The length L along the x direction was ranged from 5 nm to 10 μm.

The number of MC particles (per unit cross-sectional area) in the i-th section is then given by

$$\tilde{N}_i^* = \frac{\tilde{U}(T_i)}{E^*} \tag{4.28}$$

where T_i is the local temperature defined in the i-th section. As the simulation proceeds, the particles are redistributed and the local thermal energy may be changed. The local temperature is then updated by counting the changed number of MC particles $N_{i,\text{new}}^{-1}$, and calculated as:

$$T_{i,\text{new}} = \tilde{U}^{-1}\left(\tilde{N}_{i,\text{new}}^* E^*\right) \tag{4.29}$$

where $\tilde{U}^{-1}$ denotes the inverse function of Equation (4.27).

In the heat source region, $\tilde{Q}\Delta t/E^*$ particles are generated at random positions every time step, where $\tilde{Q}$ is the power density and Δt is the time step, while **u** was isotropically selected. In the reservoir region, the particle number was controlled to be constant ($\tilde{N}_0^* = \tilde{U}(T_0)/E^*$), where T_0 is the temperature assigned to the reservoir.

4.3.2 Simple 1-D Structure

Figure 4.18 shows the steady state temperature distribution in the Si structures. In this simulation, $\tilde{Q}$ was set to the values that cause a small temperature rise (5–15 K) across the structure to obtain the thermal resistances at 300 K. Note that linear temperature distributions were found in the long channel case (Figure 4.18 (a)), suggesting the diffusive heat conduction obeying the Fourier law. On the other hand, in the case of small channel length (Figure 4.18 (b)), nonlinear profiles, with a discontinuous temperature change at the heat sink, were confirmed.

It has been already reported that such behaviors are caused by the ballistic phonon emanating from the heat source, when the device dimension is significantly smaller than λ [4.26, 4.30]. By using the maximum temperature rise $T_{\max}$ observed at $x = 0.1\,L$ (the right edge of the heat source) under the steady state condition, we evaluated the equivalent thermal resistance R_{th} between the heat source and the sink from $T_{\max}/\tilde{Q}$. Figure 4.19 shows the calculated R_{th} (per unit cross sectional area) plotted as a function of L. Here, we also investigated the effect of the phonon dispersion models (see Figure 4.1) implemented in the MC simulator.

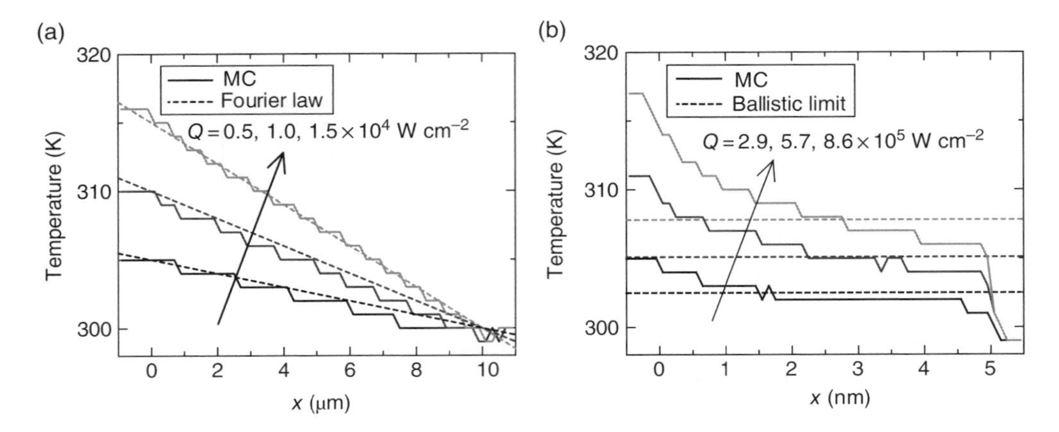

Figure 4.18 Steady state temperature distribution observed in the simulated structures presented in Figure 4.17 with (a) $L = 10\,\mu\text{m}$; and (b) 5 nm. Various power densities Q were assigned to the heat source. In this simulation, the realistic full band dispersion model was used.

Note that the results are not dependent on the dispersion model in the case of long L ($> \approx 1\,\mu\text{m}$), and well fitted by the formula based on the Fourier law [4.62]:

$$R_{\text{th}}^{F} = \frac{L}{\kappa} \tag{4.30}$$

Considering that the MC simulators with two different dispersion models are both calibrated to yield the correct κ, this coincidence can be understood straightforwardly. On the other hand, with decreasing L, the simulation results deviate from Equation (4.30), and exhibit dependence on the dispersion model. This suggests transition from the diffusive to the ballistic phonon transport. In the ballistic transport limit, the heat flux is limited by the finite phonon density and group velocity, and then the thermal resistance is not dependent on L. In Figure 4.19, the thermal resistances that are expected from the ballistic transport theory [4.20]:

$$R_{\text{th}}^{b} = \frac{4}{C_V \langle v \rangle} \tag{4.31}$$

are also plotted. In this formula, $\langle v \rangle$ is the average phonon velocity, and we evaluated it from:

$$v = \frac{\sum_p \int d\omega\, \overline{v}_p(\omega) c_v(\omega, p)}{C_V} \tag{4.32}$$

where the summation was carried out over the acoustic modes. Note that Equation (4.31) well explains the asymptotic values of the simulation results depending on the dispersion model. In Figure 4.20, the temperature dependence of $C_V\langle v \rangle$ are presented. Since this value is calculated only from the information about the phonon dispersion relation, it is essential to use the appropriate dispersion model. The smaller $C_V\langle v \rangle$ is obtained from the approximated [100] model, which is mainly due to the underestimation of c_v (ω, p), confirmed especially in the low

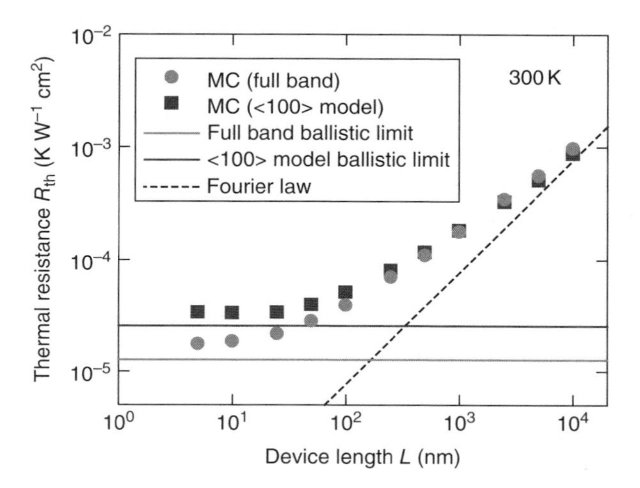

Figure 4.19 Thermal resistance R_{th} calculated from the MC simulation results with two different phonon dispersion models (symbols). The characteristics expected from the Fourier law (Equation (4.30); dashed line) and the ballistic transport theory (Equation (4.31); solid lines) are also plotted.

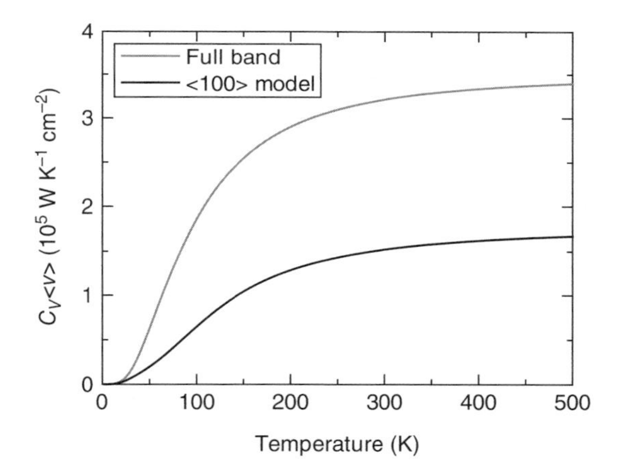

Figure 4.20 Calculated $C_V\langle v\rangle$ plotted as a function of temperature using the two different phonon dispersion models.

frequency region in Figure 4.3. However, for calculating κ, this discrepancy was canceled by using the different phonon scattering rates, as shown in Figure 4.7.

Considering the analytical expression $\kappa = C_V v\lambda/3$, the point of intersection between $R^F_{th}(L)$ and $R^b_{th}(L)$ is given by $L = 4\lambda/3$. Thus, as is well known, the dividing point between the ballistic and diffusive transport regime can be estimated by comparing the characteristic dimension of the system and λ. However, as is confirmed in Figure 4.19, the transition between the two extreme transport mechanisms is not sharp, but occurs progressively with varying L, which comes from the long-tailed nature of the phonon mean free path distribution [4.35, 4.36, 4.48].

Recently, the typical dimensions of the electronic devices are scaled down into the nanoscale regime, and such quasi-ballistic phonon transport effect is considered to become important. In FETs, when the distance from the heat source in the drain to the contact becomes very close, then R_{th} limited by the ballistic transport would become a bottleneck for heat dissipation [4.36], even if high thermal conductivity metals (e.g., carbon nanotube [4.63, 4.64]) are used as a thermal via. In order to accurately simulate such quasi-ballistic phonon transport, we believe that the MC method, taking account of the appropriate dispersion model, provides a useful tool.

4.3.3 FinFET Structure

The phonon MC simulator was modified to simulate the three-dimensional structure, and the heat conduction in FinFET was investigated. Figure 4.21 shows a schematic view of the simulated device structure. Bulk FinFET structure, with a gate length of 22 nm and a Fin thickness of 8 nm [4.65], was considered. MC particles were assumed to be confined only inside the Si region – that is, the perfectly reflecting boundary condition (assuming purely diffuse boundary scattering) was enforced at the Si/insulator interfaces. A heat source with a power density of $Q = 7.1$ TW/cm^3 mimicking the hot spot (total input power = 34 μW) was placed at the drain edge. In addition, the constant-temperature reservoirs at 300 K were set on the source/drain contacts and below the Si substrate, which act as heat sinks to remove the accumulated heat in the device.

Figure 4.22 shows the net dissipated heat in each reservoir calculated by averaging over the simulation time after turning on the heat source. The steady state was reached after ≈ 10 ns, and the drain and substrate were found to be the main thermal contacts to remove the heat. The MC simulation results are compared to those obtained by solving the heat conduction equation based on the Fourier's law, as we will see below.

In Figures 4.23 and 4.24, the comparison between the MC and Fourier-law based simulations are plotted for the steady state condition. Note that a significant difference in the

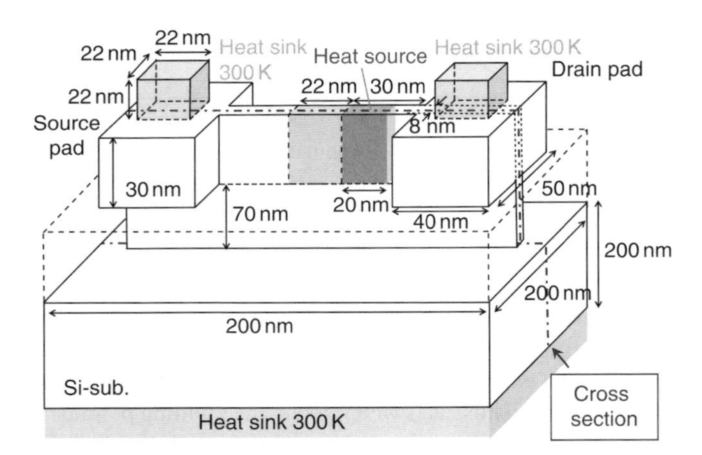

Figure 4.21 Schematic view of the FinFET structure simulated in this study. The heat source was placed at the drain edge, with a heat density of 7.1 TW/cm^3, and the constant-temperature reservoirs at 300 K were set on the top of the source/drain pads and below the bottom of the Si substrate.

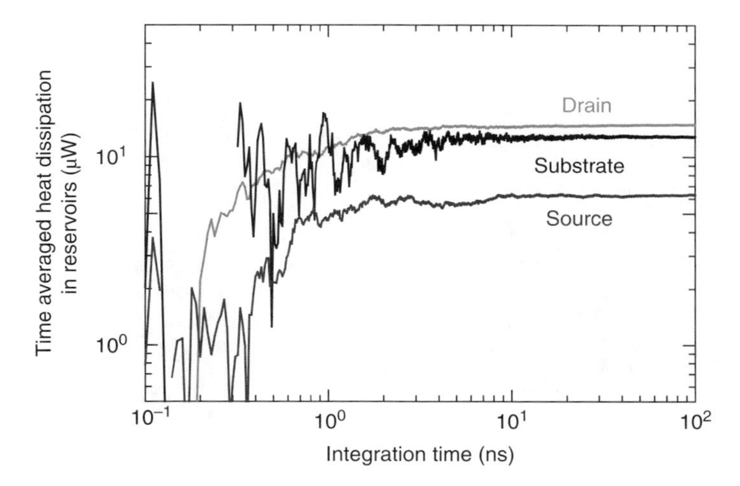

Figure 4.22 Time-averaged heat dissipation in reservoirs placed at the source, drain, and substrate contacts. The net removal of MC particles was counted in each reservoir region.

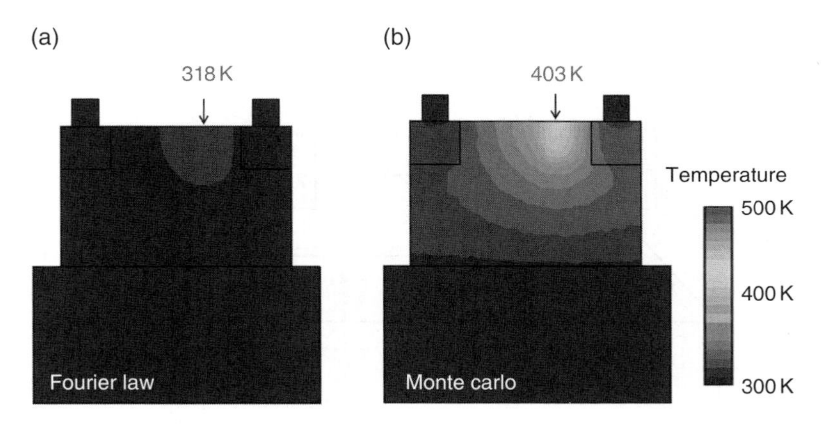

Figure 4.23 Steady-state temperature distribution on the vertical cross-section indicated in Figure 4.21. The simulation results, without taking account of the diffusive boundary scattering effect, are obtained by (a) the heat conduction equation based on Fourier's law; and (b) the present MC method.

maximum temperature at the hot spot was observed between the two methods in the case without the boundary scattering. The temperature rise ΔT at the hot spot is only ≈ 18 K in the case of Fourier law, while a much larger ΔT of ≈ 103 K was observed in the result of the MC simulator. This is due to the quasi-ballistic transport effect, especially remarkable just around the hot spot. On the other hand, when the boundary scattering was turned on, the discrepancy of ΔT was suppressed between the two methods; ΔT are ≈ 98 K and ≈ 193 K for the Fourier-law and MC simulations, respectively. This suggests that the heat conduction inside the Fin region becomes diffusive, and well described by the Fourier law.

To understand the mechanisms behind the simulation results, considerations based on the equivalent thermal circuit model were carried out. As shown in Figure 4.25, the thermal paths

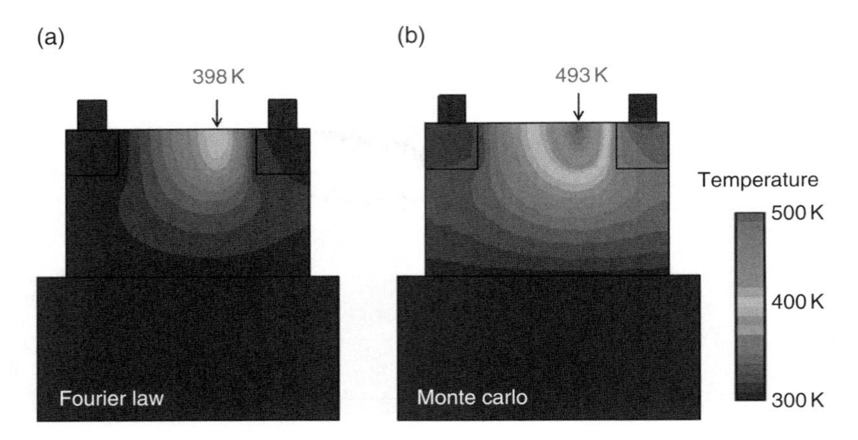

Figure 4.24 Steady-state temperature distribution on the vertical cross-section indicated in Figure 4.21. The simulation results, taking account of the diffusive boundary scattering effect, are obtained by (a) the heat conduction equation based on Fourier's law; and (b) the present MC method.

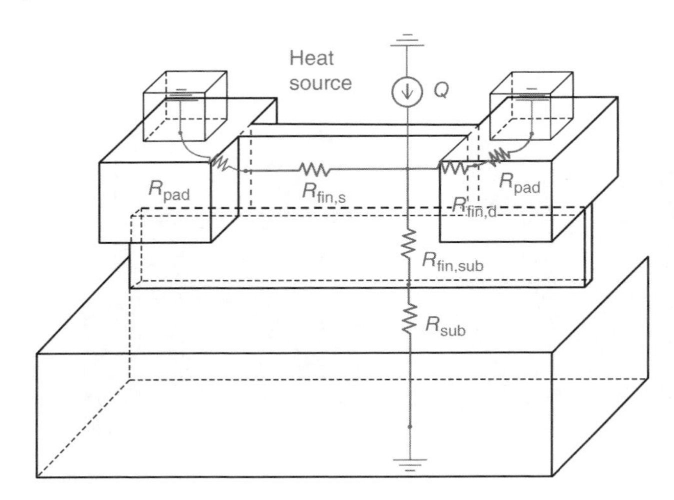

Figure 4.25 Equivalent thermal circuit model for the simulated FinFET structure.

from the heat source to each reservoir are divided into two resistance components in series (i.e., the inside Fin region (R_{fin}) and the residual part (R_{pad} or R_{sub})). The magnitudes of thermal resistances were then roughly estimated from $R = \Delta T/Q$, where ΔT is the temperature difference across the heat path and Q is the heat flux. The calculated results are shown in Figure 4.26.

The MC results exhibit larger thermal resistances compared to the Fourier law, particularly in the paths from the Fin exit to the heat sink. This is due to the quasi-ballistic transport effect of phonons, which becomes significant when the system size is comparable or less than the phonon mean free path. Also note that, even inside the Fin, the MC results show a larger R_{fin}, indicating the quasi-ballistic transport effect. Although λ is considered to be significantly shortened in the Fin, due to the frequent phonon boundary scattering, the phonon conduction nature is not fully diffusive in the simulated system assumed in this study.

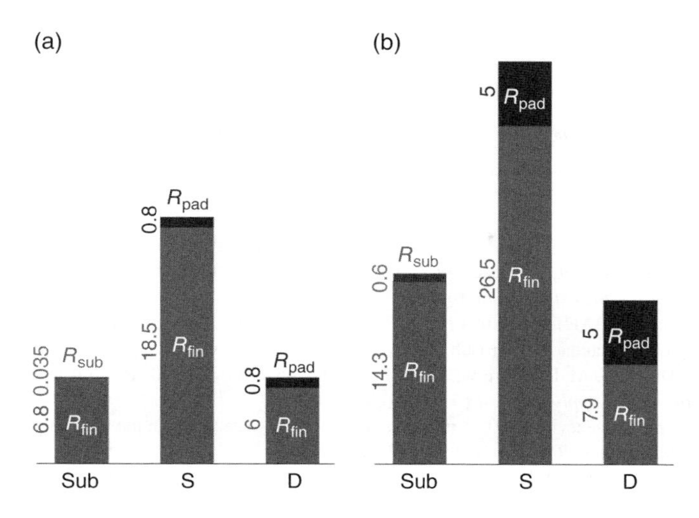

Figure 4.26 Magnitudes for the thermal resistances (MK/W) in Figure 4.25, evaluated from the simulations using (a) Fourier law; and (b) MC method.

References

[4.1] Y. Wang, K. P. Cheung, A. S. Oates and P. Mason (2007). Ballistic phonon enhanced NBTI. *Procedures of the IEEE International Reliability Physics Symposium*, 258–263.

[4.2] J.-P. Colinge, *FinFETs and Other Multi-Gate Transistors* (Springer, Berlin, 2007).

[4.3] E. Pop, S. Sinha and K. E. Goodson, Heat generation and transport in nanometer-scale transistors, *Proc. IEEE* 94, **8**, 1587-1601, Aug. 2006.

[4.4] R. Wang, J. Zhunge, C. Liu, R. Huang, D.-W. Kim, D. Park and Y. Wang, Exprimental study on quasi-ballistic transport in silicon nanowire transistors and the impact of self-heating effects. *IEDM Technical Digest* 2008, 753–756.

[4.5] J. A. Rowlette and K. E. Goodson, Fully coupled nonequilibrium electron-phonon transport in nanometer-scale silicon FETs, *IEEE Transactions on Electron Devices* **55**,1, 220–232, Jan. 2008.

[4.6] T. Takahashi, N. Beppu, K. Chen, S. Oda and K. Uchida (2011). Thermal-aware device design of nanoscale bulk/SOI FinFETs: Suppression of operation temperature and its variability. *IEDM Technical Digest*, 809–812.

[4.7] F. X. Alvarez, D. Jou and A. Sellitto (Feb. 2011). Phonon boundary effects and thermal conductivitiy of rough concentric nanowires. *ASME Journal of Heat Transfer* **133**, p. 022402.

[4.8] A. J. H. McGaughey, E. S. Landry, D. P. Sellan and C. H. Amon (2011). Size-dependent model for thin film and nanowire thermal conductivity. *Applied Physics Letters* **99**(13), p. 131904.

[4.9] A. I. Boukai, Y. Buninovich, J. Tahir-Kheli, J.-K. Yu, W. A. Goddard III and J. R. Heath (Jan. 2008). Silicon nanowires as efficient thermoelectric materials. *Nature* **451**, 168–171.

[4.10] A. I. Hochbaum, R. Chen, R. D. Delgado, W. Liang, E. C. Garnett, M. Najarian, A. Majumdar and P. Yang (Jan. 2008). Enhanced thermoelectric performance of rough silicon nanowires. *Nature* **451**, 163–167.

[4.11] Y. Li, K. Buddharaju, B. C. Tinh, N. Singh and S. J. Lee (May 2012). Improved vertical silicon nanowire based thermoelectric power generator with polyimide filling. *IEEE Electron Device Letters* **33**(5), 715–717.

[4.12] S. V. J. Narumanchi, J. Y. Murthy and C. H. Amon (Dec. 2004). Submicron heat transport model in silicon accounting for phonon dispersion and polarization. *ASME Journal of Heat Transfer* **126**, 946–955.

[4.13] X. W. Zhou, R. E. Jones and S. Aubry (2010). Molecular dynamics prediction of thermal conductivity of GaN films and wires at realistic length scales. *Physical Review B* **81**(15), p. 155321.

[4.14] T. Zushi, Y. Kamakura, K. Taniguchi, I. Ohdomari and T. Watanabe (2010). Molecular dynamics simulation of heat transport in silicon nano-structures covered with oxide films. *Japanese Journal of Applied Physics* **49**, p. 04DN08.

[4.15] T. Yamamoto and K. Watanabe (2006). Nonequilibrium Green's function approach to phonon transport in defective carbon nanotubes. *Physical Review Letters* **96**(25), p. 255503.
[4.16] N. Mingo (2006). Anharmonic phonon flow through molecular-sized junctions. *Physical Review B* **74**(12), p. 125402.
[4.17] Y. Xu, J.-S. Wang, W. Duan, B.-L. Gu and B. Li (2008). Nonequilibrium Green's function method for phonon-phonon interactions and ballistic-diffusive thermal transport. *Physical Review B* **78**(22), p. 224303.
[4.18] M. Luisier (2012). Atomistic modeling of anharmonic phonon-phonon scattering in nanowires. *Physical Review B* **86**(24), p. 245407.
[4.19] M. G. Holland (Dec. 1963). Analysis of lattice thermal conductivity. *Physical Review* **132**(6), 2461–2471.
[4.20] P. Chantrenne, J. L. Barrat, X. Blase and J. D. Gale (2005). An analytical model for the thermal conductivity of silicon nanostructures. *Journal of Applied Physics* **97**(10), p. 104318.
[4.21] C. Jeong, S. Datta and M. Lungstrom (2011). Full dispersion versus Debye model evaluation of lattice thermal conductivity with a Landauer approach. *Journal of Applied Physics* **109**(7), p. 073718.
[4.22] C. Jeong, S. Datta and M. Lungstrom (2012). Thermal conductivity of bulk and thin-film silicon: A Landauer approach. *Journal of Applied Physics* **111**(9), p. 093708.
[4.23] Z. Aksamija and I. Knezevic (2010). Anisotropy and boundary scattering in the lattice thermal conductivity of silicon nanomembranes. *Physical Review B* **82**(4), p. 045319.
[4.24] R. B. Peterson (Nov. 1994). Direct simulation of phonon-mediated heat transfer in a Debye crystal. *ASME Journal of Heat Transfer* **116**, 815–822.
[4.25] S. Mazumder and A. Majumdar (Aug. 2001). Monte Carlo study of phonon transport in solid thin films including dispersion and polarization. *ASME Journal of Heat Transfer* **123**, 749–759.
[4.26] D. Lacroix, K. Joulain and D. Lemonnier (2005). Monte Carlo transient phonon transport in silicon and germanium at nanoscales. *Physical Review B* **72**(6), p. 064305.
[4.27] Y. Chen, D. Li, J. R. Lukes and A. Majumdar (Oct. 2005). Monte Carlo simulation of silicon nanowire thermal conductivity. *ASME Journal of Heat Transfer* **127**, 1129–1137.
[4.28] M.-S. Jeng, R. Yang, D. Song and G. Chen (Apr. 2008). Modeling the thermal conductivity and phonon transport in nanoparticle composites using Monte Carlo simulation. *ASME Journal of Heat Transfer* **130**, p. 042410.
[4.29] A. Mittal and S. Mazumder (May 2010). Monte Carlo study of phonon heat conduction in silicon thin films including contributions of optical phonons. *ASME Journal of Heat Transfer* b, p. 052402.
[4.30] B. T. Wong, M. Francoeur and M. P. Mengüç (2011). A Monte Carlo simulation for phonon transport within silicon structures at nanoscales with heat generation. *International Journal of Heat and Mass Transfer* **54**, 1825–1838.
[4.31] C. Bera (2012). Monte Carlo simulation of thermal conductivity of Si nanowire: An investigation on the phonon confinement effect on the thermal transport. *Journal of Applied Physics* **112**(7), p. 074323.
[4.32] A. J. H. McGaughey and A. Jain (2012). Nanostructure thermal conductivity prediction by Monte Carlo sampling of phonon free paths. *Applied Physics Letters* **100**(6), p. 061911.
[4.33] E. B. Ramayya, L. N. Maurer, A. H. Davoody and I. Knezevic (2012). Thermoelectric properties of ultrathin silicon nanowires. *Physical Review B* **86**(11), p. 115328.
[4.34] K. Kukita and Y. Kamakura (2013). Monte Carlo simulation of phonon transport in silicon including a realistic dispersion relation. *Journal of Applied Physics* **114**(15), p. 154312.
[4.35] K. Kukita, I. N. Adisusilo and K. Kamakura (2014). Monte Carlo simulation of diffusive-to-ballistic transition in phonon transport. *Journal of Computational Electronics* **13**, 264–270.
[4.36] K. Kukita, I. N. Adisusilo and K. Kamakura (2012). Impact of quasi-ballistic phonon transport on thermal properties in nanoscale devices: A Monte Carlo approach. *IEDM Technical Digest*, 411–414.
[4.37] K. Kukita, I.N. Adisusilo and K. Kamakura (2014). Monte Carlo simulation of thermal conduction in silicon nanowires including realistic phonon dispersion relation. *Japanese Journal of Applied Physics* **53**, p. 015001.
[4.38] J.E. Turney, A.J.H. McGaughey and C.H. Amon (2010). In-plane phonon transport in thin films. *Journal of Applied Physics* **107**(2), p. 024317.
[4.39] W. Weber (May 1977). Adiabatic bond charge model for the phonons in diamond, Si, Ge and α-Sn. *Physical Review B* **15**(10), 4789–4803.
[4.40] G. Dolling (1963). *Inelastic Scattering of Neutrons in Solids and Liquids*, I, p. 37. IAEA, Vienna.
[4.41] G. Nilsson and G. Nelin (Nov. 1972). Study of the homology between silicon and germanium by thermal-neutron spectrometry. *Physical Review B* **6**(10), 3777–3786.

[4.42] O. Jepson and O.K. Anderson (1971). The electronic structure of h.c.p. Ytterbium. *Solid State Communications* **9**(20), 1763–1767.
[4.43] J. Callaway (Feb. 1959). Model for lattice thermal conductivity at low temperatures, *Physical Review* **113**(4), 1046–1051.
[4.44] C. Kittel and H. Kroemer (1980). *Thermal Physics*. W H Freeman and Co., New York.
[4.45] P. Flubacher, A. J. Leadbetter and J. A. Morrison (1959). The heat capacity of pure silicon and germanium and properties of their vibrational frequency spectra. *Philosophical Magazine* **4**(39), 273–294.
[4.46] K. Yamaguchi and K. Itagaki (2002). Measurement of high temperature heat content of silicon by drop calorimetry. *Journal of Thermal Analysis and Calorimetry* **69**, 1059–1066.
[4.47] A. S. Henry and G. Chen (Feb. 2008). Spectral phonon transport properties of silicon based on molecular dynamics simulations and lattice dynamics. *Journal of Computational and Theoretical Nanoscience* **5**(2), 141–152.
[4.48] D. M. Rowe (2005). *Thermoelectrics Handbook: Macro to Nano*. CRC Press, New York.
[4.49] M.-J. Huang, T.-M. Chang, W.-Y. Chong, C.-K. Liu and C.-K. Yu (2007). A new lattice thermal conductivity model of a thin-film semiconductor. *International Journal of Heat and Mass Transfer* **50**, 67–74.
[4.50] D. Marx and W. Eisenmenger (1982). Phonon scattering at siliconcrystal surfaces. *Zeitschrift für Physik B Condensed Matter* **48**, 277–291.
[4.51] H. G. Lipinski and A. Struppler (1989). New trends in computer graphics and computer vision to assist functional neurosurgery. *Stereotactic and Functional Neurosurgery* **52**(2–4), 234–241.
[4.52] W.-J. Yang, H. Taniguchi and K. Kudo (1995). *Advances in Heat Transfer*, **27**. Academic Press, San Diego.
[4.53] M. Asheghi, M. N. Touzelbaev, K. E. Goodson, Y. K. Leung and S. S. Wong (Feb. 1998). Temperature-dependent thermal conductivity of single-crystal silicon layers in SOI substrates. *ASME Journal of Heat Transfer* **120**, 30–36.
[4.54] Y. S. Ju (2005). Phonon heat transport in silicon nanostructures. *Applied Physics Letters* **87**(15), p. 153106.
[4.55] Y. S. Ju and K. E. Goodson (May 1999). Phonon scattering in silicon films with thickness of order 100 nm. *Applied Physics Letters* **74**(20), 3005–3007.
[4.56] W. Liu and M. Asheghi (May 2004). Phonon-boundary scattering in ultrathin single-crystal silicon layers. *Applied Physics Letters* **84**(19), 3819–3821.
[4.57] J. C. Duda, T. E. Beechem, J. L. Smoyer, P. M. Norris and P. E. Hopkins (2010). Role of dispersion on phononic thermal boundary conductance. *Journal of Applied Physics* **108**(7), p. 073515.
[4.58] J. Zou and A. Balandin (Mar. 2001). Phonon heat conduction in a semiconductor nanowire. *Journal of Applied Physics* **89**(5), 2932–2938.
[4.59] D. Li, Y. Wu, P. Kim, L. Shi, P. Yang and A. Majumdar (Oct. 2003). Thermal conductivity of individual silicon nanowires. *Applied Physics Letters* **83**(14), 2934–2936.
[4.60] D. P. Sellan, J. E. Turney, A. J. H. McGaughey and C. H. Amon (2010). Cross-plane phonon transport in thin films. *Journal of Applied Physics* **108**(11), p. 113524.
[4.61] M. E. Siemens, Q. Li, R. Yang, K. A. Nelson, E. H. Anderson, M. M. Murnane and H. C. Kapteyn (Jan. 2010). Quasi-ballistic thermal transport from nanoscale interfaces observed using ultrafast coherent soft X-ray beams. *Nature Materials* **9**, 26–30.
[4.62] M. Kaviany (2011). *Essentials of Heat Transfer: Principles, Materials and Applications*. Cambridge University Press, Cambridge.
[4.63] J. Hone, M. Whitney, C. Piskoti and A. Zettle (Jan. 1999). Thermal conductivity of single-walled carbon nanotubes. *Physical Review B* **59**(4), R2514–R2516.
[4.64] M. Nihei, A. Kawabata, T. Murakami, M. Sato and N. Yokoyama (2012). Improved thermal conductivity by vertical graphene contact formation for thermal TSV. *IEDM Technical Digest*, 797–800.
[4.65] M. Shrivastava, M. Agrawal, S. Mahajan, H. Gossner, T. Schulz, D. Sharma and V. R. Rao (May 2012). Physical insight toward heat transport and an improved electrothermal modeling framework for finfet architectures. *IEEE Transactions on Electron Devices* **59**(5), 1353–1363.

5

Carrier Transport in High-mobility MOSFETs

The remarkable advancement of LSI circuit technology has been primarily based on the downsizing of MOSFETs. However, in the present post-scaling era, the downsizing has become less effective in improving the device performance, because we need to suppress leakage current, minimize short channel effects, and maintain high drive current at the same time. This means that conventional Si CMOS scaling has been reaching its fundamental limits. As a new challenge, the application of high mobility materials, such as III-V semiconductors and Ge, to MOSFET channels is expected to achieve better device performance without miniaturization. In this chapter, we examine practical advantages of the new channel materials using semi-classical MC and quantum mechanical Wigner MC simulations, where scattering effects, quantum mechanical effects and new device structure are considered.

5.1 Quantum-corrected MC Simulation of High-mobility MOSFETs

To achieve a high current drive with a low power supply voltage, channel materials with high mobility and (more essentially) low effective mass are preferable, even in the ultrashort channel regime, since low-field mobility is still a good indicator for a high-current drive under quasi-ballistic transport, while the effective mass is the essential parameter under full ballistic transport [5.1]. In this section, we discuss the quasi-ballistic transport in high-mobility MOSFETs – that is, the channel materials are III-V semiconductors, Ge and strained-Si, using the quantum-corrected MC simulation [5.2–5.5], as explained in Section 3.1.

5.1.1 Device Structure and Band Structures of Materials

Figure 5.1 shows the device model used in this study, where UTB structure with a channel thickness of T_{ch} = 5 nm is employed. When the high-mobility materials are introduced into the

---

Carrier Transport in Nanoscale MOS Transistors, First Edition. Hideaki Tsuchiya and Yoshinari Kamakura.

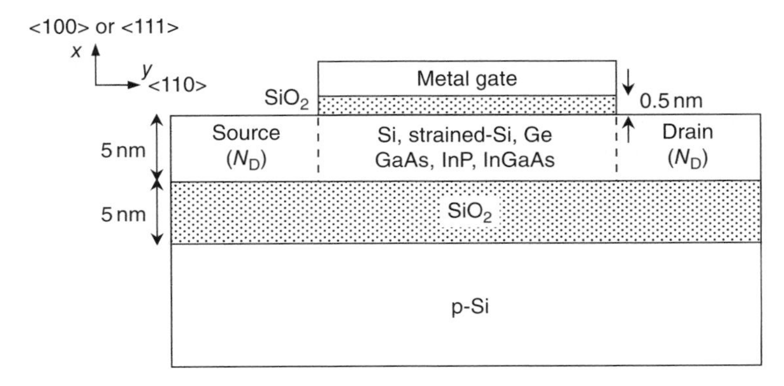

Figure 5.1 Device model used in the simulation, where a UTB structure is employed. The channel length L_{ch} is varied from 70 to 20 nm, and the channel region is assumed to be undoped. The channel thickness T_{ch} is 5 nm, and the gate oxide is SiO_2 with a thickness of 0.5 nm.

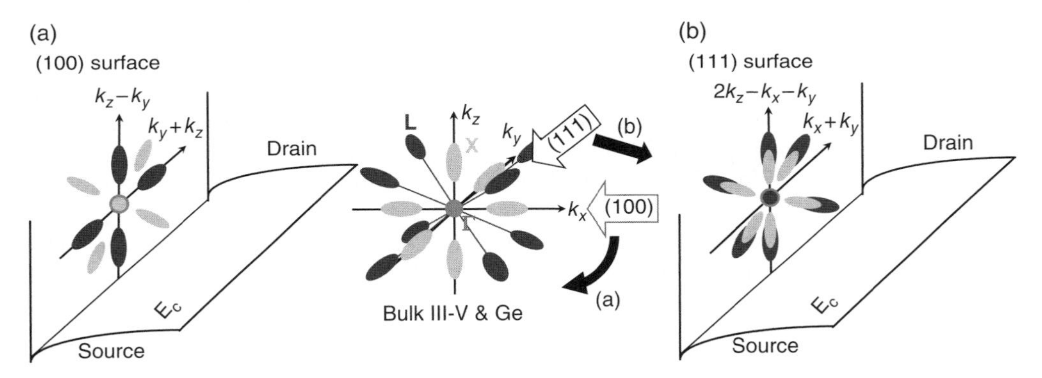

Figure 5.2 Schematics of equi-energy surfaces at Γ, L, and X valleys in band structures for III-V materials and Ge, and their 2-D projections onto (a) (100); and (b) (111) surfaces, respectively. The channel direction is taken as <110>.

channel, the usage of UTB (or multigate) III-V-on-insulator (III-V-OI) and UTB (or multigate) Ge-on-insulator (GOI) structures are required to minimize the short channel effects, because electron wave functions spread deeply into the substrate region, due to the small effective mass [5.6, 5.7].

The practical advantage of the UTB structure will be discussed in Section 5.1.4. The UTB channels are formed onto the SiO_2 layer, and the bottom substrate is assumed to be Si. The surface orientation is (100) for Si, GaAs and InP, and (100) and (111) for Ge, because the smaller transport effective mass is expected for Ge (111) surface orientation [5.7, 5.8]. Strained-Si channels are also considered. Figure 5.2 shows the schematics of equi-energy surfaces at Γ, L, and X valleys in the band structures for III-V materials and Ge, and their 2-D projections onto (100) and (111) surfaces. The channel direction is taken as <110>.

In this study, we used bulk band structures and parameters, as also explained in the next section. We have derived the electron effective masses in each direction using the transformation matrix from the principal axis of a constant-energy ellipsoid of the materials to the coordinate system we have chosen, by following the Stern and Howard theory [5.9]. The channel length L_{ch} is varied from 70–20nm, and the channel region is assumed to be undoped. The gate oxide is SiO_2 with a thickness of 0.5 nm.

Here, we pay attention to the physical limits of solid solubility of donors in III-V semiconductors – namely, that activated donor concentrations larger than 2×10^{19} cm^{-3} cannot be obtained in III-V semiconductors [5.10]. Therefore, the donor concentration of 2×10^{19} cm^{-3} was assumed in the source and drain (S/D) regions for GaAs, InP and InGaAs, while a higher donor concentration of 1×10^{20} cm^{-3} was used for Si and Ge. While implant-free devices, such as metal S/D electrodes are of current interest, implanted devices, including regrown S/D, are also subject to active research [5.11–5.14].

5.1.2 Band Parameters of Si, Ge, and III-V Semiconductors

We have extended the quantum-corrected MC simulator to the study of III-V and Ge MOSFETs, as described below. Table 5.1 shows the fundamental physical properties and band parameters of the main semiconductor materials, including Si, Ge, and III-V materials.

It is well known that, in III-V semiconductors and Ge, electron transfer to the higher valleys with a heavier transport mass degrades the device performance significantly [5.6, 5.7]. Therefore, the energy differences between the lowest and higher valleys ($\Delta E_{\Gamma L}$, $\Delta E_{\Gamma X}$, $\Delta E_{L\Gamma}$ and ΔE_{LX}) are important physical parameters. Note also that the nonparabolicity of the Γ valley is the largest for $In_{0.53}Ga_{0.47}As$, which may be disadvantageous for achieving a higher carrier velocity.

Strained-Si channels under 1% biaxial and uniaxial tensile strains both in the channel and S/D regions are also considered in this study. The band-edge energy variation due to the biaxial (uniaxial) strain was given by ΔE_{2f}=–0.25 eV (–0.125 eV) and ΔE_{4f}=–0.04 eV (–0.016 eV) for

Table 5.1 Fundamental physical properties and band parameters of the main semiconductor materials including Si, Ge, and III-V materials.

		Si	Ge	GeAs	InP	$In_{0.53}Ga_{0.47}As$
mobility (cm^{2}V^{-1}s^{-1})		1600	3900	9200	5400	25 000
mass (Γ)		–	0.037	0.063	0.082	0.046
mass (X)	$m_t(m_0)$	0.19	0.29	0.229	0.273	0.251
	$m_l(m_0)$	0.98	1.35	1.987	1.321	2.852
mass (L)	$m_t(m_0)$	0.126	0.082	0.127	0.153	0.125
	$m_l(m_0)$	1.634	1.59	1.538	1.878	1.552
nonparabolicity		0.5 (X)	0.65	1.16 (Γ)	0.61 (Γ)	1.18 (Γ)
α (ev^{-1})		0.3 (L)	(L, Γ, X)	0.40 (L)	0.49 (L)	0.43 (L)
				0.55 (X)	0.12 (X)	0.33 (X)
ΔE_{XL} (eV)		1.049	–	–	–	–
$\Delta E_{L\Gamma}/\Delta E_{LX}$ (eV)		–	0.14/0.18	–	–	–
$\Delta E_{\Gamma L}/\Delta E_{\Gamma X}$ (eV)		–	–	0.323/0.447	0.832/1.492	0.723/1.062
Band gap (eV)		1.12	0.66	1.42	1.34	0.86
Dilectric const. (ε_0)		11.9	16.0	12.9	12.6	14.1

the two-fold and four-fold valleys, respectively [2.18], and the effective mass variation due to the uniaxial tensile strain was taken into account based on a first-principles bandstructure calculation [2.14]. The scattering processes considered are impurity and phonon scatterings, where polar optical phonon (POP) scattering is also considered for III-V materials. The phonon scattering parameters are taken from [3.12] for Si, from [5.15] for Ge, and from [5.16] for III-V materials, of which parameters are explained in detail in the next section.

The roughness and electron-electron scatterings are ignored in this study, to evaluate the device performances governed by the intrinsic transport properties of each material. For impurity scattering, we considered Fermi-Dirac statistics to calculate the screening length, which is particularly important for III-V materials [5.11]. The enhanced scattering probabilities due to the confinement of wave-functions are approximately considered, by adding the quantum correction energy into the electron kinetic energy [3.2]. In the III-V MOSFET simulations, only the Γ and L valleys were considered, since the energy difference between the Γ and X valleys, $\Delta E_{\Gamma X}$, is substantially larger than the kinetic energy increase of electrons, due to the supply voltage considered in this study. We further point out that the Fermi energies of high-mobility materials in the S/D regions are comparable to $\Delta E_{\Gamma L}$ in III-V materials, and $\Delta E_{L\Gamma}$ and ΔE_{LX} in Ge [5.2–5.4] and, therefore, a certain number of electrons already populate higher valleys inside the S/D region and affect the device performance under ballistic transport [5.3].

5.1.3 Polar-optical Phonon (POP) Scattering in III-V Semiconductors

In polar materials such as III-V and II-VI compound semiconductors, the POP scattering plays an important role in carrier transport analysis. The POP scattering is treated as an inelastic process, and its scattering rate for 3-D electron gas is given by [3.11]

$$W(\mathbf{k}) = \frac{e^2 \omega_{op}}{8\pi\varepsilon_p} \frac{k}{E_{\mathbf{k}}} \left(N_q + \frac{1}{2} \mp \frac{1}{2} \right) \text{In}\left(\frac{q_{\max}}{q_{\min}} \right) \tag{5.1}$$

where

$$\frac{1}{\varepsilon_p} = \left(\frac{1}{\varepsilon_\infty} - \frac{1}{\varepsilon_0} \right) \tag{5.2}$$

$$q_{\min} = k\left| 1 - \sqrt{1 \pm \frac{\hbar\omega_{op}}{E_{\mathbf{k}}}} \right|, \; q_{\max} = k\left(1 + \sqrt{1 \pm \frac{\hbar\omega_{op}}{E_{\mathbf{k}}}} \right) \tag{5.3}$$

$\hbar\omega_{op}$ is the POP energy, and ε_∞ and ε_0 are the optical and static dielectric constants, respectively. Equation (5.2) comes from the polar coupling constant given by the usual Fröhlich expression [5.16]. As found from Equation (5.1), the POP scattering rate decreases with the electron kinetic energy $E_{\mathbf{k}}$, in contrast to the AP and IP scattering rates, which increase with $E_{\mathbf{k}}$, due to the DOS for carriers at the final state after scattering, as described in Section 3.1.2. In addition, the POP scattering is anisotropic, and the scattering angle θ is determined in the same way as used in the electron-plasmon scattering presented in Section 3.1.2 [3.11]. For the II, AP, and non-POP (or IP) scattering, the same scattering rates as those in Section 3.1.2 were used, except that the inverse of Debye length q_D defined in Equation (3.6) was replaced by using the Thomas-Fermi screening length with Fermi energy E_F, as:

$$q_D = \sqrt{\frac{3e^2 n}{2\varepsilon E_F}} \tag{5.4}$$

The phonon scattering parameters for GaAs, InP and InGaAs used in the simulations are listed in Table 5.2, and those for Ge in Table 5.3, which are taken from [5.15] and [5.16], respectively. For Si, the phonon scattering parameters listed in Table 3.1 was used. Note that in III-V semiconductors, the POP, AP, and II scattering causes an intravalley transition such as $\Gamma \rightarrow \Gamma$ and $L_1 \rightarrow L_1$, whereas the non-POP scattering causes an intervalley transition such as $\Gamma \rightarrow L_1$ and $L_1 \rightarrow L_2$.

5.1.4 Advantage of UTB Structure

We demonstrate the advantage of the UTB structure for the III-V MOSFETs. Figure 5.3 shows the $I_D - V_G$ characteristics computed for the bulk and UTB structures with unstrained-Si and

Table 5.2 Phonon scattering parameters for GaAs, InP and $In_{0.53}Ga_{0.47}As$ used in the simulation.

Material	GaAs	
Crystal density ρ	5360 (kg/m^3)	
Sound velocity v_s	5240 (m/s)	
Optical dielectric constant ε_∞	10.92 ε_0 (F/m)	
Phonon Type	Deformation potentials & fields	Phonon energy
Acoustic phonon	5.0 eV (Γ), 5.0 eV (L)	–
Polar optical phonon	–	35.36 meV
Non-polar optical phonon	(inter Γ – L) 5.25×10^8 eV/cm	22.69 meV
	(inter Γ – L) 5.94×10^8 eV/cm	24.97 meV
Material	**InP**	
Crystal density ρ	4810 (kg/m^3)	
Sound velocity v_s	5130 (m/s)	
Optical dielectric constant ε_∞	9.61 ε_0 (F/m)	
Phonon Type	Deformation potentials & fields	Phonon energy
Acoustic phonon	5.0 eV (Γ), 5.0 eV (L)	–
Polar optical phonon	–	42.40 meV
Non-polar optical phonon	(inter Γ – L) 5.06×10^8 eV/cm	21.15 meV
	(inter Γ – L) 5.75×10^8 eV/cm	24.27 meV
Material	**$In_{0.53}Ga_{0.47}As$**	
Crystal density ρ	5523 (kg/m^3)	
Sound velocity v_s	4731 (m/s)	
Optical dielectric constant ε_∞	11.89 ε_0 (F/m)	
Phonon Type	Deformation potentials & fields	Phonon energy
Acoustic phonon	5.0 eV (Γ), 5.0 eV (L)	–
Polar optical phonon	–	32.56 meV
Non-polar optical phonon	(inter Γ – L) 5.43×10^8 eV/cm	19.90 meV
	(inter Γ – L) 6.15×10^8 eV/cm	21.90 meV

Table 5.3 Phonon scattering parameters for Ge used in the simulation.

Material	Ge	
Crystal density ρ	5327 (kg/m^3)	
Sound velocity v_s	5400 (m/s)	
Phonon Type	Deformation potentials & fields	Phonon energy
Acoustic phonon	11.0 eV (Γ), 5.0 eV (L), 9.0 eV (X)	–
Non-polar optical phonon	(intra-L valley) 5.5×10^8 eV/cm	21.15 meV
	(inter L – L) with LA 3.0×10^8 eV/cm	24.60 meV
	(inter L – L) with LO 3.0×10^8 eV/cm	27.60 meV
	(inter L – L) with TA 2.0×10^8 eV/cm	10.30 meV
	(inter L – X) with LA 4.1×10^8 eV/cm	24.60 meV
	(inter L – Γ) with LA 2.0×10^8 eV/cm	27.60 meV
	(inter X – Γ) with TA 1.0×10^8 eV/cm	27.60 meV
	(inter X – X) with LA 7.9×10^8 eV/cm	8.60 meV
	(inter X – X) with LO 9.5×10^8 eV/cm	37.10 meV

GaAs channels, where $L_{ch}=50$ nm and threshold voltage V_{th} is set at 0.3 V. The drain voltage V_D is given as 0.5 V. For the bulk MOSFETs, the acceptor density in the substrate was given as $N_A=1\times10^{18}$ cm^{-3} in the simulation. Although the acceptor in the substrate increases the threshold voltage, the impurity scattering due to the acceptor is considered to be negligible for the drive current estimation at on-state.

From Figure 5.3, it is found that the drive current enhancement in the GaAs channel is considerably improved by using the UTB structure. As predicted in [5.6], the performance improvement in the GaAs-UTB structure should be due to the smaller inversion-layer thickness, which is owing to the physical confinement of electron wave-functions. To demonstrate this, Figure 5.4 shows the electron density distributions in the inversion-layers computed for each device structure, where the distributions in each valley are separately plotted. Note that these distributions are extracted in the middle of the channel, so the profiles are influenced by the applied drain voltage.

As shown in Figure 5.4 (b), the electrons in the Γ valley of the GaAs bulk MOSFET spread deeply into the substrate, due to the quantum confinement effect, while those in the L valleys are confined within a region a few nm from the Si/SiO$_2$ interface ($x=0$). This is due to the difference in the effective masses between the Γ and L valleys in the confinement direction. Note that, as the gate bias increases, more electrons occupy the L valleys in GaAs. This is due to the small $\Delta E_{\Gamma L}$, and also the preferential increase in the quantized subband energy of the Γ valley. This causes the saturation in the drain current, as shown in Figure 5.3 [5.6].

Compared to the bulk structure, the Γ valley electrons in the UTB structure are also confined in the ultrathin channel region by the buried oxide layer, as shown in Figure 5.4(d). As a result, the inversion-layer thickness decreases and then the inversion-layer capacitance increases, which leads to the increase of gate capacitance in the UTB MOSFETs. Therefore, the drive current in the GaAs UTB MOSFET significantly enhances, as shown in Figure 5.3. The advantage of the UTB structure also applies to other III-V materials, such as InP and InGaAs. On the other hand, the electron density distributions in the unstrained-Si channels are almost the same in the two structures as shown in Figures 5.4(a) and (c), because the inversion-layer thickness is less than 5 nm for the present Si MOSFETs. The above results suggest that the UTB structure is indispensable to realize high-performance III-V MOSFETs.

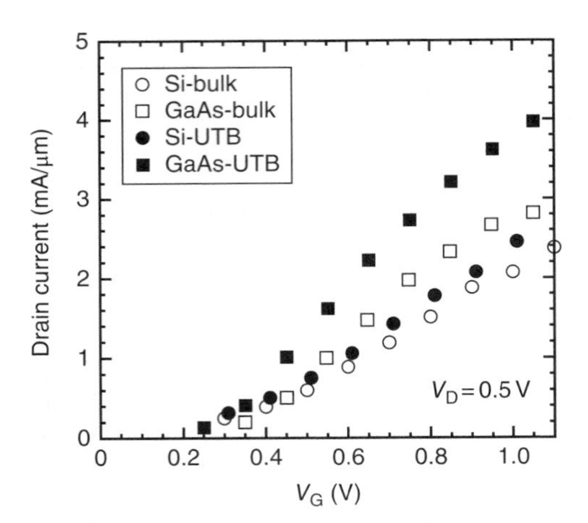

Figure 5.3 $I_D - V_G$ characteristics computed for bulk and UTB structures with unstrained-Si and GaAs channels. $L_{ch} = 50$ nm, $V_{th} = 0.3$ V and $V_D = 0.5$ V.

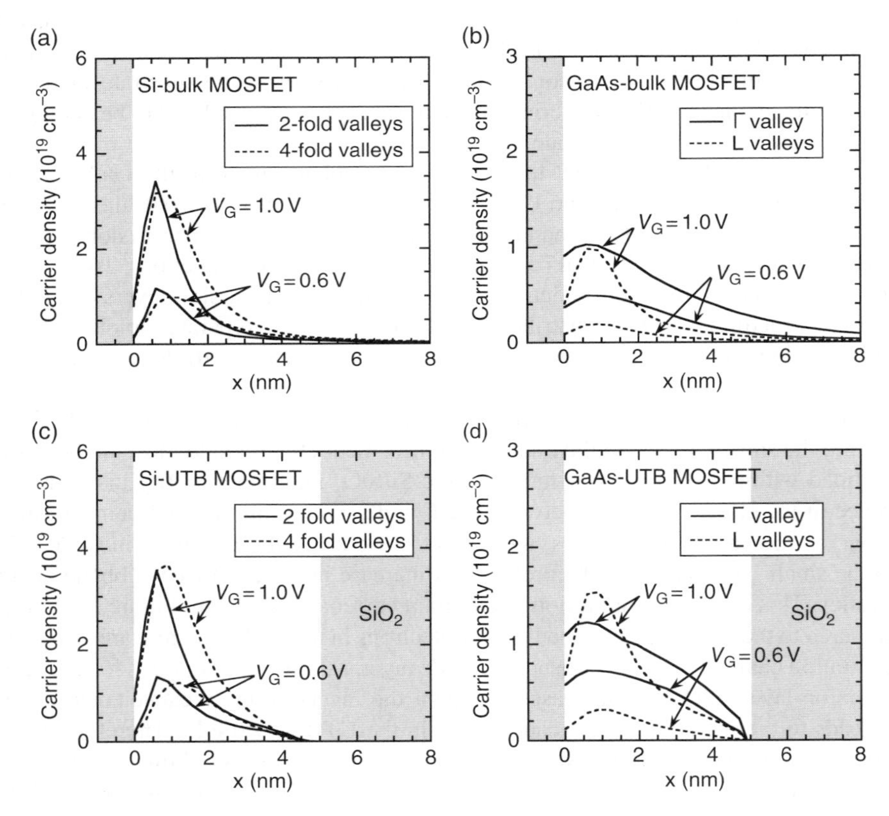

Figure 5.4 Electron density distributions in the inversion-layers computed for bulk and UTB structures with unstrained-Si and GaAs channels, where: (a) Si-bulk; (b) GaAs-bulk; (c) Si-UTB; and (d) GaAs-UTB MOSFETs. The distributions in each valley are separately plotted. $V_G = 0.6$ and 1.0V, and $V_D = 0.5$V. Note that these profiles are extracted in the middle of the channel region.

5.1.5 *Drive Current of III-V, Ge and Si* n-*MOSFETs*

We compare the drive current of Si, strained-Si, Ge, GaAs and InP MOSFETs, all with the UTB structure shown in Figure 5.1. First, the inversion-layer electron density distributions computed for Ge (100), Ge (111), 1% biaxially-tensile strained-Si and InP MOSFETs are shown in Figure 5.5, where the bias conditions are the same as in Figure 5.4.

For the InP channel, there is no electron population in the L valleys, even at high gate voltage, which is due to the large $\Delta E_{\Gamma L}$ shown in Table 5.1. On the other hand, the Ge channels have comparable electron populations in the L and X valleys for the both surface orientations, which is due to the much smaller ΔE_{LX} and, more importantly, a high transition rate from the L to X valleys, due to the inter-valley phonon scattering [5.3]. Furthermore, the electron population in the Γ valley of Ge is vanishingly small, because a much higher subband level is formed in the Γ valley, due to the smaller effective mass shown in Table 5.1. For the biaxially-tensile strained-Si channel in Figure 5.5 (c), the electron population in the two-fold valleys is found to be dominant compared with the unstrained Si channel shown in Figure 5.4 (c).

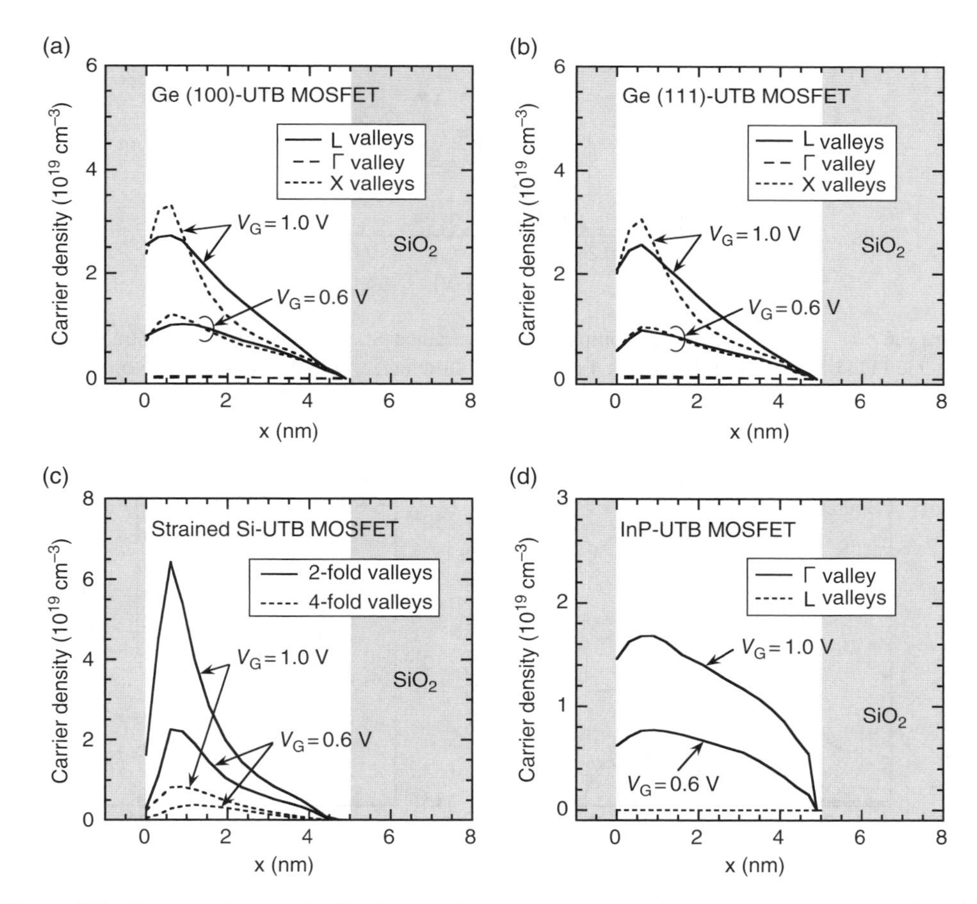

Figure 5.5 Electron density distributions in the inversion-layers computed for: (a) Ge (100); (b) Ge (111); (c) 1% biaxially-tensile strained-Si; and (d) InP with UTB structure. The bias conditions are the same as in Figure 5.4.

Figure 5.6 shows the $I_D - V_G$ characteristics computed for unstrained-Si, 1% biaxially-tensile strained-Si, Ge (100), Ge (111), GaAs and InP MOSFETs with the UTB structure. It is found that the III-V materials provide higher on-current, I_{ON}, than the group IV materials in the present long channel devices. We also found that I_{ON} of the Ge (111)-MOSFET is the highest among the group IV materials.

To understand the mechanism on such a large current enhancement in the III-V MOSFETs, we computed the averaged electron velocities and sheet electron densities, as shown in Figures 5.7(a) and (b), respectively. It is found that the III-V materials have velocities about three times higher than those of the group IV materials at the source-end of channel. On the

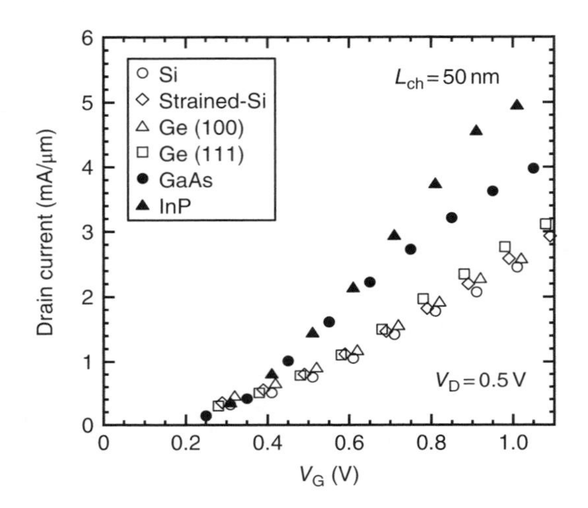

Figure 5.6 $I_D - V_G$ characteristics computed for unstrained-Si, 1% biaxially-tensile strained-Si, Ge (100), Ge (111), GaAs and InP MOSFETs with UTB structure, where the solid and the open symbols represent the group III-V and the group IV materials, respectively. $L_{ch} = 50$ nm and $V_{th} = 0.3$V.

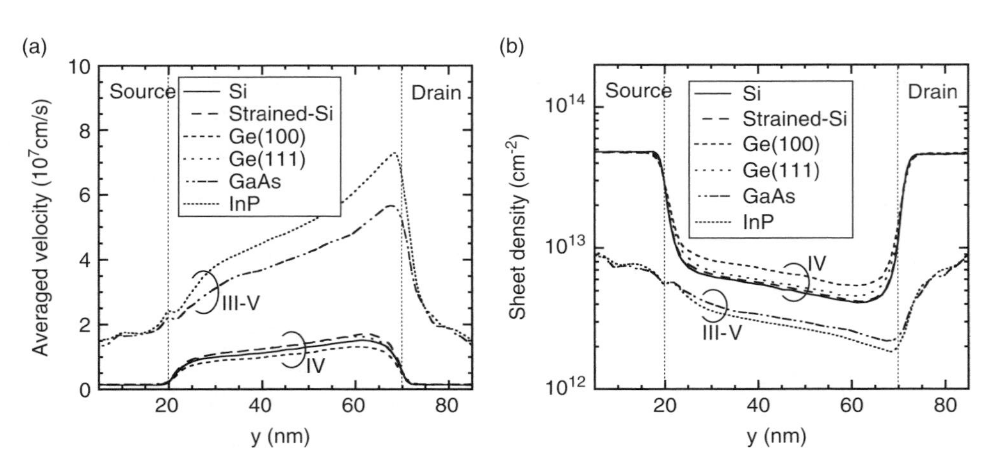

Figure 5.7 (a) Average electron velocities; and (b) sheet electron densities computed for unstrained-Si, 1% biaxially-tensile strained-Si, Ge (100), Ge (111), GaAs and InP MOSFETs with UTB structure, where $V_G = 0.6$ V and $V_D = 0.5$ V. The channel region extends from $y = 20$–70 nm.

other hand, the sheet electron density decreases to about a half of those in the group IV materials, due to the S/D parasitic resistance, but it is not crucial, owing to the UTB structure. As a result, the higher velocities effectively enhance I_{ON} of the III-V long channel MOSFETs, as shown in Figure 5.6.

Figure 5.8 shows the channel length dependences of I_{ON} for all channel materials, including 1% uniaxially-tensile strained-Si and $In_{0.53}Ga_{0.47}As$ channels. The drain voltage V_D is set sufficiently small (0.5 V), so band-to-band tunneling, considered in [5.17], was ignored in this simulation.

In Figure 5.8 (b), the GaAs channel degrades due to the Gunn effect. In the figures, ballistic-limit data for each material are also plotted as B.L., and were obtained by Monte Carlo simulations, ignoring any scattering processes in the channel region. This gives the ideal device performance of the MOSFET, which has been shown to yield an equivalent result to Natori's model [3.2]. Although some revisions may be needed for nanoscale devices [5.12], we use here the above qualitative guideline to understand the ballistic limit.

First, as Fischetti and Laux pointed out for sub-0.1-μm devices [5.10], the current drive of a device appears to be largely independent of the material, with the exception of In-based materials, even for the channel lengths of several tens of nanometers. However, it should be noted that the advantage of In-based materials decreases as the ballistic limit approaches, because ballistic transport is more effective in group IV materials. In other words, III-V MOSFETs are already quasi-ballistic in the present channel lengths. Incidentally, the calculated current drive in Figure 5.8 is higher than that of the corresponding experimental results, because roughness scattering is ignored in the present calculation.

To understand such a limited performance enhancement in III-V MOSFETs, we discuss the role of carrier transport in the source and drain electrodes. First, to recognize the influence of the lower S/D donor concentration in III-V MOSFETs, we plot potential energy profiles in Figure 5.9 for Si, biaxial strained-Si, Ge (111) and InP channels, where the results for each valley are also plotted. As shown by the solid lines in Figure 5.9(d), a large potential drop

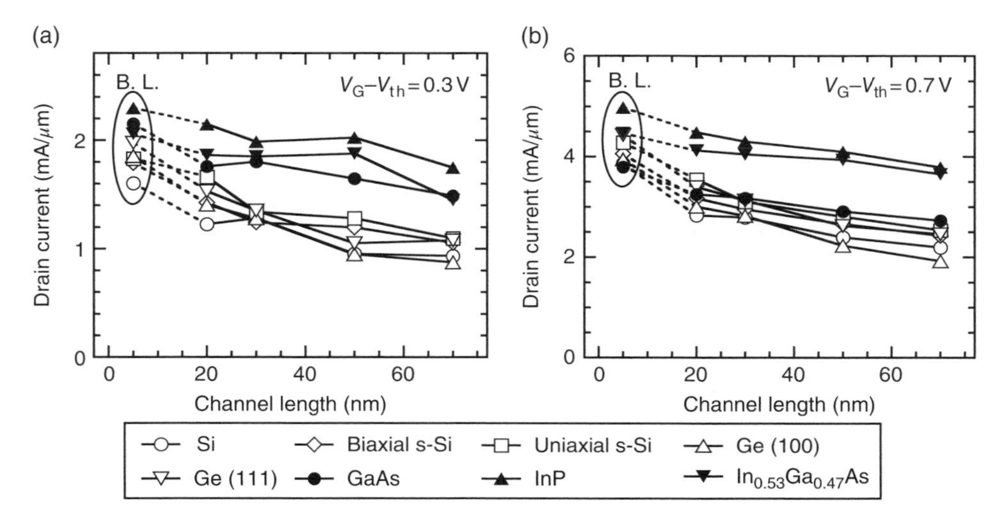

Figure 5.8 Channel length dependences of I_{ON} for all channel materials, including 1% uniaxially-tensile strained-Si and $In_{0.53}Ga_{0.47}As$ channels, where: (a) V_G-V_{th}=0.3 V; and (b) 0.7 V. V_D=0.5 V. B.L. represents ballistic-limit data. The degradation of the GaAs channel in (b) is due to the Gunn effect.

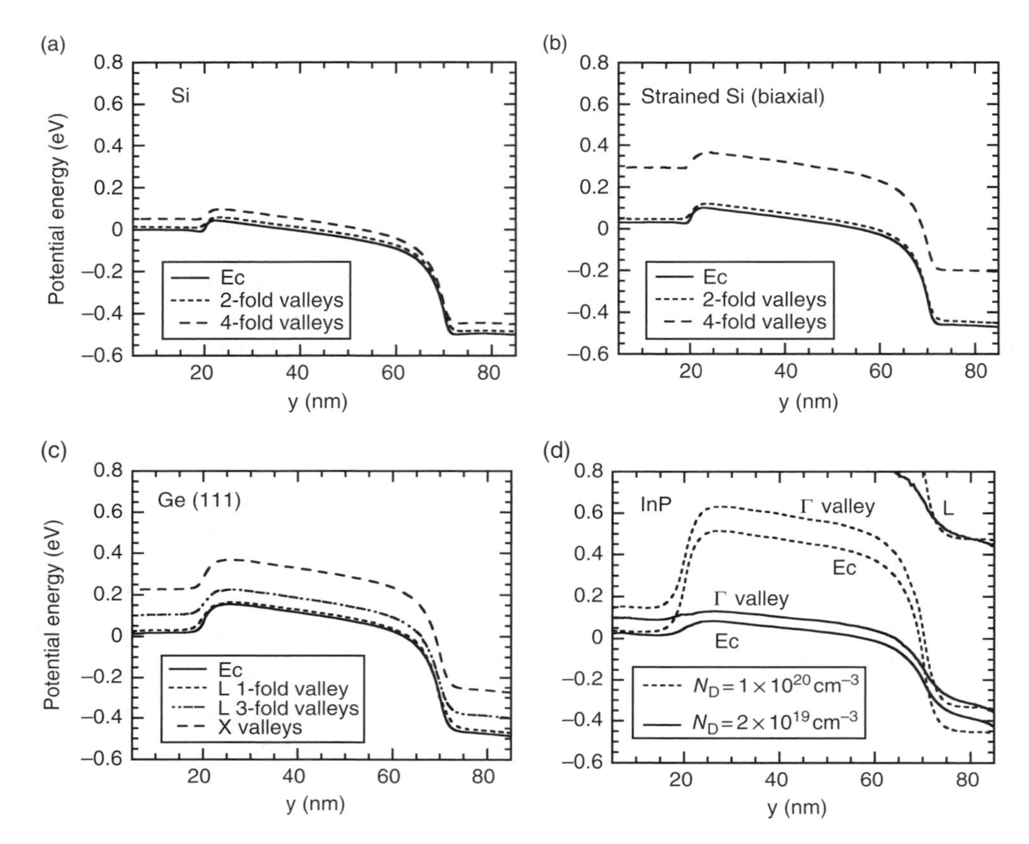

Figure 5.9 Potential energy profiles for each valley in: (a) Si; (b) biaxial strained-Si; (c) Ge (111); and (d) InP channels at $V_G - V_{th} = 0.3$ V and $V_D = 0.5$ V, where $L_{ch} = 50$ nm. Note that the quantum correction of the potential is included for each valley, and thus the lowest valley is located above the conduction band-edge E_c. In (d), the results for heavily-doped S/D are also plotted by dashed lines.

appears in the drain region of the InP-MOSFET, which represents significant parasitic resistance owing to the lower donor concentration and intensified optical phonon scattering in the drain region. As a result, the channel electrical field decreases.

We next increased the S/D donor concentration in the III-V MOSFETs up to $N_D = 1 \times 10^{20}$ cm^{-3}, as a trial. Such a high donor concentration has been reported to be possible for InP using Te-doping [5.18] and Si-doping [5.19]. The potential profiles for such heavily-doped S/D are also plotted in Figure 5.9(d) by the dashed lines. Although the potential barrier height at the source junction increases due to the increased Fermi energy of the source, the parasitic resistance is found to be ideally reduced by heavy doping. Further, Figure 5.10 compares average electron velocity and sheet electron density between the lightly doped ($N_D = 2 \times 10^{19}$ cm^{-3}) and heavily doped S/D ($N_D = 1 \times 10^{20}$ cm^{-3}).

Unexpectedly, the averaged velocity inside the channel is almost unchanged, as shown in Figure 5.10(a). This is due to the fact that acceleration due to the enhanced channel field is offset by a reduction in the injection velocity from the source due to the increased impurity

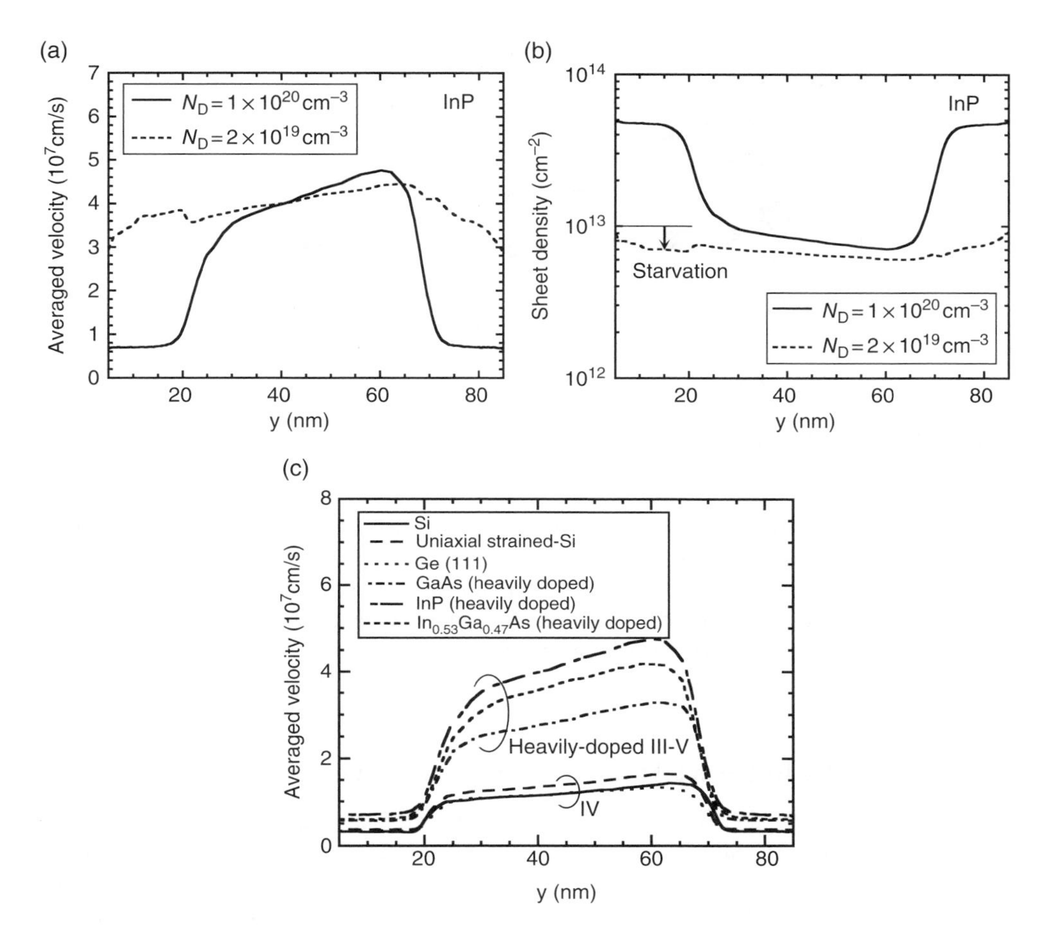

Figure 5.10 Comparison of: (a) average electron velocity; and (b) sheet electron density profiles of InP-MOSFET between lightly-doped and heavily-doped S/Ds, where $L_{ch}=50$ nm, V_G-$V_{th}=0.7$ V and $V_D=0.5$ V. "Source starvation" appears in the lightly-doped source in (b). In (c), the average velocities are compared between various semiconductor channels.

scattering. On the other hand, the sheet electron density inside the channel increases, as shown in Figure 5.10(b). This is because the inability of the lightly-doped source to sustain a large flow of ballistic carriers heading into the channel, termed "source starvation" [5.11, 5.12], is avoided in the heavily-doped source, which supplies more carriers that can be scattered, due mainly to impurities and redistributed into the channel direction.

In Figure 5.10(c), we compare average electron velocities between unstrained-Si, uniaxial strained-Si, Ge (111) and III-V MOSFETs, highlighting the high-speed capability of III-V MOSFETs – In-based materials in particular. Due mainly to the lack of source starvation in heavily-doped S/D [5.11, 5.12], the current drive of III-V MOSFETs significantly increases, as shown in Figure 5.11. In particular, heavily-doped III-V MOSFETs can provide about 1.5 times the current drivability of Si-based MOSFETs even at the ballistic limit.

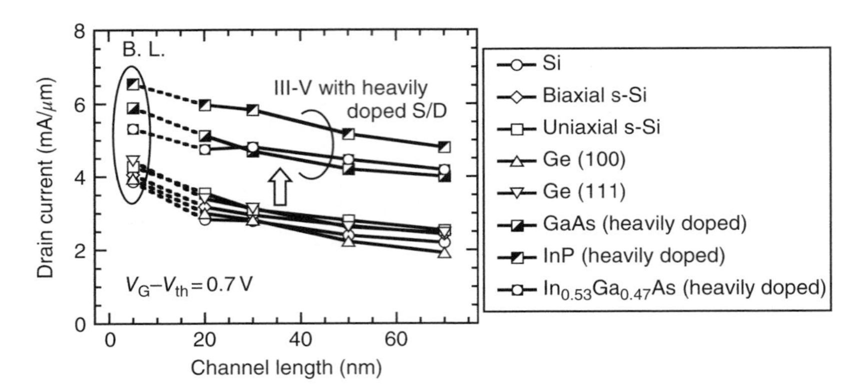

Figure 5.11 Drive current enhancement due to heavily-doped S/D in III-V MOSFETs. $V_G-V_{th}=0.7$ V and $V_D=0.5$ V.

In conclusion, for III-V MOSFETs definitely to outperform Si and Ge MOSFETs under quasi-ballistic transport, a high source and drain-doping concentration is required, because the parasitic resistance in drain and source starvation impede the high-mobility materials providing a higher current drive. Our Monte Carlo simulation showed that if a heavily-doped source and drain can be achieved, III-V MOSFETs can provide at least 1.5 times the current drivability of Si-based MOSFETs, even for the channel lengths of shorter than several 10 nm.

5.2 Source-drain Direct Tunneling in Ultrascaled MOSFETs

As discussed in Section 5.1, III-V compound semiconductors are expected to replace Si as the channel material in *n*-MOSFETs, because of their higher electron mobility and lower effective mass. In fact, the high performance of MOSFETs with InGaAs [5.1, 5.14, 5.20–5.23] and InP [5.13, 5.24] channels has already been experimentally demonstrated. However, since III-V compound semiconductors have a lower transport effective mass and an enhanced quasi-ballistic transport as compared with Si, their use in MOSFETs may lead to more serious quantum transport effects along the channel direction, such as quantum reflection and tunneling [5.25–5.28].

According to previous studies on Si-MOSFETs, their subthreshold current properties are considered to suffer from source-drain direct tunneling (SDT) for channel lengths smaller than 6–8 nm [5.29–5.33]. For instance, the channel length dependences of the $I_D - V_G$ characteristics and the subthreshold swings computed for Si-nanowire MOSFETs with a gate-all-around architecture are shown in Figure 5.12, where the computed results using both quantum and classical models are plotted so that the effect of SDT can be directly identified [5.32]. The subthreshold current and the subthreshold slope begin to increase due to SDT when the channel length becomes smaller than about 8 nm, so there are concerns that the performance of III-V MOSFETs may be degraded due to such enhanced quantum transport effects, even in devices with longer channels than those used in Si-MOSFETs.

In this section, the influence of quantum transport effects in ultrashort-channel III-V MOSFETs is investigated, based on a comparison between Wigner MC (WMC) simulations [5.34–5.41], in which both quantum transport and carrier scattering effects can be fully incorporated, and conventional Boltzmann MC (BMC) simulation, in which no quantum transport effects are considered.

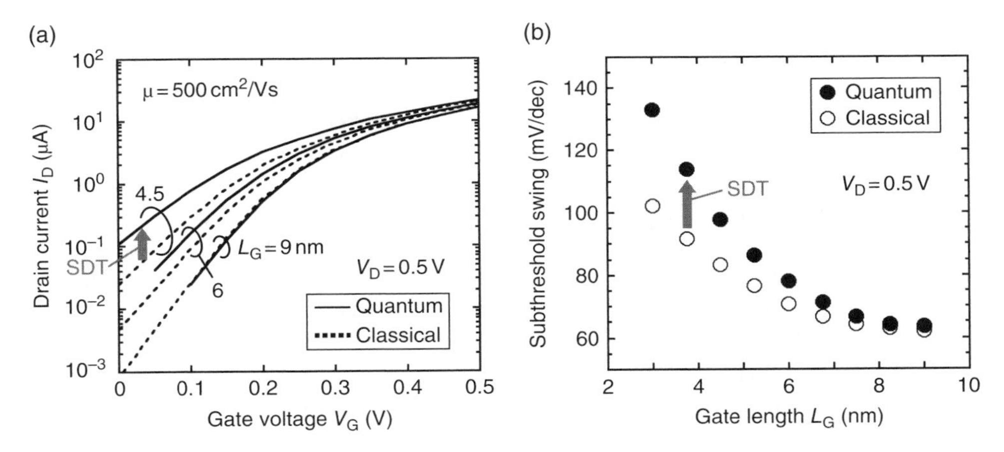

Figure 5.12 Channel length dependences of: (a) $I_D - V_G$ characteristics; and (b) subthreshold swings computed for Si-nanowire MOSFETs with a gate-all-around architecture [5.32]. The computed results using both quantum and classical models are plotted so that the effect of SDT can be directly identified. It is clearly shown that the subthreshold current and the subthreshold slope begin to increase due to SDT when the channel length becomes smaller than about 8 nm.

5.3 Wigner Monte Carlo (WMC) Method

To accurately predict the electrical characteristics of nanoscale devices at normal conditions, device simulation must reliably consider both quantum and scattering effects in carrier transport. The particle-based MC solution of the Boltzmann transport equation (BTE) is acknowledged as a powerful method for accurately describing carrier transport in semiconductor devices within the semiclassical approximation [3.12]. However, as quantum effects become more and more important with continued downscaling, the semiclassical approach, based on the BTE, fails to describe the carrier transport accurately. Quantum effects are often incorporated in conventional MC simulation by considering quantum corrections [3.4, 3.8, 5.42, 5.43], which represent repulsive force from interface and tunneling through potential barrier, in terms of "smoothed effective potential", while they keep 3-D description of particles.

On the other hand, most quantum simulations are based on the nonequilibrium Green's function (NEGF) method, which is the most fundamental theoretical model of quantum transport, and has proved to be robust and versatile in ballistic transport simulation [5.44]. However, in the case of incoherent transport with realistic scattering, the practical solution of the NEGF model requires a large computational effort. Therefore, when scattering effects are included, 2-D or 3-D simulation of actual devices such as MOSFET still remains a difficult problem.

Alternatively, the Wigner function formalism seems very appropriate to deal with realistic problems. It is based on the Wigner function defined in the phase-space, and allows us a rigorous description of quantum transport [5.45]. The well-known strong analogy between Wigner and Boltzmann formalisms makes it possible to use similar numerical techniques, such as the particle MC method, by representing the Wigner function as an ensemble of pseudo-particles and by making use of the same scattering probabilities as in the Boltzmann collision operator [5.34, 5.35]. The major difference between the BTE and the Wigner transport equation (WTE) is the non-local potential term of the WTE. While the BTE treats the potential as a localized force term, the WTE treats the potential term non-locally. Thus, the WTE fully incorporates quantum effects and the non-local term causes the Wigner function to take negative values [5.46–5.48].

The negative parts of the Wigner function cannot be accommodated in a normal MC technique so, in order to account for the negative parts of the Wigner function, the introduction of a new property – particle affinity – has been proposed [5.34, 5.35]. The affinity is a weighting given to each particle to represent its contribution to the total charge distribution of the system. The affinity evolves with time, and its magnitude is updated according to the quantum evolution term governed by the non-local potential of the WTE.

Such a Wigner function-based MC technique is called the Wigner Monte Carlo (WMC) method. The WMC method based on the affinity technique has been proposed by Shifren *et al.* [5.34] and, lately, a fully self-consistent WMC method, in which the handling of the self-consistent Poisson potential and particle injection conditions is revised, has been reported by Querlioz *et al.* [5.35]. In this book, we apply the approach by Querlioz *et al.* to quantum transport simulation in several nanoscale devices and structures, and demonstrate the ability of the fully self-consistent WMC method for studying quantum and dissipative transport of carriers.

5.3.1 Wigner Transport Formalism

The Wigner function was first derived by E. Wigner [5.45] and was successfully extended for analyzing quantum transport phenomena in a resonant-tunneling diode by W. Frensley [5.46]. The Wigner function has a strong connection with density-matrix operator and strong analogy with the classical BTE, as described below.

The quantum Boltzmann transport equation (QBTE), including an explicit expression of collisional term, has been formulated by Kadanoff and Baym [5.49], Keldysh [5.50], and Mahan [5.51], starting from the Dyson's equation for the double-time correlation functions of the field operator of a single fermion based on NEGF formalism. The QBTE is a differential kinetic equation for the Green's function $G^{<}(\mathbf{k}, \omega; \mathbf{R}, T)$, which can be formulated by using standard techniques (Equation (31) in [5.52]). The Wigner function $f_w(\mathbf{R}, \mathbf{k}, T)$ is defined as the energy-integral of $G^{<}(\mathbf{k}, \omega; \mathbf{R}, T)$ as:

$$f_w(\mathbf{R},\mathbf{k},T) = \int \frac{d\omega}{2\pi}\left[-iG^{<}(\mathbf{k},\omega;\mathbf{R},T)\right] \tag{5.5}$$

The electron density $n(\mathbf{R}, T)$ and the current density $j(\mathbf{R}, T)$ are represented using the Wigner function as follows:

$$n(\mathbf{R},T) = \int \frac{d\mathbf{k}}{(2\pi)^3} f_w(\mathbf{R},\mathbf{k},T) \tag{5.6}$$

$$j(\mathbf{R},T) = \int \frac{d\mathbf{k}}{(2\pi)^3} \frac{\hbar\mathbf{k}}{m^*} f_w(\mathbf{R},\mathbf{k},T) \tag{5.7}$$

where $\hbar$ and m^* are Planck's constant divided by 2π and the effective mass, respectively. In this book, we employ the WTE with respect to $f_w(\mathbf{R}, \mathbf{k}, T)$ for nanoscale device simulation, and formulate it starting from the following effective mass Hamiltonian:

$$H(\mathbf{r},t) = -\frac{\hbar^2}{2m^*}\nabla^2 + U(\mathbf{r},t) \tag{5.8}$$

$U(\mathbf{r}, t)$ is the potential energy distribution. For a one-electron system, the time-dependent Schrödinger equation is expressed as:

$$i\hbar\frac{\partial\varphi_j(\mathbf{r},t)}{\partial t} = H(\mathbf{r},t)\varphi_j(\mathbf{r},t) \tag{5.9}$$

Here, we define a density matrix $\rho(\mathbf{r}, \mathbf{r}', t)$ using wave functions at different spatial points – that is, $\varphi_j(\mathbf{r}, t)$ and $\varphi_j(\mathbf{r}', t)$, as follows:

$$\rho(\mathbf{r},\mathbf{r}',t) = \sum_j P_j\varphi_j(\mathbf{r},t)\varphi_j^*(\mathbf{r}',t) \tag{5.10}$$

P_j stands for the occupation probability of a state j. The density matrix $\rho(\mathbf{r}, \mathbf{r}', t)$, which is represented in the coordinate space, describes the non-local correlation of wave function, which is an essential feature of quantum mechanics. By differentiating Equation (5.10) with respect to time and using Equation (5.9) (and also its complex conjugate), time evolution of the density matrix (quantum mechanical Liouville equation) is obtained as:

$$\begin{aligned}\frac{\partial\rho(\mathbf{r},\mathbf{r}',t)}{\partial t} &= \sum_j P_j\left[\frac{\partial\varphi_j(\mathbf{r},t)}{\partial t}\varphi_j^*(\mathbf{r}',t) + \varphi_j(\mathbf{r},t)\frac{\partial\varphi_j^*(\mathbf{r}',t)}{\partial t}\right]\\ &= \sum_j P_j\left[\frac{1}{i\hbar}H(\mathbf{r},t)\varphi_j(\mathbf{r},t)\varphi_j^*(\mathbf{r}',t) + \varphi_j(\mathbf{r},t)\left(\frac{-1}{i\hbar}\right)H^*(\mathbf{r}',t)\varphi_j^*(\mathbf{r}',t)\right]\\ &= \frac{1}{i\hbar}\left[H(\mathbf{r},t) - H^*(\mathbf{r}',t)\right]\rho(\mathbf{r},\mathbf{r}',t)\end{aligned} \tag{5.11}$$

Substituting Equation (5.8) into Equation (5.11), we obtain:

$$\frac{\partial\rho(\mathbf{r},\mathbf{r}',t)}{\partial t} = \frac{1}{i\hbar}\left[-\frac{\hbar^2}{2m^*}\left(\nabla_{\mathbf{r}}^2 - \nabla_{\mathbf{r}'}^2\right) + U(\mathbf{r},t) - U(\mathbf{r}',t)\right]\rho(\mathbf{r},\mathbf{r}',t) \tag{5.12}$$

Here, to emphasize the non-local nature of the system, we introduce the center of mass coordinate vector $\mathbf{R} = (\mathbf{r} + \mathbf{r}')/2$ and relative coordinate vector $u = \mathbf{r} - \mathbf{r}'$. Then, the differential operators are transformed as:

$$\nabla_{\mathbf{r}} = \frac{1}{2}\nabla_{\mathbf{R}} + \nabla_{\mathbf{u}} \quad \rightarrow \quad \nabla_{\mathbf{r}}^2 = \frac{1}{4}\nabla_{\mathbf{R}}^2 + \nabla_{\mathbf{R}}\cdot\nabla_{\mathbf{u}} + \nabla_{\mathbf{u}}^2 \tag{5.13}$$

$$\nabla_{\mathbf{r}'} = \frac{1}{2}\nabla_{\mathbf{R}} - \nabla_{\mathbf{u}} \quad \rightarrow \quad \nabla_{\mathbf{r}'}^2 = \frac{1}{4}\nabla_{\mathbf{R}}^2 - \nabla_{\mathbf{R}}\cdot\nabla_{\mathbf{u}} + \nabla_{\mathbf{u}}^2 \tag{5.14}$$

and the quantum mechanical Liouville equation is rewritten using the **R** and **u** vectors as:

$$i\hbar\frac{\partial\rho\left(\mathbf{R}+\frac{\mathbf{u}}{2},\mathbf{R}-\frac{\mathbf{u}}{2}\right)}{\partial t}=-\frac{\hbar^2}{m^*}\nabla_{\mathbf{R}}\cdot\nabla_{\mathbf{u}}\rho\left(\mathbf{R}+\frac{\mathbf{u}}{2},\mathbf{R}-\frac{\mathbf{u}}{2}\right)+\left[U\left(\mathbf{R}+\frac{\mathbf{u}}{2}\right)-U\left(\mathbf{R}-\frac{\mathbf{u}}{2}\right)\right]\rho\left(\mathbf{R}+\frac{\mathbf{u}}{2},\mathbf{R}-\frac{\mathbf{u}}{2}\right) \tag{5.15}$$

where the time variable t is omitted for simplicity.

The Wigner function is defined as a Fourier transform of the density matrix with respect to the relative coordinate vector **u** as [5.45]:

$$f_w(\mathbf{R},\mathbf{k})=\int_{-\infty}^{\infty}d\mathbf{u}\ \rho\left(\mathbf{R}+\frac{\mathbf{u}}{2},\mathbf{R}-\frac{\mathbf{u}}{2}\right)e^{-i\mathbf{k}\cdot\mathbf{u}} \tag{5.16}$$

On the other hand, the density matrix is represented by the inverse Fourier transform of the Wigner function with respect to the wavenumber **k**:

$$\rho\left(\mathbf{R}+\frac{\mathbf{u}}{2},\mathbf{R}-\frac{\mathbf{u}}{2}\right)=\int_{-\infty}^{\infty}\frac{d\mathbf{k}}{(2\pi)^3}f_w(\mathbf{R},\mathbf{k})\ e^{i\mathbf{k}\cdot\mathbf{u}} \tag{5.17}$$

The Wigner function is real-valued, but can have negative value. In the classical picture, "distribution function" never takes negative value and, hence, the presence of a Wigner function with negative value is interpreted as the signature of the system being in a quantum state (e.g., exhibiting tunneling and interference effects). In addition, since the Wigner function has negative value, it cannot be viewed as the existence probability of electrons in the phase space. Accordingly, it should be called "Wigner quasi-distribution function", or simply "Wigner function". In this book, we call it Wigner function for simplicity.

Performing the Fourier transformation of the quantum mechanical Liouville equation (5.15), the following WTE is obtained:

$$\frac{\partial f_w}{\partial t}+\frac{\hbar\mathbf{k}}{m^*}\cdot\nabla_{\mathbf{R}}f_w=Qf_w(\mathbf{R},\mathbf{k})+Cf_w(\mathbf{R},\mathbf{k}) \tag{5.18}$$

where $Qf_w(\mathbf{R},\mathbf{k})$ is called the quantum evolution term, and given by:

$$Qf_w(\mathbf{R},\mathbf{k})=-\frac{1}{\hbar}\int_{-\infty}^{\infty}\frac{d\mathbf{k}'}{(2\pi)^3}V(\mathbf{R},\mathbf{k}-\mathbf{k}')f_w(\mathbf{R},\mathbf{k}') \tag{5.19}$$

$V(\mathbf{R},\mathbf{k}-\mathbf{k}')$ is the non-local potential term expressed in the following equation:

$$\begin{aligned}V(\mathbf{R},\mathbf{k}-\mathbf{k}')&=i\int_{-\infty}^{\infty}d\mathbf{u}\ e^{-i(\mathbf{k}-\mathbf{k}')\cdot\mathbf{u}}\left[U\left(\mathbf{R}+\frac{\mathbf{u}}{2}\right)-U\left(\mathbf{R}-\frac{\mathbf{u}}{2}\right)\right]\\&=2\int_{0}^{\infty}d\mathbf{u}\ \sin\left[(\mathbf{k}-\mathbf{k}')\cdot\mathbf{u}\right]\left[U\left(\mathbf{R}+\frac{\mathbf{u}}{2}\right)-U\left(\mathbf{R}-\frac{\mathbf{u}}{2}\right)\right]\end{aligned} \tag{5.20}$$

As can be seen in Equation (5.20), electrons evolve under not only by the local potential $U(\mathbf{R})$, but also by spatially-separate potential $U(\mathbf{R}\pm\mathbf{u}/2)$. It is for this reason that Equation (5.20) is called the non-local potential term.

$Cf_w(\mathbf{R}, \mathbf{k})$ in Equation (5.18) represents the collision term. Accordingly, the WTE is very similar to the classical BTE. If quantum collision effects are neglected and the same collision operator as in the classical BTE is used, the only difference comes from the quantum evolution term $Qf_w(\mathbf{R}, \mathbf{k})$.

In this study, we will use the same scattering probabilities as in the classical BTE, which enable us clear comparison between the classical MC and the WMC formalisms. In addition, both the potential barrier and the Poisson potential are considered in $U(\mathbf{R})$ of Equation (5.20), in order to incorporate quantum effects not only in quantum regions, but also in external electrodes [5.35]. Alternatively, the quantum evolution term Qf_w can be represented in powers of $\hbar$ and higher-order spatial derivatives of the potential energy [5.45]. The $\hbar^2$ – order term gives a quantum correction of potential [3.2, 3.4–3.7, 3.10, 3.18, 5.53] or quantum diffusion current [5.54] and quantum hydrodynamic models [5.55, 5.56], with the form of density-gradient.

5.3.2 *Relation with Quantum-corrected MC Method*

WTE in Equation (5.18) can be rewritten in the form of a modified BTE as:

$$\frac{\partial f_w}{\partial t} + \frac{\hbar\mathbf{k}}{m^*}\cdot\nabla_{\mathbf{R}} f_w - \frac{1}{\hbar}\nabla_{\mathbf{R}} U\cdot\nabla_{\mathbf{k}} f_w + C_q(\mathbf{R},\mathbf{k}) = Cf_w \tag{5.21}$$

where $C_q(\mathbf{R}, \mathbf{k})$ denotes the quantum correction due to the spatially varying potential energy $U(\mathbf{R})$, which is expressed by using Taylor expansion in the non-local potential term (5.20) as follows [5.45]:

$$C_q(\mathbf{R},\mathbf{k}) = \sum_{\alpha=1}^{\infty}\frac{(-1)^{\alpha+1}}{4^{\alpha}\hbar(2\alpha+1)!}(\nabla_{\mathbf{R}}\cdot\nabla_{\mathbf{k}})^{2\alpha+1}Uf_w \tag{5.22}$$

The quantum correction accounts for various quantum effects. Here, using the relation of $\mathbf{p}=\hbar\mathbf{k}$, Equation (5.21) yields:

$$\frac{\partial f_w}{\partial t} + \frac{\mathbf{p}}{m^*}\cdot\nabla_{\mathbf{R}} f_w - \nabla_{\mathbf{R}} U\cdot\nabla_{\mathbf{p}} f_w + \sum_{\alpha=1}^{\infty}\frac{(-1)^{\alpha+1}\hbar^{2\alpha}}{4^{\alpha}(2\alpha+1)!}(\nabla_{\mathbf{R}}\cdot\nabla_{\mathbf{p}})^{2\alpha+1}Uf_w = Cf_w \tag{5.23}$$

In the classical limit of $\hbar\to 0$, the fourth term of the left-hand side in Equation (5.23) vanishes, and we obtain the conventional BTE as:

$$\frac{\partial f}{\partial t} + \frac{\mathbf{p}}{m^*}\cdot\nabla_{\mathbf{R}} f - \nabla_{\mathbf{R}} U\cdot\nabla_{\mathbf{p}} f = Cf \tag{5.24}$$

Next, we consider the lowest-order quantum correction to the BTE by taking only the $\alpha = 1$ term in Equation (5.22), that is:

$$\frac{\partial f_w}{\partial t} + \frac{\hbar \mathbf{k}}{m^*} \cdot \nabla_{\mathbf{R}} f_w - \frac{1}{\hbar} \nabla_{\mathbf{R}} U \cdot \nabla_{\mathbf{k}} f_w + \frac{1}{24\hbar} \nabla_{\mathbf{R}}^3 U \cdot \nabla_{\mathbf{k}}^3 f_w = C f_w \tag{5.25}$$

This lowest-order term induces a major contribution in the quantum correction components. When the system is close to equilibrium, the following approximate relations are derived using displaced Maxwellian statistics:

$$\nabla_{\mathbf{R}}^3 U \approx \nabla_{\mathbf{R}} \left(-k_B T \nabla_{\mathbf{R}}^2 \ln(n) \right), \nabla_{\mathbf{k}}^3 f_w \approx -\frac{2\hbar^2}{m^* k_B T} \nabla_{\mathbf{k}} f \tag{5.26}$$

From these approximations, the following quantum-corrected BTE is obtained [3.4, 3.7, 5.53]:

$$\frac{\partial f}{\partial t} + \frac{\hbar \mathbf{k}}{m^*} \cdot \nabla_{\mathbf{R}} f - \frac{1}{\hbar} \nabla_{\mathbf{R}} \left(U + U^{QC} \right) \cdot \nabla_{\mathbf{k}} f = C f \tag{5.27}$$

where U^{QC} is the quantum correction of potential introduced in Section 3.1.1 and, under an isotropic effective mass, it is expressed by:

$$U^{QC} = -\frac{\hbar^2}{12 m^*} \nabla_{\mathbf{R}}^2 \ln(n) \tag{5.28}$$

The quantum effects are incorporated in terms of quantum mechanically corrected potential in the driving term. The advantage of the quantum-corrected MC method is that only a little modification is needed to introduce into conventional MC simulator (i.e., just adding U^{QC} to the potential energy). As illustrated in Figure 5.13, quantum tunneling and confinement effects are described by potential lowering and potential rising phenomena due to U^{QC}, respectively. However, the quantum-corrected MC method cannot rigorously handle quantum mechanical effects, because the lowest-order approximation in quantum correction terms, and also the analytical distribution function at quasi-equilibrium state, are assumed in the derivation of U^{QC}.

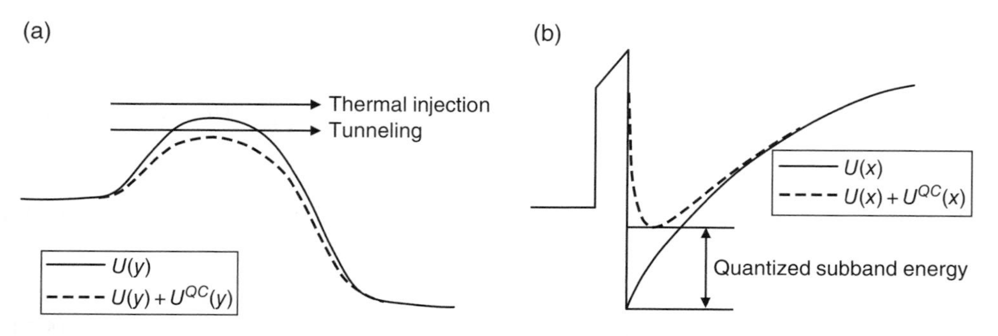

Figure 5.13 Roles of U^{QC} in a MOSFET along: (a) transport direction; and (b) confinement direction.

5.3.3 WMC Algorithm

We consider 1-D transport system for sake of simplicity. In this case, the WTE is represented by:

$$\frac{\partial f_w}{\partial t}+\frac{\hbar k}{m^*}\frac{\partial f_w}{\partial \chi}=Qf_w(\chi,k)+Cf_w(\chi,k) \tag{5.29}$$

where:

$$Qf_w(\chi,k)=-\frac{1}{\hbar}\int_{-\infty}^{\infty}\frac{dk'}{2\pi}V(\chi,k-k')f_w(\chi,k') \tag{5.30}$$

and:

$$V(\chi,k-k')=2\int_{0}^{\infty}du\ \sin\left[(k-k')u\right]\left[U\left(\chi+\frac{u}{2}\right)-U\left(\chi-\frac{u}{2}\right)\right] \tag{5.31}$$

According to Shifren *et al.* [5.34] and Querlioz *et al.* [5.35], we use the affinity technique to extend the particle MC algorithm to WTE. In this technique, the Wigner function is defined by the position and the wave vector, and then described as an ensemble of pseudo-particles weighted by the affinity. Hence, the Wigner function takes the form:

$$f_w(\chi,k,t)=\sum_i A_i(t)\delta(\chi-\chi_i(t))\delta(k-k_i(t)) \tag{5.32}$$

where χ_i, k_i and A_i are the position, the wavenumber and the affinity, respectively, of the *i*-th particle. Note that the affinity technique incorporates the wave properties of particles in the simulation. Such quantum mechanical particles behave and scatter as classical particles, except that the potential no longer influences the wavenumber, but only the affinity through the quantum evolution term $Qf_w(\chi, k)$. This means that the wavenumber can be changed only by scattering and, as a result, the equations of motion during a free flight are given by [5.35–5.39]:

$$\frac{d\chi_i}{dt}=\frac{\hbar k_i}{m} \tag{5.33}$$

$$\frac{dk_i}{dt}=0 \tag{5.34}$$

$$\sum_{i\in M(\chi,k)}\frac{dA_i}{dt}=Q f_w(\chi_i,k_i) \tag{5.35}$$

As shown in Equation (5.34), the wavenumber k_i is constant with time during a free flight, because the classical drift term is already incorporated in Equation (5.35) in terms of the affinity change owing to $Qf_w(\chi, k)$. Here, $M(\chi, k)$ represents a mesh of the phase-space that the *i*-th particle belongs to. In other words, in the WMC approach, quantum transport of carriers is described by temporal change in the affinity of pseudo-particles moving with a constant velocity. It is also found from Equation (5.35) or Equation (5.32) that the affinity can take negative values. As Querlioz *et al.* pointed out, each mesh of the phase-space must always contain at least one pseudo-particle to ensure the conservation of the total affinity of particles. Therefore, we should inject particles with zero-affinity to every empty mesh where the quantum evolution term is not null. However, by injecting particles to every empty mesh, computational effort drastically increases. To reduce the number of pseudo-particles to be simulated, and to ensure the computational accuracy simultaneously, we introduce a criterion for the zero-affinity particle injection as follows:

$$\left|Qf_w(\chi,k)\right| \rangle \frac{0.01}{\Delta t}\left(\mathrm{s}^{-1}\right) \tag{5.36}$$

Here, Δt is the time step of the MC simulation. This criterion was derived using the following simple argument. From Equation (5.35), the time evolution of the affinities is calculated as:

$$A_i(t+\Delta t)-A_i(t)=\frac{1}{n}\Delta t\times Qf_w(\chi,k,t+\Delta t) \tag{5.37}$$

where n is the number of pseudo-particles in the mesh $M(\chi, k)$ of the phase-space. If the right-hand side quantity of Equation (5.37) is smaller than 0.01, the difference is negligible since A_i usually takes a value close to 1. In a word, particles with zero-affinity are not necessary to be injected into such a trivial mesh.

Consequently, $\Delta t\times|Qf_w(\chi,k,t+\Delta t)|>0.01n>0.01$ can be used as a guideline as to whether to inject the zero-affinity particle into empty mesh. We have carefully checked its validity by changing the threshold value, and have verified Equation (5.36) to work very well. We also add here that a predictor technique of fourth order was adopted to update the pseudo-particle affinities following Equation (5.37), which makes the solution stable [5.38, 5.39].

To accurately describe the time evolution of pseudo-particles, the discretization scheme for the phase-space must be also stable. Since Equation (5.31) is periodic in k-space with a period of $2\pi/\Delta_u$, a mesh spacing Δ_u for the relative coordinate u should be sufficiently small so that π/Δ_u must always be larger than the maximum k value that carriers can reach. In the WMC method, Δ_u can be chosen independently of the real-space mesh spacing Δ_χ.

On the other hand, k-space meshing determines the number of pseudo-particles and, hence, the computational time, so its mesh spacing Δ_k with an arbitrary number of mesh points N_k, that is $\Delta_k=2\pi/(N_k\Delta_u)$, should be adjusted not to reduce the computational accuracy. Details of other numerical techniques for the fully self-consistent WMC method, including boundary conditions, are found in [5.35–5.39].

Since the WMC technique is a quantum ensemble Monte Carlo (EMC) based on the full particle nature of the EMC technique, we can utilize ensemble statistics by representing that any ensemble average takes the form:

$$\langle Q\rangle = \frac{\sum_i A_i Q_i}{\sum_i A_i} \tag{5.38}$$

where Q is the quantity of interest, such as velocity and energy. The current flowing through the device is calculated by averaging the electron velocities based on the Ramo-Shockley theorem [5.57] as:

$$I(t) = \frac{q}{L}\sum_i A_i(t) v_i(t) \tag{5.39}$$

where L is the device length.

5.3.4 Description of Higher-order Quantized Subbands

First, we examined the ability of the WMC method to describe higher-order quantized subbands by simulating quantum well structures. Figure 5.14 shows the electron density distributions computed for single quantum wells with a well depth of 0.2 eV and well widths of 4 nm, 10 nm and 14 nm. The barrier region consists of AlGaAs doped to 10^{18} cm^{-3} and the well region GaAs slightly doped to 10^{16} cm^{-3}. The temperature is 300 K, and the scattering mechanisms considered are polar optical phonons, elastic acoustic phonons and impurities. In this study, we considered only transport within the Γ valley, and thus non-polar optical phonons are disregarded.

It is found that the electron distribution in the quantum well has single-peaked pattern in (a), double-peaked pattern in (b), and triple-peaked pattern in (c), which can easily be assumed to be due to the change of electron population into higher-order quantized subbands. Note that Poisson's equation was not solved in the calculation, which means that the potential energy profile is frozen to be flat, except at the heterojunctions as shown in the insets of Figure 5.14. This is because the self-consistent Poisson potential induces band-bending inside the quantum well, and drastically modifies electron density profile, even when only a single quantized subband exists.

To check the actual number of quantized subbands, we have numerically solved the Schrödinger equation to calculate the energy eigenvalues, as shown in Table 5.4. Comparing them to Figure 5.14, the number of peaks in the WMC density distributions has proved to be completely consistent with the total number of eigenvalues obtained from the Schrödinger equation. Consequently, the WMC technique is found to be a more accurate quantum tool than quantum-corrected MC techniques, which can represent only the lowest-quantized subband via smoothed effective potential [3.4 - 3.9, 5.42, 5.43, 5.53 - 5.56].

5.3.5 Application to Resonant-tunneling Diode

Next, we apply the WMC method to a resonant-tunneling diode (RTD), which is one of the appropriate devices to demonstrate the ability of the WMC method to describe quantum tunneling and scattering effects. The RTD structure used in the simulation is shown in Figure 5.15, where a GaAs quantum well of 5 nm is sandwiched between two AlGaAs barriers of 0.3 eV high and 3 nm wide, which is similar to that computed in [5.35]. The quantum well, the barriers and 10 nm thick spacer regions adjacent to the barriers are slightly doped to 10^{16} cm^{-3}, while the 60 nm long access regions are doped to 10^{18} cm^{-3}. As in the previous section, the temperature is 300 K, and the scattering mechanisms considered are polar optical

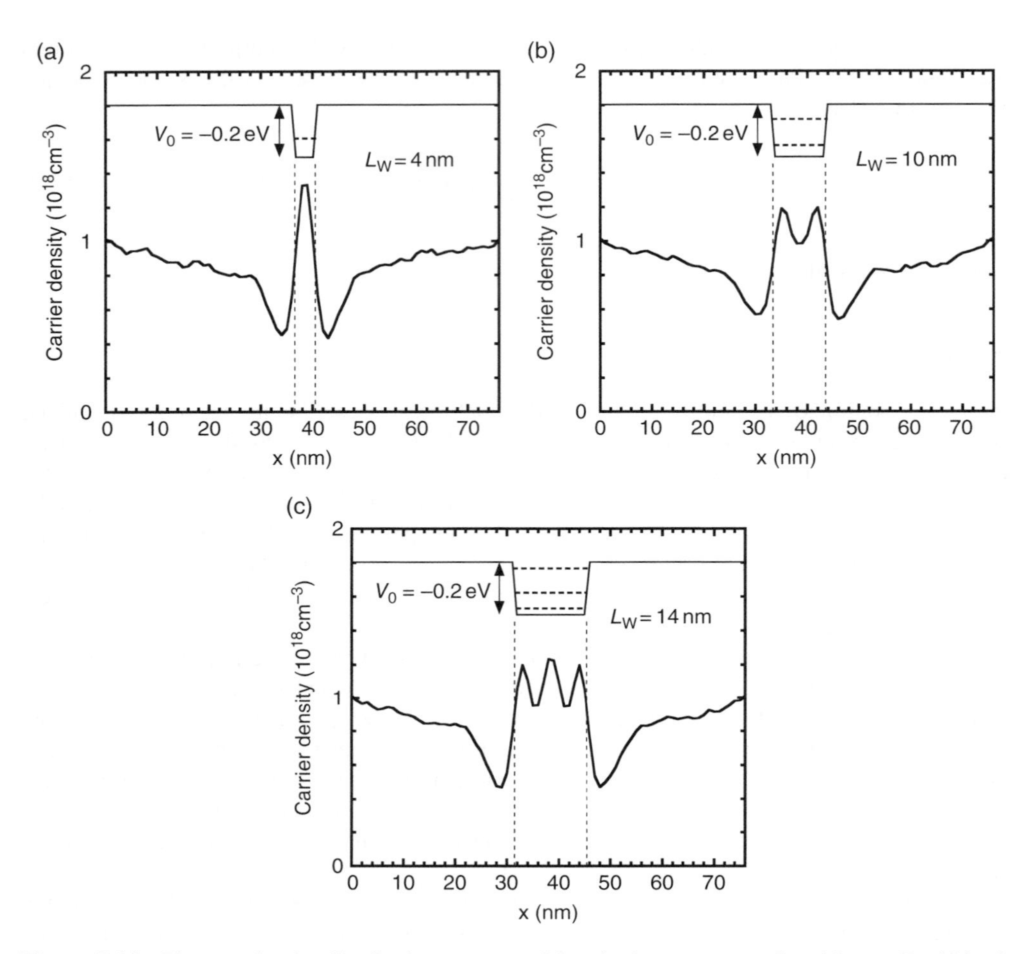

Figure 5.14 Electron density distributions computed for single quantum wells with a well width of: (a) 4 nm; (b) 10 nm; and (c) 14 nm, where the well depth is assumed to be 0.2 eV. The barrier is AlGaAs doped to 10^{18} cm^{-3} and the well is GaAs slightly doped to 10^{16} cm^{-3}. The Poisson equation is not solved in the simulation.

Table 5.4 Energy eigenvalues estimated for single quantum wells with three different well widths. They were obtained by numerically solving the Schrödinger equation.

Well width (nm)	E_1 (eV)	E_2 (eV)	E_3 (eV)
4	0.094	–	–
10	0.031	0.120	–
14	0.018	0.073	0.150

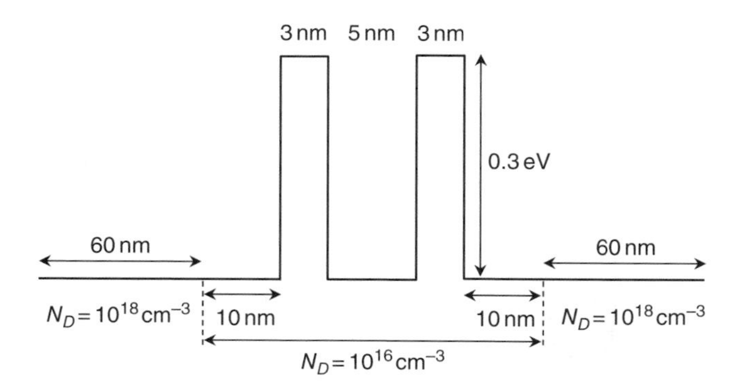

Figure 5.15 RTD structure used in the simulation.

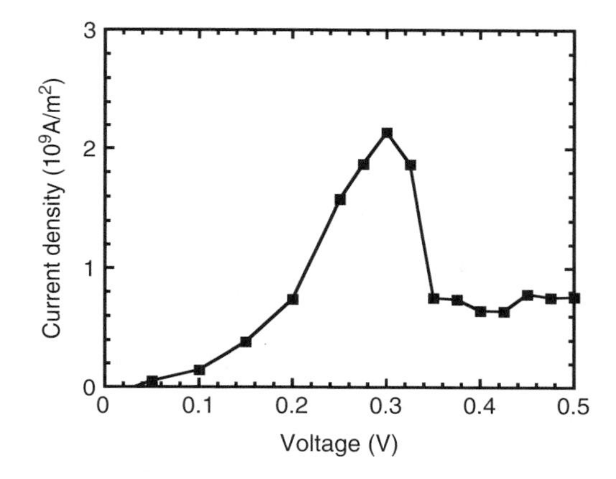

Figure 5.16 Computed current-voltage characteristics of RTD.

phonons, elastic acoustic phonons and impurities; only the transport within the Γ valley is considered. Here, we should state that Poisson's equation was self-consistently solved in the present RTD simulation, and both the potential barriers and the Poisson potential are treated in the affinity evolution.

The computed current-voltage characteristics are shown in Figure 5.16, which corresponds well to the result reported in [5.35], in terms of peak and valley current densities, and also their occurring bias voltages. In Figure 5.17, we plot electron densities and self-consistent potentials for the resonant and the non-resonant biases. On resonance, an obvious peak of electron density is found in the central quantum well while, on non-resonance, an accumulation peak is formed in the left electrode, which is caused by the electrons bounced off the barriers. These results indicate that the WMC method can handle quantum mechanical resonant-tunneling accurately.

Next, we examine the ability of handling scattering effects [5.38, 5.58]. In Figure 5.18, we plot three kinds of current-voltage characteristics computed for no scattering inside the double-barrier region, standard scattering rate and scattering rates multiplied by 5. As expected, the peak current density decreases and the valley current density increases with scattering rates, and thus the peak-to-valley current ratio reduces by scattering.

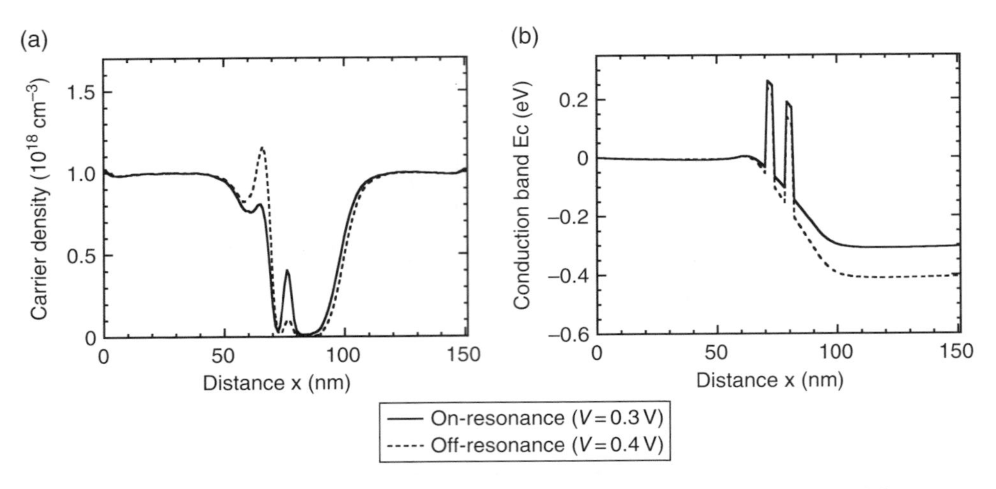

Figure 5.17 (a) Electron density; and (b) self-consistent potential distributions computed for resonant and non-resonant biases.

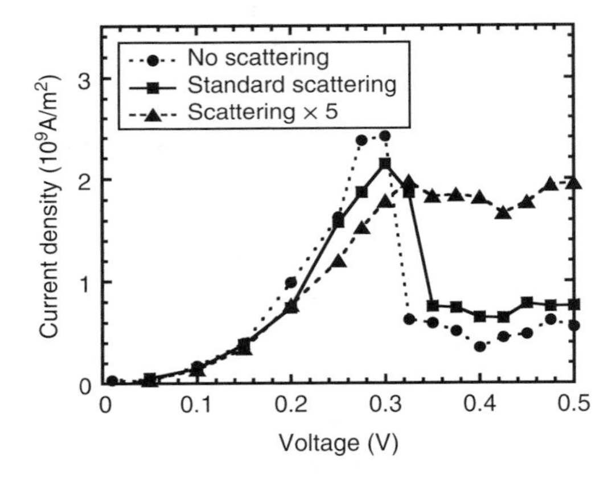

Figure 5.18 Influences of scattering on current-voltage characteristics of RTD. Three results computed for no scattering inside the double-barrier region, standard scattering rate and scattering rates multiplied by 5 are plotted.

Such decoherence effects due to scattering can also be verified by rendering the Wigner distribution function in the phase-space, as shown in Figure 5.19, with no scattering inside the double-barrier region, standard scattering rate, and scattering rates multiplied by 5. The bias voltage is all set at $V = 0.3$ V. In the right-hand side electrode of Figure 5.19(a) and (b), we can see the signature of electron waves tunneling through the double-barrier. In addition, a quantum interference pattern inside the central quantum well is clearly visible in Figure 5.19(a) and (b). On the other hand, such quantum mechanical behaviors almost disappear when the scattering rates are increased by five times, as shown in Figure 5.19(c). This scattering-induced decoherence tends to degrade the resonant tunneling property as shown in Figure 5.18, by

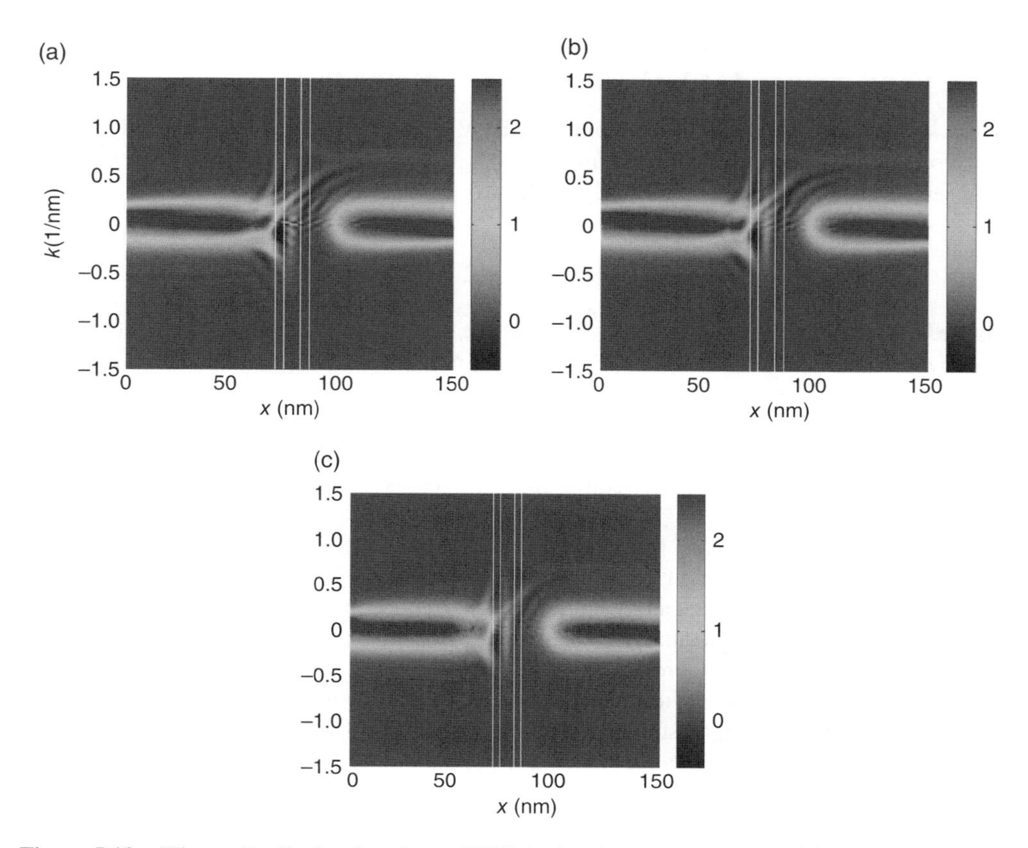

Figure 5.19 Wigner distribution functions of RTD in the phase-space computed for: (a) no scattering inside the double-barrier region; (b) standard scattering rate; and (c) scattering rates multiplied by 5. The bias voltage is all set at $V = 0.3$ V.

increasing the valley current to a current level such that the negative resistance becomes almost unobservable.

As for computational efforts, a large number of simulated particles (i.e., typically between 65 000 and 110 000, depending on bias voltage) are required. The computational time also increases by a factor of about two when compared with the classical MC simulation. Even so, the WMC technique has the ability to handle all quantum effects, such as resonant tunneling and higher-order quantization, not to mention that all relevant scattering mechanisms found in the standard MC procedure can be handled.

In conclusion, we have shown that the WMC approach can describe higher-order quantized subbands by simulating electron injection processes into single quantum wells. This is one of the advantages over the quantum correction approaches, which can produce spatial carrier distribution of only the lowest-quantized subband via smoothed effective potential. We have also demonstrated that the WMC approach can quite clearly simulate tunneling and decoherence processes occurring in nanoscale semiconductor devices.

5.4 Quantum Transport Simulation of III-V *n*-MOSFETs with Multi-subband WMC (MSB-WMC) Method

In this section, the WMC approach is applied to III-V MOSFETs, and the influence of quantum transport on the subthreshold current properties is investigated on the basis of a comparison between WMC and conventional BMC simulations [5.26, 5.28].

5.4.1 Device Structure

Figure 5.20 shows the device structure used in the simulation. A double-gate structure was employed with a channel thickness (T_{ch}) of 5 or 3 nm, and a SiO_2 gate oxide thickness (T_{ox}) of 0.5 nm. The channel, source, and drain materials were $In_{0.53}Ga_{0.47}As$ or InP, and the band parameters listed in Table 5.1 were used. The bulk band parameters for both materials were used, and hence the present simulation might overestimate the effect of SDT in UTB III-V MOSFETs, because the actual effective mass increases owing to quantum confinement [5.59–5.67].

In the simulation, only the Γ and L valleys are considered, since electrons are never distributed in X valleys with an energy gap $\Delta E_{\Gamma X}$ sufficiently larger than the supply voltage considered in this study. The source and drain donor concentrations used were 2×10^{19} cm^{-3} [5.2, 5.4, 5.16, 5.17], and the channel was undoped, where we assumed that the donor distribution abruptly changes at source-channel and drain-channel junctions. The channel length (L_{ch}) was varied from 5 to 40 nm. The electrical characteristics were calculated using the WMC device simulator, and also using the BMC simulator [3.12], in both of which electron transport is simulated along the channel direction y for each valley and each quantized subband obtained with the Schrödinger-Poisson solver. Therefore, we can assess quantum transport effects, including SDT, by comparing the two MC results. The scattering processes considered in this study are acoustic phonon, non-polar and polar optical phonon, and impurity scatterings [5.2, 5.4, 5.16].

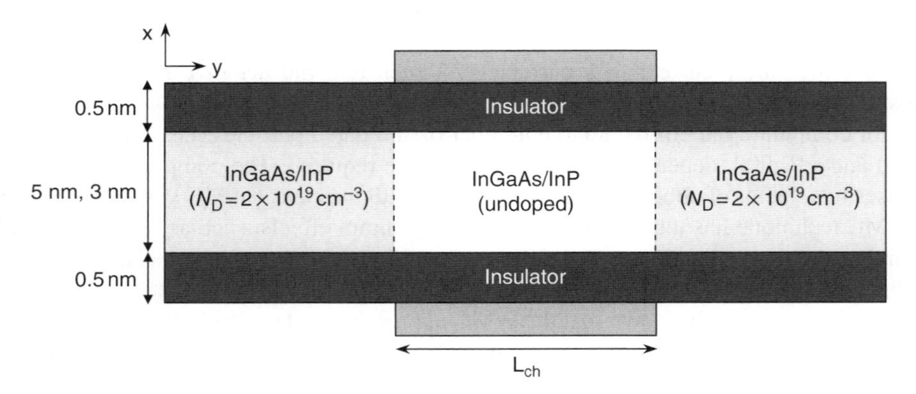

Figure 5.20 Device structure of III-V *n*-MOSFET used in this study. A double-gate structure was employed with a channel thickness (T_{ch}) of 5 or 3 nm and a SiO_2 gate oxide thickness (T_{ox}) of 0.5 nm. A (001) surface was assumed and the channel direction was set as <110>throughout the present study.

5.4.2 POP Scattering Rate for 2-D Electron Gas

In the WMC and BMC simulations of UTB III-V MOSFETs, we use the POP scattering rate for 2-D electron gas in the inversion-layer as follows:

$$W(\mathbf{k}) = \frac{m_d^*}{\hbar^2}\frac{e^2\omega_{op}}{8\pi^2\varepsilon_p}\left(N_q + \frac{1}{2} \mp \frac{1}{2}\right)\int_0^{2\pi}\int_{-\infty}^{\infty}\frac{1}{k_{\parallel}^2 - 2k_{\parallel}k_{\parallel}^{'}\cos\theta + k_{\parallel}^{'2} + q_x^2}\left|G_{m,n}(q_x)\right|^2 dq_x d\theta \quad (5.40)$$

where:

$$k_{\parallel}^{'} = \sqrt{k_{\parallel}^2 - \frac{2m_d^*}{\hbar^2}\left(E_n - E_m \pm \hbar\omega_{op}\right)} \quad (5.41)$$

$$G_{m,n}(q_x) = \int_0^{T_{Si}} \varphi_n^*(x)e^{iq_x x}\varphi_m(x)dx \quad (5.42)$$

As can be seen in Equation (5.40), the POP scattering is anisotropic and the scattering angle θ is determined using Equations (3.13) and (3.14), presented in Section 3.1.2. For the II, AP, and non-POP scattering, the same scattering rates as those in Section 3.2.1 were used. Surface roughness scattering is not considered here, because a reliable model for the surface roughness scattering in III-V MOSFETs has not been established yet. This will not be a major drawback, because subthreshold current properties of MOSFETs, which is the main subject of this section, are not influenced much by carrier scattering [5.32, 5.68].

Figure 5.21 shows all the partial scattering rates calculated for the lowest subband in Γ and L valleys in the source region of InGaAs n-MOSFET with $T_{\rm ch}=3$ nm and $L_{\rm ch}=7$ nm, as a function of the electron kinetic energy. The zero in the horizontal axis represents the lowest quantized subband energy in each valley, and the nonparabolicity of the conduction band in the x and z directions is taken into account. The II scattering rate corresponds to the electron density of $n=N_{\rm D}=1\times10^{20}$ cm^{-3}. In the Γ valley, the intravalley POP emission scattering and II scattering in the SD electrodes are found to have an important role in determining the transport properties of electrons with kinetic energies smaller than 0.5 eV.

5.4.3 $I_D - V_G$ Characteristics for InGaAs DG-MOSFETs

Figure 5.22 shows the drain current versus gate voltage ($I_{\rm D}$ – $V_{\rm G}$) characteristics of $In_{0.53}Ga_{0.47}As$ n-MOSFETs with $T_{\rm ch}=5$ nm computed at $V_{\rm D}=0.5$ V for three channel lengths [5.28], where the WMC and BMC results are plotted as solid and dashed lines, respectively. We add that that similar results were obtained for InP n-MOSFETs, as reported in Reference [5.26].

It can be seen that the subthreshold curves in both sets of MC results are almost identical for $L_{\rm ch}=30$ nm. However, as the channel length decreases below 15 nm, the subthreshold current determined by the WMC simulation rapidly becomes larger than that determined by the BMC simulation. This is the effect of SDT, as will be shown later, based on phase-space distribution function. It should also be noted that the drain current at high gate voltages is

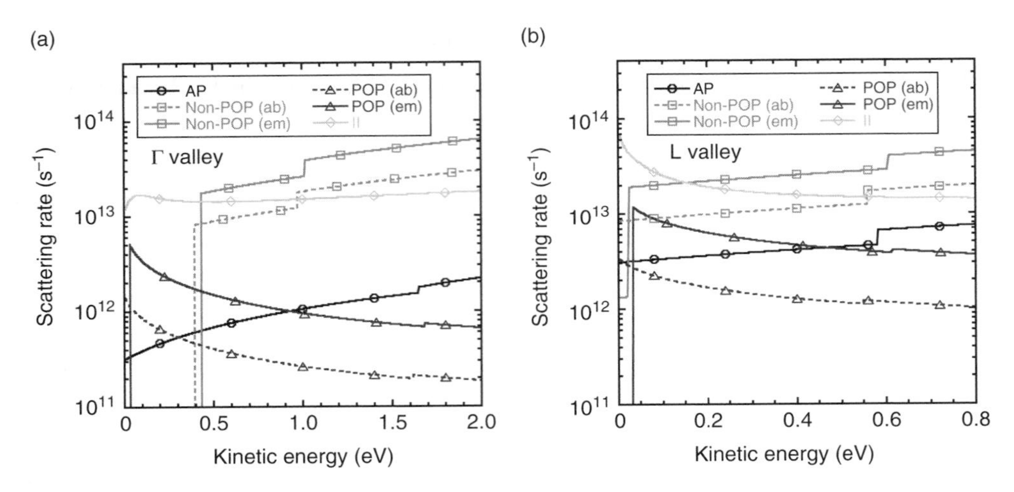

Figure 5.21 Partial scattering rates calculated for the lowest subband in: (a) Γ and; (b) L valleys in the source region of InGaAs *n*-MOSFET, as a function of the electron kinetic energy. $T_{ch}=3$ nm and $L_{ch}=7$ nm. Note that the zero in the horizontal axis represents the lowest quantized subband energy in each valley. $V_G=0.2$ V and $V_D=0.5$ V. "ab" and "em" represent the absorption and emission processes in non-POP and POP scatterings, respectively.

nearly the same for both simulations, regardless of the channel length. This implies that quantum reflection is weak and leads to slightly decreased on-state drain current [5.26, 5.28].

To further elucidate the effects of SDT and quantum reflection, the distribution functions in phase-space were computed using both the WMC and BMC approaches. Figures 5.23 and 5.24 show the computed distribution functions for a gate voltage corresponding to a threshold voltage V_{th} and an on-state voltage V_{on}, respectively, for $L_{ch}=30$ and 10 nm. V_{th} and V_{on} are defined as the gate voltages corresponding to $I_D=0.03$ and 3 mA/μm, respectively, as determined by the BMC simulation. Specifically, $V_{th}=-0.3$ V for $L_{ch}=30$ nm and -0.4 V for $L_{ch}=10$ nm, whereas $V_{on}=0.25$ V for $L_{ch}=30$ nm and 0.2 V for $L_{ch}=10$ nm.

The contrast in the lower panels of the figures represents the number of electrons present in each cell in the phase-space, where Boltzmann and Wigner distribution functions are obtained by BMC and WMC simulations, respectively. The upper panels show the spatial distributions of the lowest subband energy in the Γ valley and the total sheet electron density. From Figure 5.23(a), it can be seen that the two distribution functions are remarkably similar, not only in the source and drain regions but also in the channel region, indicating that SDT is almost negligible for $L_{ch}=30$ nm. This is also confirmed by the sheet electron density distributions (i.e., both MC simulations indicate almost the same electron densities inside the channel).

On the other hand, for the shorter channel device with $L_{ch}=10$ nm, an interference pattern is observed in the Wigner distribution function inside the channel region, as shown in Figure 5.23 (b). Since this interference pattern exhibits a tunneling phenomenon of electrons [5.35–5.41], SDT actually occurs in this short-channel device. This is also confirmed by the significant increase in the channel density of the WMC simulation as seen in the upper panel of Figure 5.23(b), even though the lowest-subband energy profiles are identical for both MC simulations. Note that the interference pattern is more evident in the $In_{0.53}Ga_{0.47}As$ MOSFET than in the InP MOSFET [5.26], indicating that SDT becomes more notable in the $In_{0.53}Ga_{0.47}As$ MOSFET, with the lower effective mass.

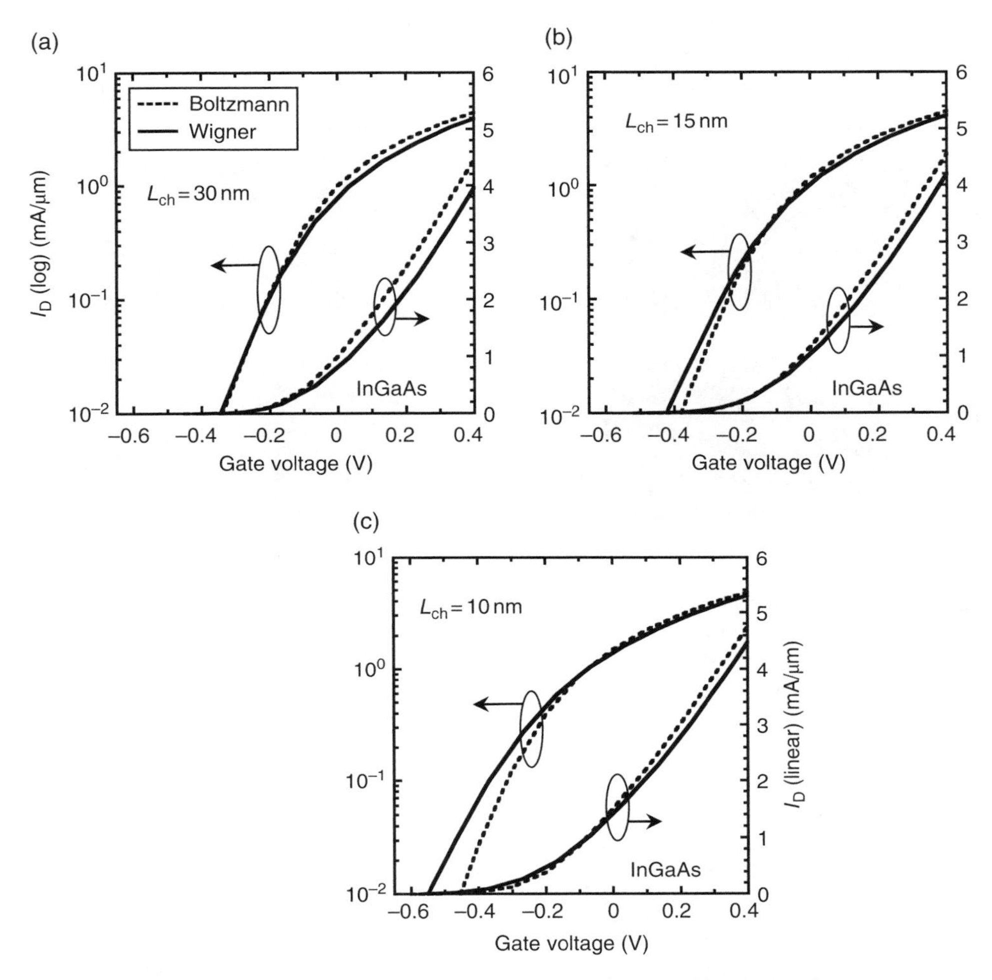

Figure 5.22 $I_D - V_G$ characteristics of $In_{0.53}Ga_{0.47}As$ n-MOSFETs with $T_{ch} = 5$ nm computed at $V_D = 0.5$ V for $L_{ch} =$ (a) 30; (b) 15; and (c) 10 nm, where the WMC and BMC results are plotted as solid and dashed lines, respectively.

As seen in Figure 5.24, for a gate voltage of V_{on}, there are only slight differences between the WMC and BMC distribution functions for either channel length. In particular, in Figure 5.24(a), the two distribution functions are almost identical. In addition, the sheet electron density distributions calculated using the two MC methods are also almost identical, regardless of the channel length. This is because the main current is governed by thermal electron emission at the source-channel junction, which is a classical mechanism, and thus the drain current for a gate voltage of V_{on} is basically determined by classical transport. In other words, SDT is insignificant in this case. In the Wigner distribution function shown in Figure 5.24(b), an interference pattern appears inside the channel, and this is mainly associated with quantum reflection [5.35–5.41]. Due to this quantum reflection, the drain current in the on-state slightly decreases, as shown in Figure 5.22.

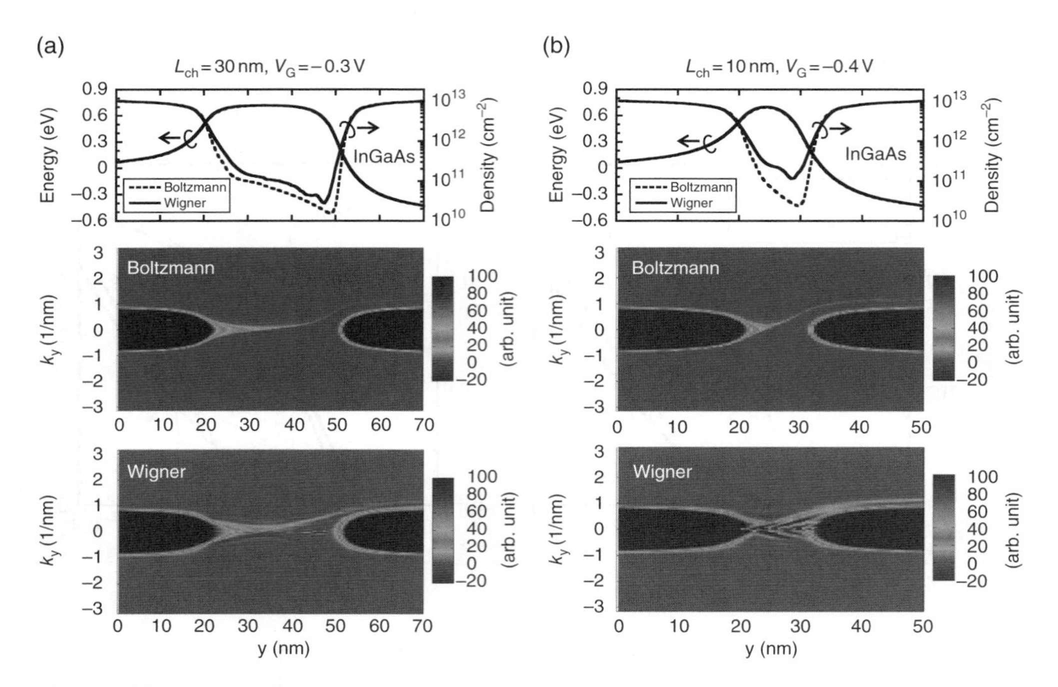

Figure 5.23 Computed phase-space distribution functions of $In_{0.53}Ga_{0.47}As$ *n*-MOSFETs with L_{ch} = (a) 30 and; (b) 10 nm at a threshold gate voltage V_{th}, defined as the gate voltage corresponding to I_D = 0.03 mA/μm, obtained by BMC simulation. T_{ch} = 5 nm and V_D = 0.5 V. The upper panels show the spatial distributions of the lowest subband energy in the Γ valley and the total sheet electron density.

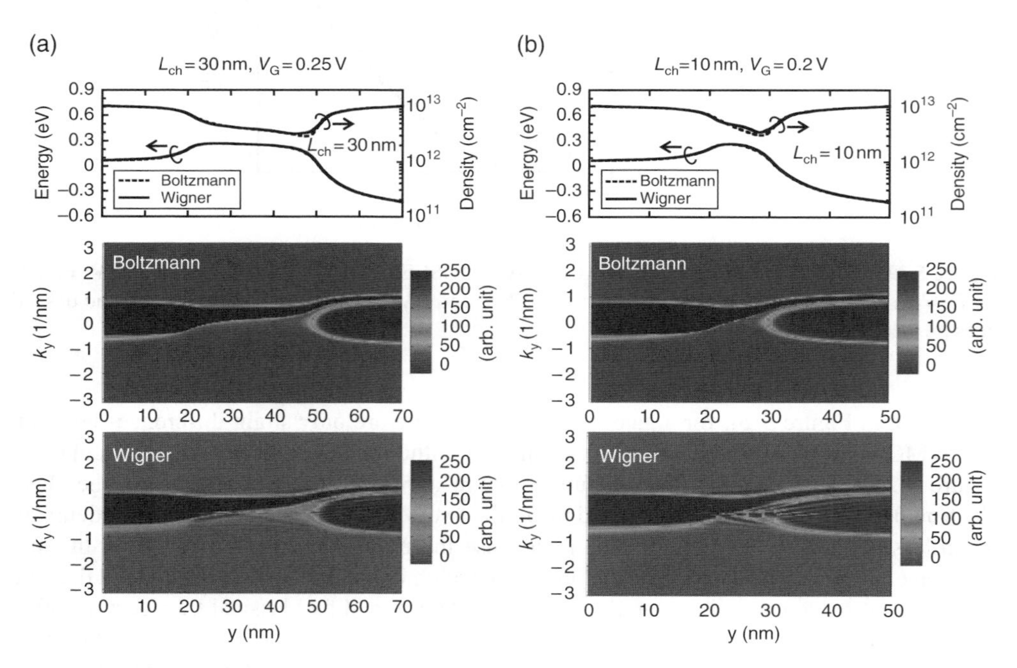

Figure 5.24 Computed phase-space distribution functions of $In_{0.53}Ga_{0.47}As$ *n*-MOSFETs with L_{ch} = (a) 30; and (b) 10 nm at an on-state gate voltage V_{on}, defined as the gate voltage corresponding to I_D = 3 mA/μm obtained by BMC simulation. T_{ch} = 5 nm and V_D = 0.5 V. The upper panels show the spatial distributions of the lowest-subband energy in the Γ valley and the total sheet electron density.

5.4.4 Channel Length Dependence of SDT Leakage Current

Here, we discuss the critical channel length for which a drastic increase in the subthreshold current occurs due to SDT. Figure 5.25(a) shows the channel length dependences of SDT current computed for $In_{0.53}Ga_{0.47}As$ and InP *n*-MOSFETs, where the method for calculating SDT current is indicated in Figure 5.25(b): it is the difference between the drain currents at $V_G = V_{th}$, determined using the WMC and BMC simulations. It can be seen that the influence of SDT becomes evident for channels shorter than about 20 nm for both III-V materials.

Furthermore, as expected, $In_{0.53}Ga_{0.47}As$ *n*-MOSFETs indicate a larger SDT current than in InP *n*-MOSFET because of the lower effective mass at the Γ valley, as shown in Table 5.1. The SDT current for Si nanowire (NW) MOSFETs is also plotted as a dashed line in Figure 5.25(a), which was obtained using a direct solution approach of the WTE and BTE [5.32].

As described in Section 5.2, SDT has an impact on Si *n*-MOSFETs with channel length smaller than 6–8 nm. Consequently, the critical channel length for both $In_{0.53}Ga_{0.47}As$ and InP *n*-MOSFETs becomes approximately three times larger than that for Si MOSFETs, suggesting that SDT can be a major obstacle in downscaling III-V *n*-MOSFETs into $L_{ch} < 20$ nm. Therefore, it will be necessary to determine methods for suppressing SDT, in order to aggressively downscale III-V MOSFETs to the deca-nanometer or nanometer scale.

As demonstrated in Reference [5.28], the gate electrostatics enhancement by reducing the gate oxide thickness is effective for controlling SDT, but it fails to suppress SDT in III-V *n*-MOSFETs deeply scaled down to the channel length of 10 nm or below. As discussed in References [5.25–5.28], the introduction of a material with a heavier transport mass should be considered to reduce SDT drastically. However, since the increase in transport mass decreases the carrier mobility, on-state device performance will be degraded using materials with heavier transport masses. On the other hand, the increase in effective mass would help to reduce the DOS bottleneck problem in the quantum capacitance limit; thus, depending on the device structure and operating condition, the use of a material with a heavier transport mass could

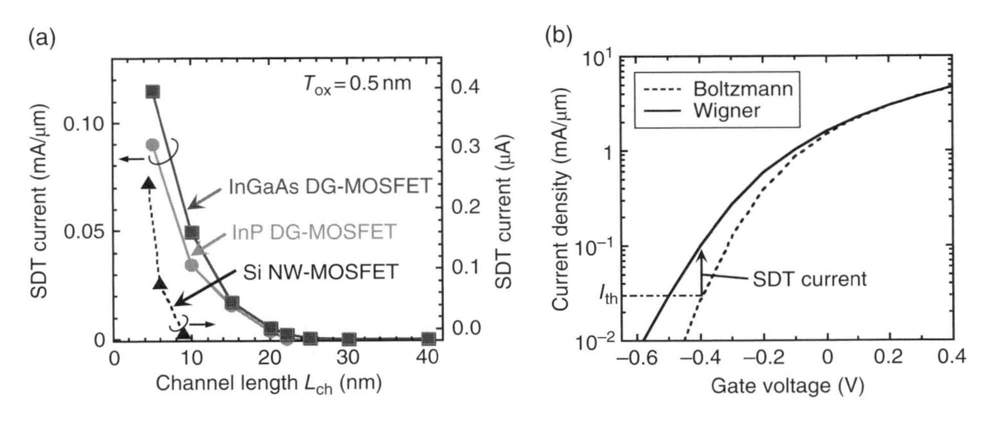

Figure 5.25 (a) Channel length dependences of SDT current computed for $In_{0.53}Ga_{0.47}As$ and InP *n*-MOSFETs with T_{ch}=5 nm. V_D=0.5 V. The results for Si nanowire (NW) MOSFETs [5.32] are also plotted for comparison. (b) Method for calculating the SDT current: the difference between the drain currents at $V_G = V_{th}$ obtained by WMC and BMC simulations. Here, V_{th} is defined as the gate voltage corresponding to I_D=0.03 mA/μm obtained by the BMC simulation.

even lead to better on-state device performance [5.66, 5.67]. Therefore, there may exist an optimum effective mass or an optimum band structure for both reducing SDT, and achieving the required on-state performance. In any case, the use of a material with a heavier transport mass might be one option to go beyond the end of the roadmap [5.27].

5.4.5 Effective Mass Dependence of Subthreshold Current Properties

As shown in Table 5.1, $In_{0.53}Ga_{0.47}As$ or InP has an electron effective mass of 0.046 m_0 or 0.082 m_0 at the Γ valley, respectively, which are smaller by about 1/4 to 1/2 of the transverse effective mass (m_t=0.19 m_0) of Si at Δ valleys. Since a tunneling probability through a finite potential barrier exponentially increases with decreasing the effective mass, the introduction of a material with a heavier transport mass is extremely effective to reduce SDT. In this section, we investigate a quantitative relationship between channel electron effective mass and SDT-induced subthreshold leakage current in ultra-scaled III-V *n*-MOSFETs.

The DG MOSFET structure shown in Figure 5.20 was used, where T_{ch}=3 nm and T_{ox}=0.5 nm. The channel, source, and drain materials were taken as InP, because the donor concentration in the source and drain can be taken to be high, $N_D = 1 \times 10^{20}$ cm^{-3}, to suppress the parasitic resistance in the source and drain regions [5.4], as described in section 5.1.5. In addition, in InP *n*-MOSFETs, almost all of electrons occupy the lowest subband at the Γ valley, because of its large energy gaps between the Γ and L/X valleys, as shown in Table 5.1, under the present bias conditions. This means that the electrical characteristics of InP *n*-MOSFETs are primarily determined by the electron transport at the Γ valley. Therefore, to derive a relationship between the channel electron effective mass and the SDT-induced leakage current in III-V *n*-MOSFETs, we changed only the effective mass at the Γ valley, m^*_Γ, from 0.05 to 0.20 m_0. Note that all other band parameters were unchanged.

Figure 5.26 shows the $I_D - V_G$ characteristics computed for III-V *n*-MOSFETs, with three different effective masses at the Γ valley.

It can be seen that as m^*_Γ decreases, the subthreshold current determined by the WMC simulation rapidly becomes larger than that determined by the BMC simulation. As discussed in Section 5.4.3, this is the effect of SDT. Furthermore, it can be speculated from Figure 5.26 that the subthreshold current increase due to SDT becomes pronounced when m^*_Γ decreases smaller than about 0.1 m_0. The m^*_Γ dependence of the subthreshold current increase will be discussed again later.

Figure 5.27 shows the computed Boltzmann and Wigner distribution functions of III-V *n*-MOSFETs.

First of all, there are few electrons inside the channel of all the Boltzmann distribution functions, which means that the devices are biased at around threshold voltage. On the other hand, in the Wigner distribution functions, an interference pattern is observed inside the channel, and it becomes more evident as m^*_Γ decreases. Since the interference pattern is a signature of tunneling, the results in Figure 5.27 indicate that SDT occurs more notably in MOSFETs with the lower effective mass. The increased SDT is also confirmed in the upper panels, by noticing that the electron density inside the channel significantly increases in the WMC results for the lower m^*_Γ.

Finally, Figure 5.28 shows the drain current slopes estimated at the smallest gate voltage, as a function of m^*_Γ.

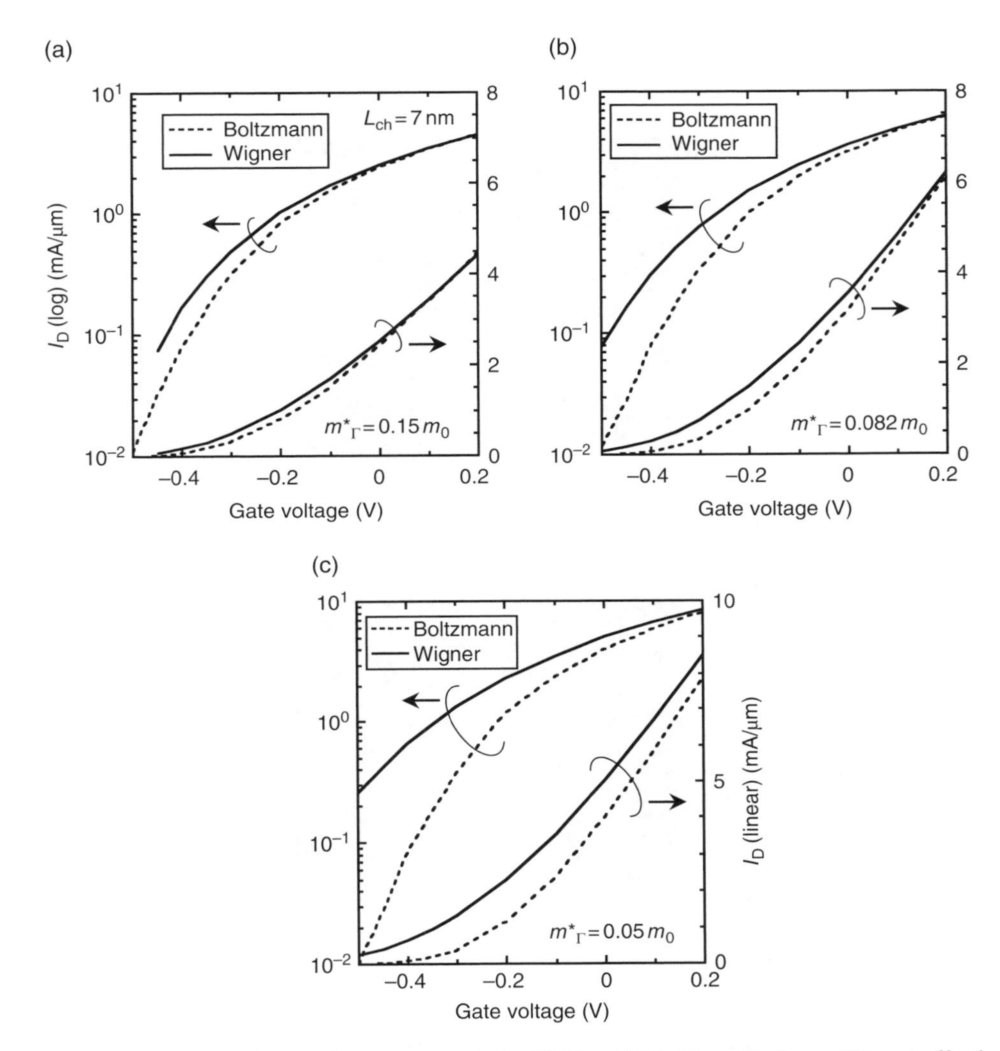

Figure 5.26 $I_D - V_G$ characteristics computed for III-V n-MOSFETs with three different effective masses at Γ valley of m^*_Γ=(a) 0.15; (b) 0.082; and (c) 0.05 m_0, where the solid and the dashed lines represent results obtained from the WMC and BMC simulations, respectively. L_{ch}=7 nm and V_D=0.5 V.

In the BMC results, the current slopes are almost constant with respect to m^*_Γ, which is reasonable because, under the classical simulation, thermal electron emission at the source-channel junction governs the subthreshold current properties. Incidentally, the current slopes are larger than 100 mV/decade, even in the BMC simulation. This is because the gate voltage could not be decreased sufficiently smaller than the threshold voltage using the particle MC technique, which is needed to estimate the subthreshold slope (SS) precisely. On the other hand, the WMC results exhibit a drastic increase in the current slope with decreasing m^*_Γ. Moreover, as L_{ch} changes from 7 to 5 nm, the current slope significantly increases for the shorter L_{ch}.

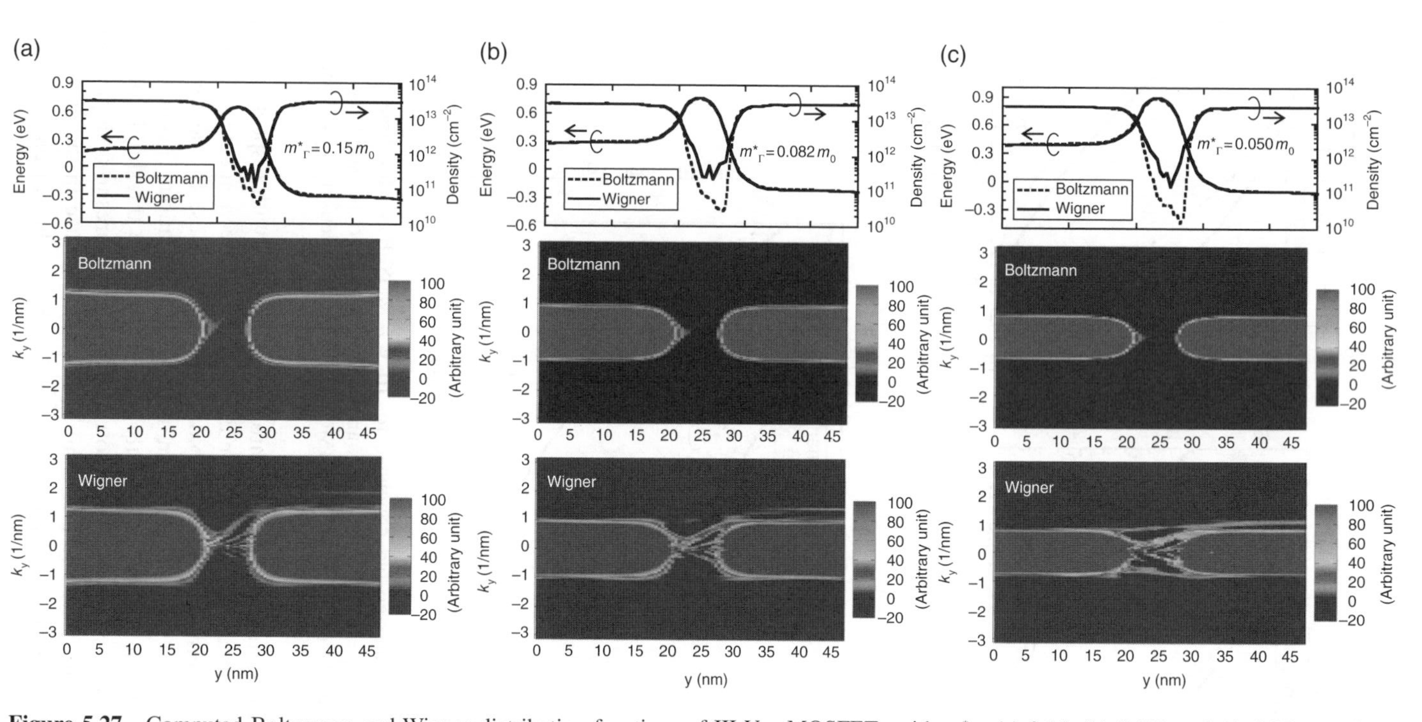

Figure 5.27 Computed Boltzmann and Wigner distribution functions of III-V *n*-MOSFETs with m^*_Γ=(a) 0.15, (b) 0.082 and (c) 0.05 m_0, where L_{ch}=7 nm, V_G=–0.45 V and V_D=0.5 V. The upper panels show the spatial variations of the lowest subband energy in the Γ valley and the total sheet electron density.

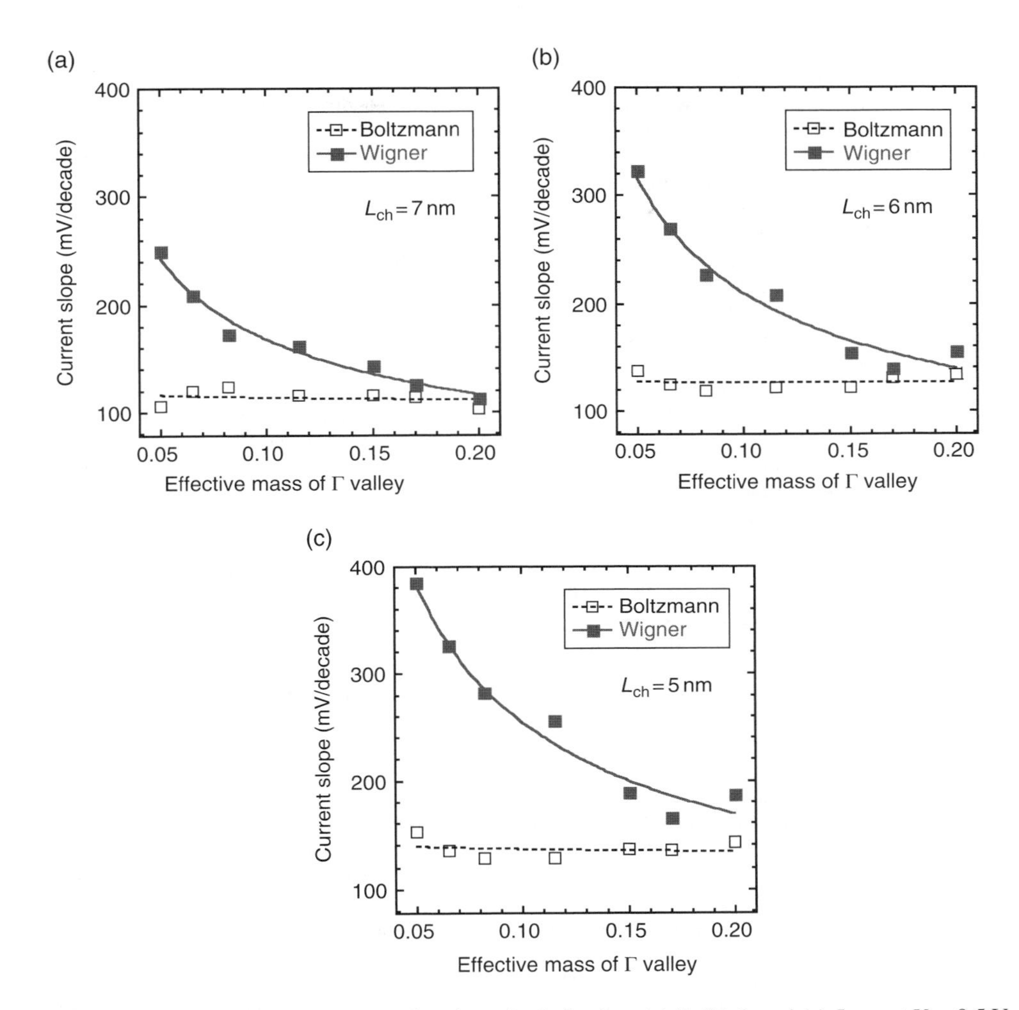

Figure 5.28 Drain current slopes as a function of m^*_Γ for L_{ch} = (a) 7; (b) 6; and (c) 5 nm at V_D = 0.5 V. The solid and open squares indicate the results obtained from the WMC and BMC simulations, respectively, and the solid and dashed lines represent the corresponding interpolation relations derived by assuming the power of m^*_Γ.

The above results indicate that the current slope increase is obviously caused by SDT. Here, it is noteworthy that the current slope increase is suppressed by using m^*_Γ equal to or larger than 0.2 m_0. This suggests that Si (m_t = 0.19 m_0) would be one of the candidates for suppressing SDT in the sub-10 nm regime and, not only that, materials with a heavier mass than Si should be required as L_{ch} decreases less than 5 nm.

References

[5.1] S. Takagi, R. Zhang, J. Suh, S.-H. Kim, M. Yokoyama, K. Nishi and M. Takenaka (May 2015). III-V/Ge channel MOS device technologies in nano CMOS era. *Japanese Journal of Applied Physics* **54**, p. 06FA01.

[5.2] T. Mori, Y. Azuma, H. Tsuchiya and T. Miyoshi (Mar. 2008). Comparative study on drive current of III-V semiconductor, Ge and Si channel n-MOSFETs based on quantum-corrected Monte Carlo simulation. *IEEE Transactions on Nanotechnology* **7**(2), 237–241.

[5.3] Y. Azuma, T. Mori and H. Tsuchiya (Mar. 2008). Drive current of ultrathin Ge-on-insulator n-channel MOSFETs. *Physica Status Solidi (c)* **5**(9), 3153–3155.

[5.4] H. Tsuchiya, A. Maenaka, T. Mori and Y. Azuma (Apr. 2010). Role of carrier transport in source and drain electrodes of high-mobility MOSFETs. *IEEE Electron Device Letters* **31**(4), 365–367.

[5.5] Y. Maegawa, S. Koba, H. Tsuchiya and M. Ogawa (Aug. 2011). Influence of source/drain resistance on device performance of ultrathin body III-V chhannel metal-oxide-semiconductor field-effect transistors. *Applied Physics Express* **4**, p. 084301.

[5.6] S. Takagi and S. Sugahara (Sep. 2006). Comparative study on influence of subband structures on electrical characteristics of III-V semiconductors, Ge and Si channel n-MISFETs. *Extended Abstracts of International Conference on Solid State Devices and Materials (SSDM)*, Yokohama, 1056–1057.

[5.7] S. Takagi, T. Irisawa, T. Tezuka, T. Numata, S. Nakaharai, N. Hirashita, Y. Moriyama, K. Usuda, E. Toyoda, S. Dissanayake, M. Shichijo, R. Nakane, S. Sugahara, M. Takenaka and N. Sugiyama (Jan. 2008). Carrier-transport-enhanced channel CMOS for improved power consumption and performance. *IEEE Transactions on Electron Devices* **55**(1), 21–39.

[5.8] S. Takagi (Sep. 2004). Physical origin of drive current enhancement in ultra-thin Ge-On-Insulator (GOI) MOSFETs under full ballistic transport. *Extended Abstracts of International Conference on Solid State Devices and Materials (SSDM)*, Tokyo, 10–11.

[5.9] F. Stern and W. E. Howard (Nov. 1967). Properties of semiconductor surface inversion layers in the electric quantum limit. *Physical Review* **163**(3), 816–835.

[5.10] M. V. Fischetti and S. E. Laux (Mar. 1991). Monte Carlo simulation of transport in technologically significant semiconductors of the diamond and zinc-blende structure–Part II: Submicrometer MOSFETs. *IEEE Transactions on Electron Devices* **38**(3), 650–660.

[5.11] M. V. Fischetti, L. Wang, B. Yu, C. Sachs, P. M. Asbeck, Y. Taur and M. Rodwell (2007). Simulation of electron transport in high-mobility MOSFETs: Density of states bottleneck and source starvation. *IEDM Technical Digest* 109–112.

[5.12] M. V. Fischetti, S. Jin, T.-W. Tang, P. Asbeck, Y. Taur, S. E. Laux, M. Rodwell and N. Sano (Jun. 2009). Scaling MOSFETs to 10nm: Coulomb effects, source starvation and virtual source model. *Journal of Computational Electronics* **8**(2), 60–77.

[5.13] R. Terao, T. Kanazawa, S. Ikeda, Y. Yonai, A. Kato and Y. Miyamoto (Apr. 2011). InP/InGaAs composite metal-oxide-semiconductor field-effect transistors with regrown source and Al_2O_3 gate dielectric exhibiting maximum drain current exceeding 1.3 mA/μm. *Applied Physics Express* **4**, p. 054201.

[5.14] X. Zhou, Q. Li, C. W. Tang and K. M. Lau (Oct. 2012). Inverted-type InGaAs metal-oxide-semiconductor high-electron-mobility transistor on Si substrate with maximum drain current exceeding 2 A/mm. *Applied Physics Express* **5**, p. 104201.

[5.15] C. Jacoboni, F. Nava, C. Canali and G. Ottaviani (July 1981). Electron drift velocity and diffusivity in germanium. *Physical Review B* **24**(2), 1014–1026.

[5.16] M. V. Fischetti and S. E. Laux (Mar. 1991). Monte Carlo simulation of transport in technologically significant semiconductors of the diamond and zinc-blende structure – Part I: Homogeneous transport. *IEEE Transactions on Electron Devices* **38**(3), 634–649.

[5.17] S. E. Laux (Sep. 2007). A simulation study of the switching times of 22- and 17-nm gate-length SOI nFETs on high mobility substrates and Si. *IEEE Transactions on Electron Devices* **54**(9), 2304–2320.

[5.18] M. J. Antonell, C. R. Abernathy, V. Krishnamoorthy, R. W. Gedridge, Jr. and T. E. Haynes (1997). Thermal stability of heavily tellurium-doped InP grown by metalorganic molecular beam epitaxy. *Journal of Electronic Materials* **26**(11), 1283–1286.

[5.19] H. Q. Zheng, K. Radahakrishnan, S. F. Yoon and G. I. Ng (Jun. 2000). Electrical and optical properties of Si-doped InP grown by solid source molecular beam epitaxy using a valved phosphorus cracker cell. *Journal of Applied Physics* **87**(11), 7988–7993.

[5.20] M. Radosavljevic, B. Chu-Kung, S. Corcoran, G. Dewey, M. K. Hudait, J. M. Fastenau, J. Kavalieros, W. K. Liu, D. Lubyshev, M. Metz, K. Millard, N. Mukherjee, W. Rachmady, U. Shah and R. Chau (2009). Advanced high-k gate dielectric for high-performance short-channel $In_{0.7}Ga_{0.3}As$ quantum well field effect transistors on silicon substrate for low power logic applications. *IEDM Technical Digest*, 319–322.

[5.21] J. A. del Alamo (Nov. 2011). Nanometer-scale electronics with III-V compound semiconductors. *Nature* **479**, 317–323.

[5.22] Y. Yonai, T. Kanazawa, S. Ikeda and Y. Miyamoto (2011). High drain current (>2A/mm) InGaAs channel MOSFET at V_D=0.5 V with shrinkage of channel length by InP anisotropic etching. *IEDM Technical Digest*, p. 307–310.

[5.23] S. H. Kim, M. Yokoyama, N. Taoka, R. Iida, S. Lee, R. Nakane, Y. Urabe, N. Miyata, T. Yasuda, H. Yamada, N. Fukuhara, M. Hata, M. Takenaka and S. Takagi (Dec. 2012). Electron mobility enhancement of extremely thin body $In_{0.7}Ga_{0.3}As$-on-insulator metal-oxide-semiconductor field-effect transistors on Si substrates by metal-oxide-semiconductor interface buffer layers. *Applied Physics Express* **5**, p. 014201.

[5.24] S. H. Kim, M. Yokoyama, N. Taoka, R. Iida, S. Lee, R. Nakane, Y. Urabe, N. Miyata, T. Yasuda, H. Yamada, N. Fukuhara, M. Hata, M. Takenaka and S. Takagi (June 2011). Self-aligned metal source/drain InP n-metal-oxide-semiconductor field-effect transistors using Ni-InP metallic alloy. *Applied Physics Letters* **98**, p. 243501.

[5.25] S. S. Sylvia, H. -H. Park, M. A. Khayer, K. Alam, G. Klimeck and R. K. Lake (Aug. 2012). Material selection for minimizing direct tunneling in nanowire transistors. *IEEE Transactions on Electron Devices* **59**(8), 2064–2069.

[5.26] S. Koba, Y. Maegawa, M. Ohmori, H. Tsuchiya, Y. Kamakura, N. Mori and M. Ogawa (May 2013). Increased subthreshold current due to source-drain direct tunneling in ultrashort channel III-V metal-oxide-semiconductor field-effect transistors. *Applied Physics Express* **6**, p. 064301.

[5.27] S. R. Mehrotra, S.G. Kim, T. Kubis, M. Povolotskyi, M. S. Lundstrom and G. Klimeck (July 2013). Engineering nanowire n-MOSFETs at $L_g < 8$ nm. *IEEE Transactions on Electron Devices* **60**(7), 2171–2177.

[5.28] S. Koba, M. Ohmori, Y. Maegawa, H. Tsuchiya, Y. Kamakura, N. Mori and M. Ogawa (Feb. 2014). Channel length scaling limits of III-V channel MOSFETs governed by source-drain direct tunneling. *Japanese Journal of Applied Physics* **53**, 04EC10.

[5.29] H. Kawaura, T. Sakamoto and T. Baba (June 2000). Observation of source-to-drain direct tunneling current in 8 nm gate electrically variable shallow junction metal-oxide-semiconductor field-effect transistors. *Applied Physics Letters* **76**(25), 3810–3812.

[5.30] J. Wang and M. Lundstrom (2002). Does source-to-drain tunneling limit the ultimate scaling of MOSFETs? *IEDM Technical Digest*, 707–710.

[5.31] H. Wakabayashi, T. Ezaki, M. Hane, T. Ikezawa, T. Sakamoto, H. Kawaura, S. Yamagami, N. Ikarashi, K. Takeuchi, T. Yamamoto and T. Mogami (2004). Transport properties of sub-10-nm planar-bulk-CMOS devices. *IEDM Technical Digest*, 429–432.

[5.32] Y. Yamada, H. Tsuchiya and M. Ogawa (July 2009). Quantum transport simulation of silicon-nanowire transistors based on direct solution approach of the Wigner transport equation. *IEEE Transactions on Electron Devices* **56**(7), 1396–1401.

[5.33] M. Luisier, M. Lundstrom, D. Antoniadis and J. Bokor (2011). Ultimate device scaling: Intrinsic performance comparisons of carbon-based, InGaAs and Si field-effect transistors for 5 nm gate length. *IEDM Technical Digest*, 251–254.

[5.34] L. Shifren, C. Ringhofer and D. K. Ferry (Mar. 2003). A Wigner function-based quantum ensemble Monte Carlo study of a resonant tunneling diode. *IEEE Transactions on Electron Devices* **50**(3), 769–773.

[5.35] D. Querlioz, P. Dollfus, V.-N. Do, A. Bournel and V. Nguyen (Dec. 2006). An improved Wigner Monte-Carlo technique for the self-consistent simulation of RTDs. *Journal of Computational Electronics* **5**(4), 443–446.

[5.36] D. Querlioz, J. Saint-Martin, V.-N. Do, A. Bournel and P. Dollfus (2006). Fully quantum self-consistent study of ultimate DG-MOSFETs including realistic scattering using a Wigner Monte-Carlo approach. *IEDM Technical Digest*, 941–944.

[5.37] D. Querlioz, J. Saint-Martin, K. Huet, A. Bournel, V. Aubry-Fortuna, C. Chassat, S. Galdin-Retailleau and P. Dollfus (Sep. 2007). On the ability of the particle Monte Carlo technique to include quantum effects in nano-MOSFET simulation. *IEEE Transactions on Electron Devices* **54**(9), 2232–2242.

[5.38] D. Querlioz, H.-N. Nguyen, J. Saint-Martin, A. Bournel, S. Galdin-Retailleau and V. Nguyen (Aug. 2009). Wigner-Boltzmann Monte Carlo approach to nanodevice simulation: from quantum to semiclassical transport. *Journal of Computational Electronics* **8**(3–4), 324–335.

[5.39] D. Querlioz and P. Dollfus (2010). *The Wigner Monte Carlo Method for Nanoelectronic Devices*. Wiley, New York.

[5.40] S. Koba, R. Aoyagi and H. Tsuchiya (Sep. 2010). Quantum transport simulation of nanoscale semiconductor devices based on Wigner Monte Carlo approach. *Journal of Applied Physics* **108**, p. 064504.

[5.41] S. Koba, H. Tsuchiya and M. Ogawa (Sep. 2011). Wigner Monte Carlo approach to quantum and dissipative transport in Si-MOSFETs. *Extended Abstracts of International Conference on Simulation of Semiconductor Processes and Devices (SISPAD)*, Osaka, 79–82.

[5.42] B. Winstead and U. Ravaioli (Feb. 2003). A quantum correction based on Schrodinger equation applied to Monte Carlo device simulation. *IEEE Transactions on Electron Devices* **50**(2), 440–446.

[5.43] B. Wu, T.W. Tang, J. Nam and J.-H. Tsai (Dec. 2003). Monte Carlo simulation of symmetric and asymmetric double-gate MOSFETs using Bohm-based quantum corrections. *IEEE Transactions on Nanotechnology* **2**(4), 291–294.

[5.44] S. Datta (1995). *Electronic Transport in Mesoscopic Systems*. Cambridge University Press, Cambridge, UK.

[5.45] E. Wigner (June 1932). On the quantum correction for thermodynamic equilibrium. *Physical Review* **40**, 749–759.

[5.46] W. Frensley (July 1987). Wigner-function model of a resonant-tunneling semiconductor device. *Physical Review B* **36**(3), 1570–1580.

[5.47] N. Kluksdahl, A. Kriman, D. Ferry and C. Ringhofer (Apr. 1989). Self-consistent study of the resonant-tunneling diode. *Physical Review B* **39**(11), 7720–7735.

[5.48] H. Tsuchiya, M. Ogawa, T. Miyoshi (June 1991). Simulation of quantum transport in quantum devices with spatially varying effective mass. *IEEE Transactions on Electron Devices* **38**(6), 1246–1252.

[5.49] L. P. Kadanoff and G. Baym (1962). *Quantum statistical mechanics*. Benjamin, New York.

[5.50] L. V. Keldysh (1965). Diagram technique for nonequilibrium processes. *Soviet Physics(JETP)* **20**,4, 1018–1026.

[5.51] G. D. Mahan (Jan. 1987). Quantum transport equation for electric and magnetic fields. *Physics Reports* **145**(5) 251–318.

[5.52] H. Tsuchiya and T. Miyoshi (Mar. 1998). Nonequilibrium Green's function approach to high-temperature quantum transport in nanostructure devices. *Journal of Applied Physics* **83**(5), 2574–2585.

[5.53] H. Tsuchiya and U. Ravaioli (Apr. 2001). Particle Monte Carlo simulation of quantum phenomena in semiconductor nanostructures. *Journal of Applied Physics* **89**(7), 4023–4029.

[5.54] M. Ancona and G. Iafrate (May 1989). Quantum correction to the equation of state of an electron gas in a semiconductor. *Physical Review B* **39**(13), 9536–9540.

[5.55] J. R. Zhou and D. K. Ferry (Mar. 1992). Simulation of ultra-small GaAs MESFET's using quantum moment equations. *IEEE Transactions on Electron Devices* **39**(3), 473–478.

[5.56] C. L. Gardner (1994). The quantum hydrodynamic model for semiconductor devices. *SIAM Journal on Applied Mathematics* **54**(2), 409–427.

[5.57] H. Kim, H. S. Min, T. W. Tang and Y. J. Park (Nov. 1991). An extended proof of the Ramo-Shockley theorem. *Solid-State Electronics* **34**(11), 1251–1253.

[5.58] D. Querlioz, J. Saint-Martin, A. Bournel and P. Dollfus (Oct. 2008). Wigner Monte Carlo simulation of phonon-induced electron decoherence in semiconductor nanodevices. *Physical Review B* **78**(16), p. 165306.

[5.59] M. Poljak, V. Jovanovic, D. Grgec and T. Suligoj (June 2012). Assessment of electron mobility in ultrathin-body InGaAs-on-insulator MOSFETs using physics-based modeling. *IEEE Transactions on Electron Devices* **59**(6), 1636–1643.

[5.60] M. A. Khayer and R. K. Lake (Nov. 2008). Performance of n-type InSb and InAs nanowire field-effect transistors. *IEEE Transactions on Electron Devices* **55**(11), 2939–2945.

[5.61] E. Lind, M. P. Persson, Y.-M. Niquet and L.-E. Wernersson (Feb. 2009). Band structure effects on the scaling properties of [111] InAs nanowire MOSFETs. *IEEE Transactions on Electron Devices* **56**(2), 201–205.

[5.62] N. Neophytou, T. Rakshit and M. S. Lundstrom (July 2009). Performance analysis of 60-nm gate-length III-V InGaAs HEMTs: Simulations versus experiments. *IEEE Transactions on Electron Devices* **56**(7), 1377–1387.

[5.63] K. Alam and R. N. Sajjad (Nov. 2010). Electronic properties and orientation-dependent performance of InAs nanowire transistors. *IEEE Transactions on Electron Devices* **57**(11), 2880–2885.

[5.64] N. Takiguchi, S. Koba, H. Tsuchiya and M. Ogawa (Jan. 2012). Comparisons of performance potentials of Si and InAs nanowire MOSFETs under ballistic transport. *IEEE Transactions on Electron Devices* **59**(1), 206–211.

[5.65] Y. Lee, K. Kakushima, K. Natori and H. Iwai (Apr. 2012). Gate capacitance modeling and diameter-dependent performance of nanowire MOSFETs. *IEEE Transactions on Electron Devices* **59**(4), 1037–1045.

[5.66] K. Shimoida, Y. Yamada, H. Tsuchiya and M. Ogawa (Jan. 2013). Orientational dependence in device performances of InAs and Si nanowire MOSFETs under ballistic transport. *IEEE Transactions on Electron Devices* **60**(1) 117–122.

[5.67] K. Shimoida, H. Tsuchiya, Y. Kamakura, N. Mori and M. Ogawa (Feb 2013). Performance comparison of InAs, InSb and GaSb n-channel nanowire metal-oxide-semiconductor field-effect transistors in the ballistic transport limit. *Applied Physics Express* **6**, p. 034301.

[5.68] S. Scaldaferri, G. Curatola and G. Iannaccone (Nov. 2007). Direct solution of the Boltzmann transport equation and Poisson-Schrödinger equation for nanoscale MOSFETs. *IEEE Transactions on Electron Devices* **54**(11), 2901–2909.

6

Atomistic Simulations of Si, Ge and III-V Nanowire MOSFETs

Si nanowire (SiNW) MOSFETs with a gate-all-around (GAA) architecture are expected to be a key device technology for future integrated circuits, because of excellent short-channel effect immunity, possible elimination of channel doping, which gives origin to characteristic fluctuation, feasibility of vertical integration, and so on. Practical fabrication of SiNW MOSFETs with diameters less than 5 nm has already been reported experimentally. In addition, the introduction of high-mobility semiconductor NWs, such as GeNW or InAsNW, has been aggressively considered toward further performance improvement. Focusing on the regime of a 10 nm or less gate length, in which NW MOSFETs will be in practical use, the number of atoms in the cross-section becomes countable and, hence, fully atomistic simulation of carrier transport is required. In this chapter, we describe atomistic approaches to simulate carrier transport in Si, Ge and III-VNW MOSFETs with GAA structure, and discuss their performance potentials by performing intercomparison.

6.1 Phonon-limited Electron Mobility in Si Nanowires

As the diameter of SiNWs shrinks down to a nanometer scale, and the number of atoms in the cross-section becomes countable, crystalline orientation, quantum confinement, and electron scattering play important roles in understanding the physical and transport characteristics of SiNWs. Also, for exploration of novel functional materials and devices in the nanometer regime, a fully atomistic simulation considering both the electron and phonon band structures and their interactions, is strongly required.

In conventional and commonly-used theoretical approaches, the transport properties of SiNWs, such as carrier mobility, have been calculated by using an analytical effective mass band structure for electrons and bulk dispersion relations for phonons, where 2-D Schrödinger-Poisson equations are solved self-consistently [3.62, 6.1–6.3]. Also, in modeling of phonon scattering processes, deformation potentials for acoustic and optical phonons are parameterized

Carrier Transport in Nanoscale MOS Transistors, First Edition. Hideaki Tsuchiya and Yoshinari Kamakura.

to reproduce the experimental data for various semiconductor materials [3.12]. Thus, atomistic details, such as crystalline orientation and cross-sectional shape, are disregarded.

However, because the lattice vibration in NWs is quite different from that in bulk materials because of the phonon confinement, as demonstrated later in this section, the deformation potentials in NWs may depend on both the diameter and the crystal orientation [6.4, 6.5]. Generally speaking, the deformation potentials depend not only on the electron wave functions at the initial and final states, but also on the relevant phonon polarization vectors [3.64, 3.65, 6.6, 6.7]. Therefore, atomistic treatment of the electron-phonon interaction is essential to predict the ultimate device performance of SiNW MOSFETs.

In this section, we address this subject and investigate the electron mobility of SiNWs with three crystalline orientations –<100>, <110>and <111>– by considering the atomistic electron-phonon interactions [6.8]. We calculate the electron band structures based on a semi-empirical $sp^3d^5s^*$ tight-binding (TB) approach [6.9, 6.10], and the phonon band structures based on the Keating potential model [6.11]. Then, by combining the electron and phonon eigenstates in SiNWs, we derive the phonon scattering rate, based on Fermi's golden rule. We then solve the Boltzmann transport equation, while considering Pauli's exclusion principle, and eventually evaluate the electron mobility of SiNWs. We show that the electron mobility of SiNWs strongly depends on the crystalline orientation and diameter, and its origin is discussed in terms of electron and phonon band structure modulation caused by the quantum confinement.

6.1.1 Band Structure Calculations

Figure 6.1 shows unit cells of SiNWs with cylindrically-shaped cross-sections, where we considered the three crystalline orientations of <100>, <110>and <111>. The NW length is assumed to be infinite in the transport direction z, which is perpendicular to the plane of the paper. The Si atoms are represented by spheres, and the surface Si atoms are passivated using an sp^3 hybridization scheme [6.12], although their terminating hydrogen atoms are not shown in Figure 6.1. The diameter of the SiNWs is determined by the formula $D = 2\sqrt{a^3 N_{\mathrm{Si}} / 8\pi L}$, where a is the Si lattice constant of 5.43 Å, N_{Si} is the number of Si atoms and L is the length of the unit cell along the transport directions, which corresponds to $L = a$, $a/\sqrt{2}$ and $\sqrt{3}a$ for the <100>, <110>and <111>orientations, respectively.

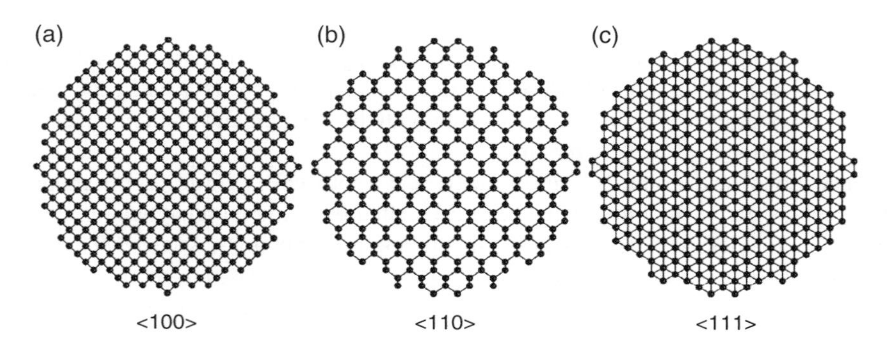

Figure 6.1 Atomic models for (a) <100>-oriented, (b) <110>-oriented, and (c) <111>-oriented SiNWs with cylindrically-shaped cross-sections. The diameter is approximately 4.3 nm.

A. Electrons

The electron band structures in SiNWs were computed using a semi-empirical TB approach, where the first nearest-neighbor and two-center orthogonal $sp^3d^5s^*$ model was used. Here, we give an outline of the computational procedure. First, the electronic wave function $\psi^n_{k,\gamma}(\mathbf{r},k)$ is expressed in the form of a linear combination of atomic orbitals using Bloch-symmetrized atomic orbitals $\chi^n_{\kappa,\gamma}(\mathbf{r},k)$, as follows:

$$\psi^n(\mathbf{r},k) = \sum_{\kappa}\sum_{\gamma} c^n_{\kappa,\gamma}(k)\,\chi^n_{\kappa,\gamma}(\mathbf{r},k) \tag{6.1}$$

where n is the band index, $\mathbf{r}$ is the position vector, k is the wave number in the transport direction, κ is the atomic index and γ represents the $sp^3d^5s^*$ TB orbitals.

Because $\chi^n_{\kappa,\gamma}(\mathbf{r},k)$ can be expressed as a Fourier transform of the atomic orbitals $\varphi^n_{\kappa,\gamma}(\mathbf{r}-\mathbf{R}_{\eta k})$, which are localized at the position (specified as $\mathbf{R}_{\eta k}$) of the k-th Si atom in the η-th unit cell, as shown in Figure 6.2, we obtain the following equation for $\chi^n_{\kappa,\gamma}(\mathbf{r},k)$:

$$\chi^n_{\kappa,\gamma}(\mathbf{r},k) = \frac{1}{\sqrt{N}}\sum_{\eta=1}^{N} e^{ik\eta L}\varphi^n_{\kappa,\gamma}(\mathbf{r}-\mathbf{R}_{\eta\kappa}) \tag{6.2}$$

where N is the number of unit cells. By substituting Equations (6.1) and (6.2) into the Schrödinger equation, we obtain an eigenvalue equation represented by the $sp^3d^5s^*$ TB orbital basis set, as follows:

$$\sum_{(\eta',\kappa')\in NN(\eta,\kappa)}\sum_{\gamma'} U^{\gamma,\gamma'}_{\eta\kappa,\eta'\kappa'} e^{ik(\eta'-\eta)L} c^n_{\kappa',\gamma'}(k) - \phi_\kappa c^n_{\kappa,\gamma}(k) = E_n(k) c^n_{\kappa,\gamma}(k) \tag{6.3}$$

where spin-orbit interactions are not considered. The interatomic potential $U^{\gamma,\gamma'}_{\eta k,\eta' k'}$ in Equation (6.3) can be formulated in terms of two-center integrals and directional cosines [6.9]. In this study, we used Boykin's parameterization [6.13, 6.14] for the onsite energies and the two-center integrals between the nearest-neighbor Si atoms. ϕ_κ is the electrostatic potential

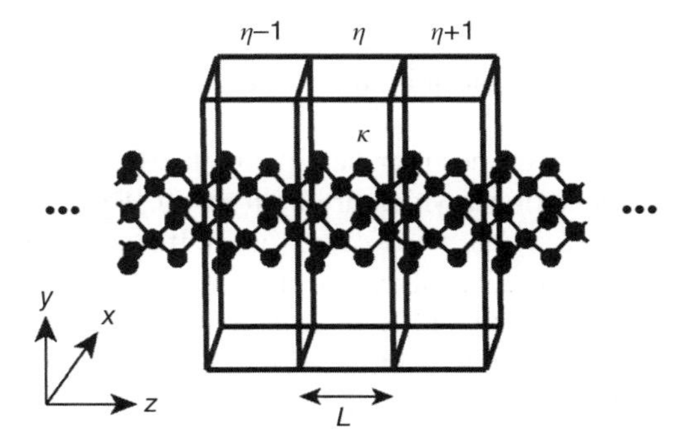

Figure 6.2 Schematic diagram representing the atomic index κ and unit cells. The position of the Si atom is identified using the atomic index κ and the unit cell index η as (η, κ). L is the length of the unit cell in the transport direction.

obtained from the 3-D Poisson equation, which is solved by using the finite volume discretization scheme described in Appendix A, until a self-consistently converged solution for ϕ_κ is obtained.

Because the main purpose of this paper is to evaluate electron mobility under a low electric field, the Fermi level of the surrounding metallic gate was fixed to the conduction band minima of bulk Si, which means a zero flat-band voltage, and the gate voltage was set to zero for all simulations. Therefore, the self-consistent solution for the electrostatic potential coupled with the Poisson equation plays only a minor role in the present mobility estimation. We confirmed this by performing a non-self-consistent calculation in solving Equation (6.3), and found that the self-consistent electrostatic potential has a negligibly small impact on determination of the electron mobility. However, the gate bias dependence of the electron mobility will be important for discussion of the on-state properties of SiNW MOSFETs, and our self-consistent approach is applicable to such simulations.

By diagonalizing the TB matrix at a given wave number k, the energy levels $E_n(k)$ and the expansion coefficients $c^n_{\kappa,\gamma}(k)$ are obtained for the electrons. As for the numerical values, the TB Hamiltonian of Equation (6.3) is a sparse matrix with a dimension of $10\,N_{\mathrm{Si}}$ for the $sp^3d^5s^*$ ten orbital basis and N_{Si} atoms, and we therefore adopted the Jacobi-Davidson method [6.15] to solve Equation (6.3) because a few eigenvalue spectra above the conduction band minima are usually relevant to the present carrier transport. Acceleration of the numerical convergence in the Jacobi-Davidson iterations was achieved by using the preconditioned generalized minimum residual solver (GMRES) [6.16], where a preconditioner was factorized using the incomplete Cholesky decomposition [6.16], and matrix elements of triangular matrices that were smaller than a predetermined threshold were omitted.

Figure 6.3 shows the electron band structures computed for <100>-oriented, <110>-oriented and <111>-oriented SiNWs with diameters of about 3 nm, where the horizontal axis denotes the wave number normalized by the Brillouin zone width in each case. The origin of the vertical axis (i.e. $E=0$), corresponds to the Fermi level. For the <100>- and <110>-oriented SiNWs, the conduction band minimum appears at the Γ point, while it appears away from the Γ point for the <111>-orientation. Also, the lowest subbands are nearly degenerate, as indicated in each figure.

The differences in the band structures among these orientations, including the number of degeneracies, can be qualitatively explained in terms of the 2-D quantum confinement of the six equivalent ellipsoidal valleys in bulk Si [2.13, 2.14]. Next, Figure 6.4 shows the transport effective mass at the conduction band minimum as a function of the diameter for the three orientations.

Note that the horizontal dashed lines represent the bulk effective masses corresponding to the <100>- and <110>-orientations, which are both $m_t=0.19\,m_0$, and corresponding to the <111>-orientation, which is $(2\,m_t+m_l)/3=0.43\,m_0$ [6.17]. It is found that the transport masses of the <100>- and <111>-orientations increase from the bulk value as the diameter decreases. In contrast, the transport mass of the <110>-orientation decreases from the bulk value. The above results for the <100>- and <110>-orientations correspond to those in Figure 2.11(b), which were obtained by the first-principles DFT calculation [2.4]. The transport effective mass behavior is caused by the 2-D quantization of the anisotropic and nonparabolic Si conduction band [2.13, 2.14]. Therefore, intuitively, higher electron mobility can be expected in <110>-oriented SiNWs with a nanometer-sized cross-section. We discuss this point later in Section 6.1.3.

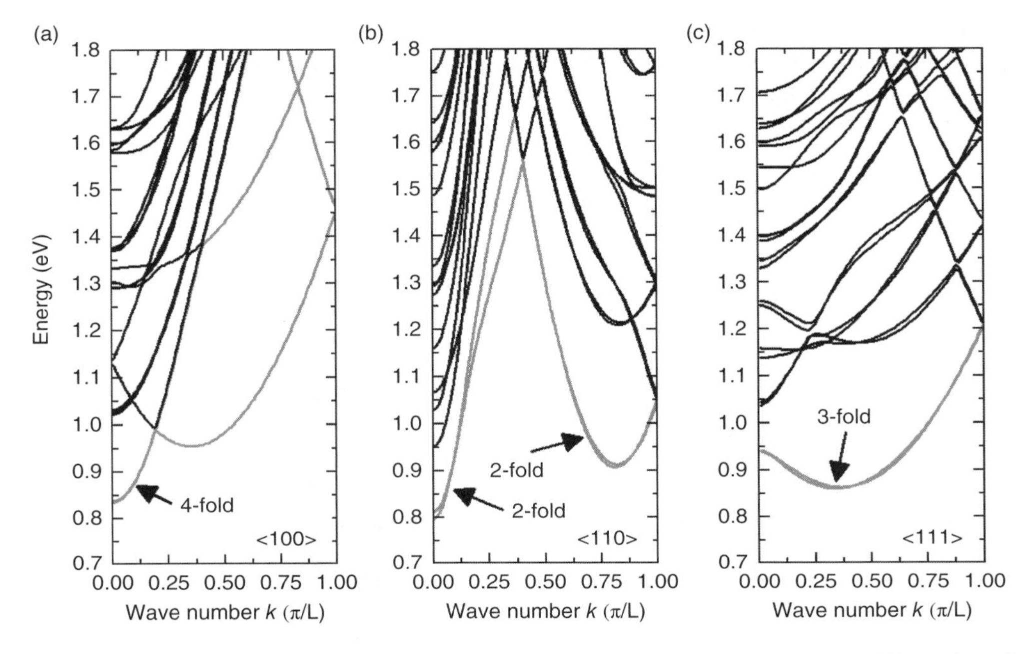

Figure 6.3 Electron band structures computed for (a) <100>-, (b) <110>- and (c) <111>-oriented SiNWs with diameters of approximately 3 nm. The gray curves denote conduction bands used in the mobility calculation.

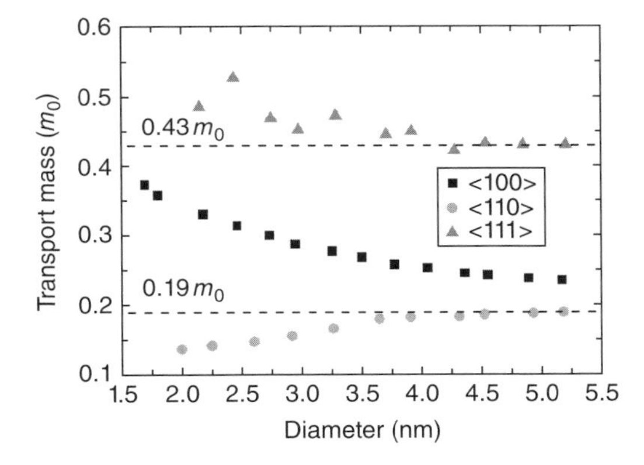

Figure 6.4 Transport effective mass at the conduction band minimum as a function of the diameter for the three crystalline orientations, where the horizontal dashed lines represent bulk effective masses corresponding to each orientation.

B. Phonons

Phonon band structures were computed using the original Keating valence force field (VFF) approach [6.11], which is known to describe the microscopic features of phonon eigenstates in nanostructures well.

The Keating potential consists of two potential energy terms. The first is the bond-stretching term, and the other is the bond-bending term. Each term is represented as a function of the atomic coordinates and, for a diamond structure, the crystal potential energy (Keating potential energy) is given by:

$$E = \frac{3}{16}\frac{\alpha}{d_1^2}\sum_{\eta,\kappa}\sum_{(\eta',\kappa')\in NN(\eta,\kappa)}\left(R_{\eta\kappa,\eta'\kappa'}^2 - d_1^2\right)^2 + \frac{3}{16}\frac{\beta}{d_1^2}\sum_{\eta,\kappa}\sum_{(\eta',\kappa')\in NN(\eta,\kappa)}\sum_{\substack{(\eta'',\kappa'')\in NN(\eta,\kappa)\\ \eta''\neq\eta' \cap \kappa''\neq\kappa'}}\left(\mathbf{R}_{\eta\kappa\eta'\kappa'}\cdot\mathbf{R}_{\eta k\eta''k''} + \frac{1}{3}d_1^2\right)^2. \tag{6.4}$$

Here, the first and the second terms correspond to the bond-stretching and bond-bending terms, respectively, and α and β are Keating's force constants. $NN(\eta, k)$ stands for a group of nearest-neighbor atoms associated with the (η, κ)-th atom. $\mathbf{R}_{\eta\kappa,\eta'\kappa'} = \mathbf{R}_{\eta'\kappa'} - \mathbf{R}_{\eta k}$ indicates the relative position vector between the two atoms and d_1 is the equilibrium bond length between nearest neighbor Si atoms. In the Keating VFF approach, the phonon band structures are calculated by introducing a harmonic perturbation approximation into Equation (6.4).

If all atoms are fixed to their equilibrium positions $(\mathbf{R}_1^0, \mathbf{R}_2^0, \cdots, \mathbf{R}_N^0)$, the potential energy and forces acting on the atoms are found to be all zero from Equation (6.4). This means that $E(\mathbf{R}_1^0, \mathbf{R}_2^0, \cdots, \mathbf{R}_N^0) \equiv E_0 = 0$ and $\nabla_{\mathbf{R}_i} E(\mathbf{R}_1, \mathbf{R}_2, \cdots, \mathbf{R}_N)\big|_{\mathbf{R}=\mathbf{R}_1^0,\mathbf{R}_2^0,\cdots,\mathbf{R}_N^0} (i \in 1, 2, \cdots, N) = 0$. Therefore, Equation (6.4) is expanded using higher-order spatial derivatives and, when we ignore the components that are higher than the third-order, the Keating potential energy is finally given by:

$$\begin{aligned} E &= E_0 + \sum_{\eta}\sum_{\kappa}\sum_{\mu\in\{x,y,z\}} \left.\frac{\partial E}{\partial R_{\eta\kappa\mu}}\right|_0 \delta R_{\eta\kappa\mu} + \frac{1}{2}\sum_{\eta}\sum_{\kappa}\sum_{\mu\in\{x,y,z\}}\sum_{\eta'}\sum_{\kappa'}\sum_{\mu'\in\{x,y,z\}} \left.\frac{\partial^2 E}{\partial R_{\eta\kappa\mu}\partial R_{\eta'\kappa'\mu'}}\right|_0 \delta R_{\eta\kappa\mu}\delta R_{\eta'\kappa'\mu'} + \cdots \\ &\approx \frac{1}{2}\sum_{\eta\kappa}\sum_{\mu\in\{x,y,z\}}\sum_{\eta'}\sum_{\kappa'}\sum_{\mu'\in\{x,y,z\}} \left.\frac{\partial^2 E}{\partial R_{\eta\kappa\mu}\partial R_{\eta'\kappa'\mu'}}\right|_0 \delta R_{\eta\kappa\mu}\delta R_{\eta'\kappa'\mu'} \end{aligned} \tag{6.5}$$

where μ denotes three Cartesian coordinates $\{x, y, z\}$, and $\delta R_{\eta\kappa\mu}$ denotes infinitesimal displacements from the equilibrium position of the (η, κ)-th atom.

By applying Equation (6.5) to Newton's equation of motion, we obtained a dynamical equation for the atomic displacement $\delta R_{\eta k\mu}$ of the (η, κ)-th atom, as follows:

$$M\frac{\partial^2 \delta R_{\eta\kappa\mu}}{\partial t^2} = -\sum_{\eta'}\sum_{\kappa'}\sum_{\mu'\in\{x,y,z\}} \left.\frac{\partial^2 E}{\partial R_{\eta\kappa\mu}\partial R_{\eta'\kappa'\mu'}}\right|_0 \delta R_{\eta'\kappa'\mu'} \tag{6.6}$$

where M denotes the mass of the Si atom. The factor 1/2 of Equation (6.5) was canceled out by the symmetry in the suffixes. Here, we assume a solution for the atomic vibration in the form of a plane wave along the transport direction such as $\delta R_{\eta\kappa\mu} \propto e^{i(qz-\omega t)}$ and, consequently, the following dynamical matrix equation has been derived:

$$\sum_{(\eta',\kappa')\in NN(\eta,\kappa)} \sum_{\mu\in\{x,y,z\}} D^1_{\eta\kappa\mu\eta'\kappa'\mu'}\left[\varepsilon^{\lambda}_{\kappa'\mu'}(q)e^{iq(\eta'-\eta)L} - \varepsilon^{\lambda}_{\kappa\mu'}(q)\right]$$
$$+ \sum_{(\eta',\kappa')\in NN(\eta,\kappa)} \sum_{\substack{(\eta'',\kappa'')\in NN(\eta',\kappa') \\ \eta''\neq\eta\cap\kappa''\neq k}} \sum_{\mu''\in\{x,y,z\}} D^2_{\eta\kappa\mu\eta'\kappa'\eta''\kappa''\mu''}\left[\varepsilon^{\lambda}_{\kappa''\mu''}(q)e^{iq(\eta''-\eta)L} - \varepsilon^{\lambda}_{k\mu''}(q)\right] = \omega^2_{\lambda}(q)\varepsilon^{\lambda}_{\kappa\mu}(q) \tag{6.7}$$

where ε and ω are eigenvectors and eigenvalues for the phonons. In Equation (6.7), $D^1_{\eta\kappa\mu\eta'\kappa'\mu'}$ and $D^2_{\eta\kappa\mu\eta'\kappa'\eta''\kappa'\mu''}$ are matrix elements between the first nearest-neighbor and the second nearest-neighbor Si atoms, respectively, which are represented by:

$$\begin{aligned} D^1_{\eta\kappa\mu,\eta'\kappa'\mu'} &= -3\frac{\alpha}{d_1^2}\frac{R^{\mu}_{\eta\kappa,\eta'\kappa'}R^{\mu'}_{\eta\kappa,\eta'\kappa'}}{\sqrt{M_\kappa M_{\kappa'}}} \\ &-\frac{3}{4}\frac{\beta}{d_1^2}\sum_{\substack{(\eta'',\kappa'')\in NN(\eta,\kappa) \\ \eta''\neq\eta'\cap\kappa''\neq\kappa'}} \frac{\left(R^{\mu}_{\eta\kappa,\eta'\kappa'} + R^{\mu}_{\eta\kappa,\eta''\kappa''}\right)R^{\mu'}_{\eta\kappa,\eta''\kappa''}}{\sqrt{M_\kappa M_{\kappa'}}} \\ &-\frac{3}{4}\frac{\beta}{d_1^2}\sum_{\substack{(\eta'',\kappa'')\in NN(\eta',\kappa') \\ \eta''\neq\eta\cap\kappa''\neq\kappa}} \frac{R^{\mu}_{\eta'\kappa',\eta''\kappa''}\left(R^{\mu'}_{\eta'\kappa',\eta\kappa} + R^{\mu'}_{\eta'\kappa',\eta''\kappa''}\right)}{\sqrt{M_\kappa M_{\kappa'}}} \end{aligned} \tag{6.8}$$

$$D^2_{\eta\kappa\mu,\eta'\kappa',\eta''\kappa''\mu''} = \frac{3}{4}\frac{\beta}{d_1^2}\frac{R^{\mu}_{\eta'\kappa',\eta''\kappa''}R^{\mu''}_{\eta'\kappa',\eta\kappa}}{\sqrt{M_\kappa M_{\kappa''}}} \tag{6.9}$$

In Equations (6.8) and (6.9), $R^{\mu}_{\eta\kappa,\eta'\kappa'}$ is the directional component of the relative position vectors from (η, k) to the (η', k')-th atoms. We note that the bond-stretching term (the first term in Equation (6.4)) is involved only in $D^1_{\eta\kappa\mu\eta'\kappa'\mu'}$, while the bond-bending term (the second term in Equation (6.4)) is involved in both $D^1_{\eta\kappa\mu\eta'\kappa'\mu'}$ and $D^2_{\eta\kappa\mu\eta'\kappa'\eta''\kappa''\mu''}$. Therefore, the dynamical matrix for a specified atom contains up to four coupling elements with the first nearest-neighbor atoms, and up to 12 coupling elements with the second nearest-neighbor atoms. As a matter of course, these matrix elements depend on the cross-sectional geometry and the crystal orientations of the nanowires. By solving Equation (6.7) using the matrix elements defined by Equations (6.8) and (6.9), the phonon band structure can be obtained.

In practical calculations, we used prescribed force constants for the bond-stretching term α and the bond-bending term β. Although the dynamical matrix is also Hermitian and positive definite, we have calculated all eigenvalues and eigenvectors using the LAPACK library [6.18], because the electron-phonon scattering rates formulated in the next section must be estimated for the entire first Brillouin zone. The phonon band structures were calculated in the absence of the gate dielectric and the metallic gate, and the surface Si atoms were assumed to be freely-vibrating.

Figure 6.5 shows the phonon band structures computed for the three orientations in the low energy regime, where the diameter is approximately 3 nm, and the horizontal axis denotes the normalized phonon wave number. Unlike in bulk phonons, four acoustic modes exist in a long wavelength regime [6.19] (i.e., $q \approx 0$). We symbolize these modes as TLA1, TLA2, TA and LA

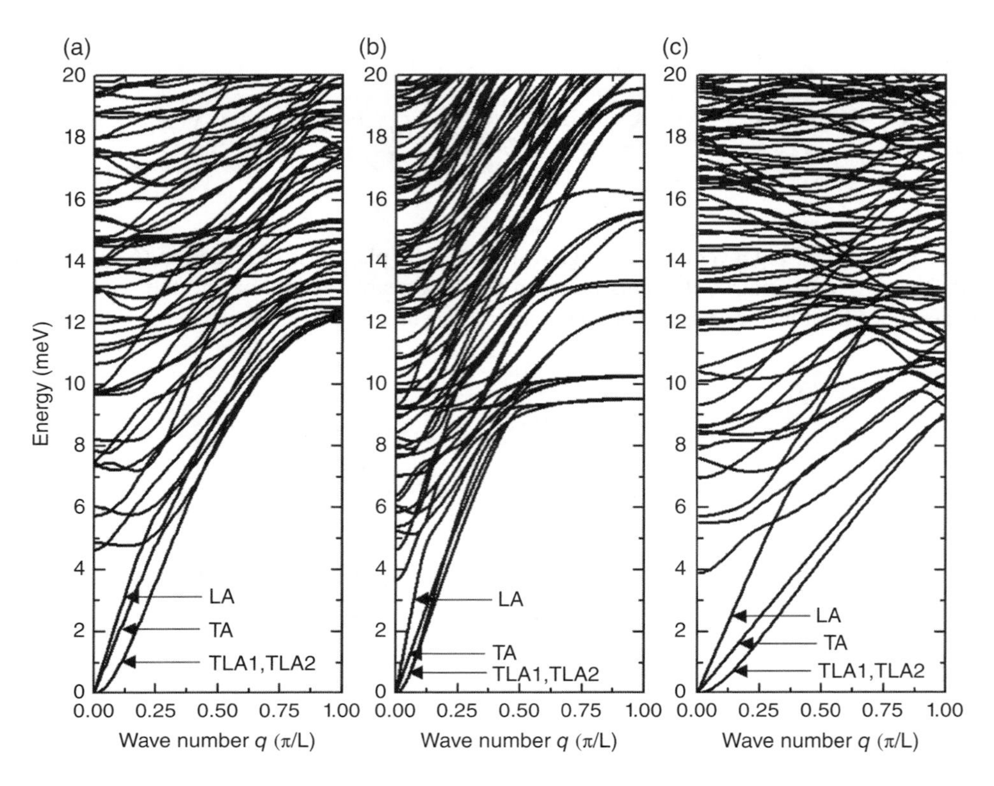

Figure 6.5 Phonon band structures computed for (a) <100>-, (b) <110>-, and (c) <111>-oriented SiNWs in a low energy regime, where the diameter is about 3 nm. Unlike bulk phonons, four acoustic modes exist in the long wavelength regime – that is, $q \approx 0$. We symbolize these modes as TLA1, TLA2, TA and LA from the bottom upwards.

from the bottom up, as shown in Figure 6.5, where TA and LA are the transverse acoustic and longitudinal acoustic modes, respectively, in common with the bulk phonon, and TLA is a mixed state of transverse and longitudinal acoustic modes.

By looking closely at these four curves, we notice that TLA1 and TLA2 are degenerate in the <100>- and <111>-orientations, but are non-degenerate in the <110>-orientation. This is because of the presence or absence of rotational symmetry. Specifically, the <100>- and <111>-oriented SiNWs have rotational symmetries of π and $2\pi/3$, respectively, but the <110>-oriented NW has no rotational symmetry. Therefore, the TLA1 and TLA2 modes become non-degenerate in the <110>-orientation. This can also be confirmed by actually plotting the atomic vibration vectors at the limit of $q = 0$. Figure 6.6 shows the atomic vibration vectors of the four acoustic phonon modes computed for the three orientations. For the <100>- and <111>-orientations, the TLA1 and TLA2 modes exhibit identical vibration patterns, considering each rotational symmetry mentioned above. However, the <110>-oriented NW clearly has different vibration patterns between the TLA1 and TLA2 modes, which means that the rotational symmetry is absent in this orientation.

Incidentally, from the vibration patterns shown in Figure 6.6, the TLA1 and TLA2 modes are named "flexural", and the TA mode is called "torsional". Also, as found in Figure 6.5, the

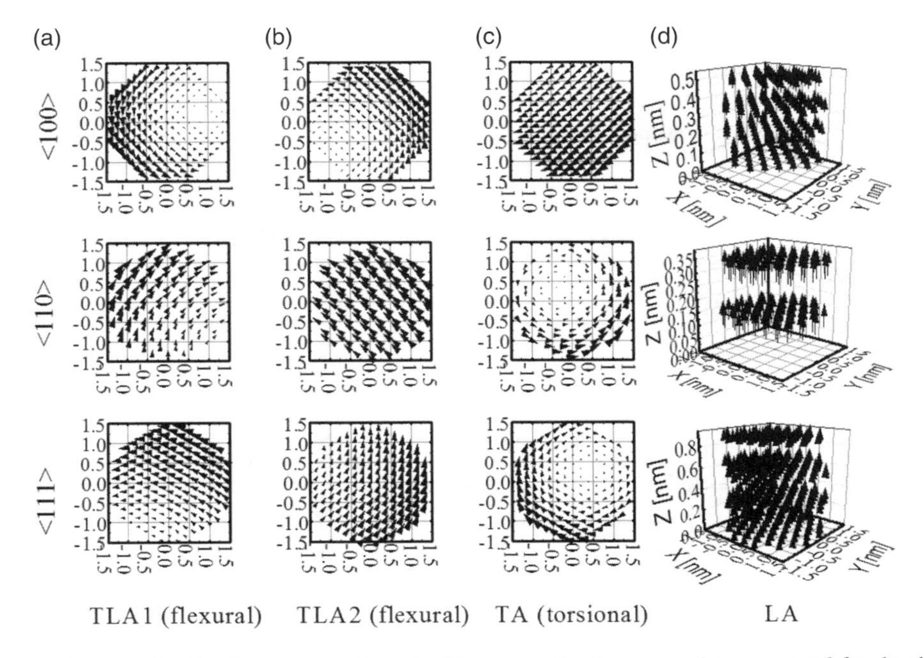

Figure 6.6 Atomic vibration vectors at $q = 0$ of four acoustic phonon modes, computed for the three crystalline orientations. (a) and (b) are flexural modes, (c) is a torsional mode, and (d) is the LA mode. The diameter is about 3 nm. Si atoms are located at the origin of each glyph.

two flexural modes exhibit $\omega \propto q^2$ dispersions, while the torsional mode has a $\omega \propto q$ dispersion, which is the nature of a purely transverse mode.

Next, we investigated another phonon property in SiNWs. Figure 6.7 shows the phonon densities-of-states (DOS) obtained for SiNWs with the three orientations and diameters of about 3 nm. First, it is found that the densities-of-states are sharply-peaked around 65 meV for all the orientations, which is because of the optical phonon modes. It is interesting that the vibrational energies of the optical phonon modes in SiNWs remain similar to those of bulk Si [3.12]. On the other hand, broad peaks, extending from $\approx$ 10 meV to $\approx$ 40 meV, result from the mixed states of the acoustic and optical phonons. We should also note that the apparent difference caused by the NW orientation is visually unrecognizable in the present DOS.

Sound velocities are evaluated from the phonon band structures at $q \approx 0$ of Figure 6.5, and are plotted as a function of diameter in Figure 6.8. The data for TA and LA represent the sound velocities for the TA and LA modes in Figure 6.5, respectively, and those of the TLA modes are not shown here. The horizontal dashed lines indicate the theoretical sound velocities for the TA and LA modes in bulk Si [6.20]. It is found that, as the diameter decreases, the sound velocities decrease because of the phonon confinement effect. Also, we note that the calculated sound velocities depend on the crystalline orientation, and their magnitude relation corresponds well to that of bulk Si, especially in a larger diameter regime. Consequently, the phonon confinement and anisotropic effects in SiNWs are considered to be successfully described by the present Keating potential approach.

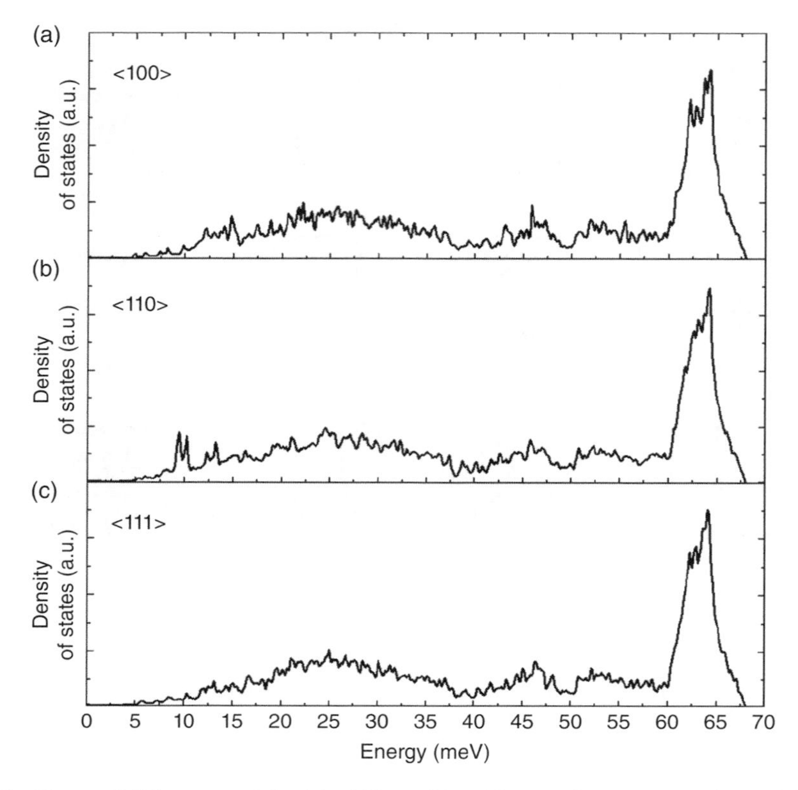

Figure 6.7 Phonon DOS computed for (a) <100>-, (b) <110>-, and (c) <111>-oriented SiNWs with $D = 3$ nm.

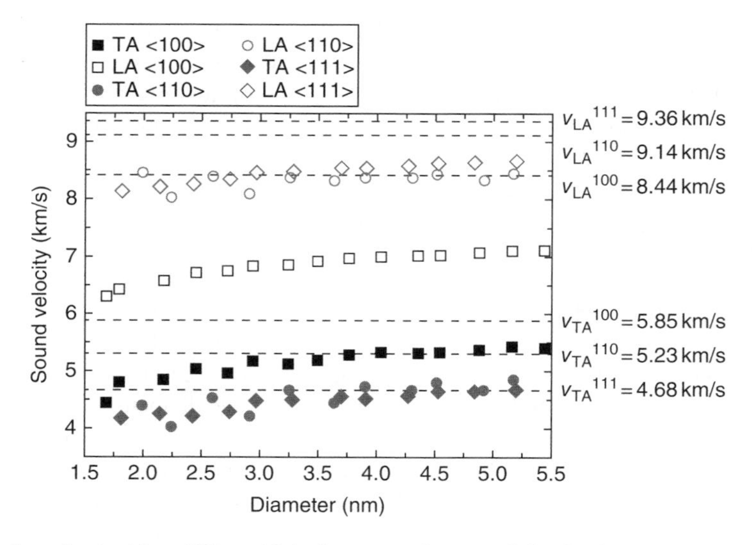

Figure 6.8 Sound velocities of TA and LA phonon modes at $q \approx 0$ for the three crystalline orientations. The horizontal dashed lines indicate theoretical sound velocities for the TA and LA modes in bulk Si.

6.1.2 Electron-phonon Interaction

In this section, we formulate the scattering rates caused by electron-phonon interaction in the SiNWs by coupling the electron and phonon eigenstates derived in Section 6.1.1.

First, we represent the atomic position vector $\mathbf{R}_{\eta\kappa}(t)$ by adding an atomic vibration vector $\delta\mathbf{R}_{\eta\kappa}(t)$ to the equilibrium position vector $\mathbf{R}^0_{\eta\kappa}$ as follows:

$$\mathbf{R}_{\eta\kappa}(t) = \mathbf{R}^0_{\eta\kappa} + \delta\mathbf{R}_{\eta\kappa}(t) \tag{6.10}$$

Here, we define the atomic vibration vector in the second quantization notation as:

$$\delta\mathbf{R}_{\eta\kappa}(t) = \sum_{q,\lambda}\sqrt{\frac{\hbar}{2MN\omega_\lambda(q)}}\left[a_{q\lambda}\varepsilon^\lambda_\kappa(q)e^{i(q\eta L-\omega_\lambda(q)t)} + a^*_{q\lambda}\varepsilon^{\lambda*}_\kappa(q)e^{-i(q\eta L-\omega_\lambda(q)t)}\right] \tag{6.11}$$

where $a_{q\lambda}$ and $a^*_{q\lambda}$ are the phonon annihilation and creation operators, respectively. $\varepsilon^\lambda_\kappa(q)$ is the eigenvector of the (η, κ)-th atom at the wavevector q and the λ-th branch. When the electron-phonon interaction Hamiltonian is represented by the $sp^3d^5s^*$ orbitals and is expanded to a first-order spatial derivative term as [6.21],

$$U^{\gamma,\gamma'}_{\eta\kappa,\eta'\kappa'}\left(\mathbf{R}_{\eta\kappa},\mathbf{R}_{\eta'\kappa'}\right) \approx U^{\gamma,\gamma'}_{\eta\kappa,\eta'\kappa'}\left(\mathbf{R}^0_{\eta\kappa},\mathbf{R}^0_{\eta'\kappa'}\right) + \left.\frac{\partial U^{\gamma,\gamma'}_{\eta\kappa,\eta'\kappa'}}{\partial\left(\mathbf{R}_{\eta'\kappa'} - \mathbf{R}_{\eta\kappa}\right)}\right|_0 \cdot \left[\delta\mathbf{R}_{\eta'\kappa'}(t) - \delta\mathbf{R}_{\eta\kappa}(t)\right] \tag{6.12}$$

then the transition matrix element from a state (n, k) to a state (n', k') is given by:

$$M^{n,n'}(k,k') = \sum_{q,\lambda}\sqrt{\frac{\hbar}{2MN\omega_\lambda(q)}}\sum_{\kappa}\sum_{(\eta',\kappa')\in NN(\eta,\kappa)}\sum_{\gamma}\sum_{\gamma'}\left.\frac{\partial U^{\gamma,\gamma'}_{\eta\kappa,\eta'\kappa'}}{\partial\left(\mathbf{R}_{\eta'\kappa'} - \mathbf{R}_{\eta\kappa}\right)}\right|_0 \times\left[\begin{array}{l}\varepsilon_{\kappa',\lambda}(\pm q)c^{n'\,*}_{\kappa,\gamma}(k')c^{n}_{\kappa',\gamma'}(k)e^{ik'(\eta'-\eta)L} \\ -\varepsilon_{\kappa,\lambda}(\pm q)c^{n'\,*}_{\kappa,\gamma}(k')c^{n}_{\kappa',\gamma'}(k)e^{ik(\eta'-\eta)L}\end{array}\right]\delta_{k-k'\pm q,\,G}\,e^{-i\left[E_n(k)/\hbar - E_{n'}(k')/\hbar \,\pm\, \omega_\lambda(q)\right]t} \tag{6.13}$$

In practical calculations, solutions for the electron wave functions $c(k)$ and the phonon eigenvalues $\omega(q)$ and polarization vectors $\varepsilon(q)$, which have been derived in Section 6.1.1, are substituted into Equation (6.13). However, the electron eigenenergies are implicitly considered in the term $\partial U^{\gamma,\gamma'}_{\eta\kappa,\eta'\kappa'} / \partial(\mathbf{R}_{\eta'\kappa'} - \mathbf{R}_{\eta\kappa})$, which is numerically calculated using analytical formulae given by the Slater-Koster table, as described in Appendix B. Note that G denotes the reciprocal lattice vector, and thus both normal ($G=0$) and Umklapp ($G\neq 0$) processes caused by the phonon scattering are automatically included in the present calculation.

Next, the scattering rates from state (n, k) to state (n', k') are represented by applying Fermi's golden rule as follows:

$$S^{n,n'}(k,k') = \frac{2\pi}{\hbar}\left[\begin{array}{l}\left|M^{n,n'}(k,k')\right|^2 g\left(\hbar\omega_\lambda(k'-k)\right)\delta\left(E_n(k) - E_{n'}(k') + \hbar\omega_\lambda(k'-k)\right) \\ +\left|M^{n,n'}(k,k')\right|^2\left[1 + g\left(\hbar\omega_\lambda(k-k')\right)\right]\delta\left(E_n(k) - E_{n'}(k') - \hbar\omega_\lambda(k-k')\right)\end{array}\right] \tag{6.14}$$

where $g(\hbar\omega_\lambda(q))$ is the equilibrium Bose-Einstein distribution function, and the Dirac delta function represents the energy conservation law. Here, as found from Equation (6.14), the final states (n', k') after the scattering events need to be searched by imposing momentum and energy conservations. To perform accurate detection of the final states over the whole of the first Brillouin zone, we introduced quadratic spline interpolations of the electron and phonon band structures between discretized points. Also, to realize a more rigorous estimation of the joint density-of-states involving the group velocities of both electrons and phonons at the final states, we transformed the Dirac delta function in Equation (6.14) by using a mathematical formula, as follows:

$$\delta\left(E_n(k)-E_{n'}(k')\pm\hbar\omega_\lambda(\pm k'\mp k)\right)=\sum_i\frac{1}{\left|\left.\frac{\partial\left[E_{n'}(k')\mp\hbar\omega_\lambda(\pm k'\mp k)\right]}{\partial k'}\right|_{k'=k_i}\right|}\delta(k'-k_i) \tag{6.15}$$

These approaches enable us to perform numerically stable and accurate estimation of the scattering rates.

6.1.3 *Electron Mobility*

Next, we describe a methodology to compute the electron mobility. By assuming that the electron distribution function $f_n(k)$ can be represented by the sum of the equilibrium Fermi-Dirac function $f_0(E_n(k))$ and a first-order perturbation component $\delta f_n(k)$ driven by an electric field F, the following equation is derived from the Boltzmann transport equation [3.40]:

$$-\frac{e}{k_BT}Fv_n(k)f_0(E_n(k))\left[1-f_0(E_n(k)\right]=\hat{C}\{\delta f_n(k)\} \tag{6.16}$$

where $v_n(k)=\partial E_n(k)/\hbar\partial k$ is the group velocity of the electrons, which is calculated using the electron band structure, and T is the temperature. For the calculations in this paper, we estimated the electron mobility at room temperature. Next, we represented the collisional integral in the right-hand side of Equation (6.16) using the scattering rate $S^{n,n'}(k,k')$ and applying detailed balance conditions. As a result, we obtained the following expression for the collisional integral:

$$\hat{C}\{\delta f_n(k)\}=\frac{NL}{2\pi}\sum_{n'}\int_{-\pi/L}^{\pi/L}dk'S^{n,n'}(k,k')\left[\delta f_{n'}(k')\frac{f_0(E_n(k))}{f_0(E_{n'}(k'))}-\delta f_n(k)\frac{1-f_0(E_{n'}(k'))}{1-f_0(E_n(k))}\right] \tag{6.17}$$

Here, L denotes the length of the unit cell in the transport direction and N is the number of the unit cells. The first and second terms correspond to the in-scattering process from (n', k')

to (n, k) and the out-scattering process from (n, k) to (n', k'), respectively. We introduce a relaxation time approximation:

$$\hat{C}\{\delta f_n(k)\} = -\frac{\delta f_n(k)}{\tau_n(k)} \tag{6.18}$$

Then, by substituting Equation (6.18) into Equation (6.17), the relaxation time $\tau_n(k)$ is calculated by numerically solving the following integral equation:

$$\frac{NL}{2\pi}\sum_{n'}\int_{-\pi/L}^{\pi/L} dk' S^{n,n'}(k,k')\frac{1-f_0(E_{n'}(k'))}{1-f_0(E_n(k))}\left[v_n(k)\tau_n(k)-v_{n'}(k')\tau_{n'}(|k'|)\right] = v_n(k) \tag{6.19}$$

It should be emphasized that Pauli's exclusion principle is strictly taken into account in Equation (6.19). Because the relaxation time is an even function of k (i.e., $\tau_n(k)=\tau_n(-k)$), we only need to solve Equation (6.19) in a half-space of the first Brillouin zone (e.g., $k \in [0, \pi / L]$).

Once the relaxation time was obtained, electronic conductivity σ can be calculated using the following equation [6.22, 6.23]:

$$\sigma = \sum_n \frac{4e^2}{2\pi k_B T}\int_0^{\pi/L} dk (v_n(k))^2 \tau_n(k) f_0(E_n(k))\left[1-f_0(E_n(k))\right] \tag{6.20}$$

Finally, we compute the electron density in the conduction band by using:

$$n = \sum_n \frac{4}{2\pi}\int_0^{\pi/L} dk f_0(E_n(k)) \tag{6.21}$$

and the electron mobility μ is evaluated from the relationship $\sigma=en\mu$. The number of conduction subbands necessary for accurate mobility estimation depends on the wire orientation and, thus, the conduction subbands used in the actual simulation are highlighted by the gray lines in Figure 6.3. The inclusion of these subbands is considered to be sufficient, because the higher subbands are located away from them by more than the thermal energy at room temperature for the present diameter range ($D<5$ nm), and also because the gate voltage was set at zero.

The computed electron mobilities are shown in Figure 6.9 as a function of the diameter. It is found that the electron mobilities significantly decrease with decreasing diameter, which suggests that the electron-phonon interaction becomes stronger.

To examine this point, we plot the total scattering rates for $D=3$ nm in Figure 6.10 and for $D=4.5$ nm in Figure 6.11, which were calculated using the following equation [6.24]:

$$\frac{1}{\tau(E)} = \left(\sum_{n,n'}\int\int_{-\pi/L}^{\pi/L} dk dk' S^{n,n'}(k,k')\frac{1-f_0(E_{n'}(k'))}{1-f_0(E_n(k))}\delta(E-E_n(k))\right) \Big/ \sum_n \int_{-\pi/L}^{\pi/L} dk \delta(E-E_n(k)) \tag{6.22}$$

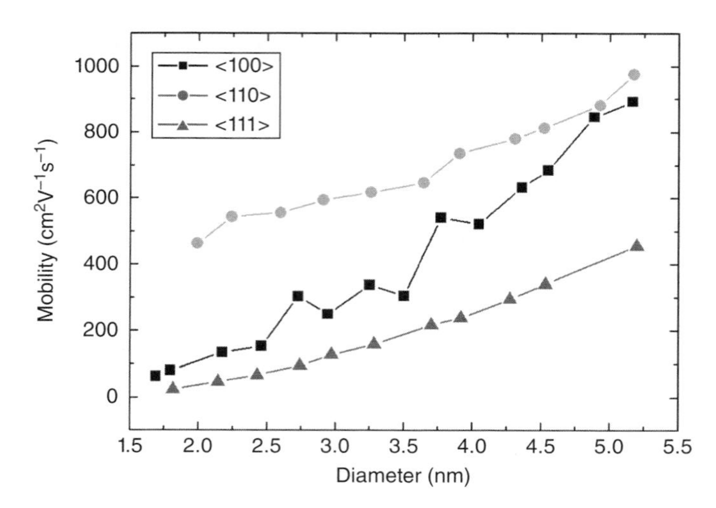

Figure 6.9 Computed electron mobilities of SiNWs for three crystalline orientations as a function of diameter.

Note that all scattering processes, including TA, LA, TLA and optical phonon modes, are considered. It was found that the total scattering rate increases with decreasing diameter for all of the orientations, as expected. To explain the complicated behavior of the scattering rates, we also plotted the density-of-states for the electrons in the dashed lines. First, a peak is found at each Van Hove singularity point, which indicates that the 1-D nature of the SiNWs is described well. Also, a small but broad peak appears right next to each singularity point, which is caused by the excitation of acoustic phonons at the Brillouin zone boundary. An extensive broad peak caused by optical phonon emission, where the electrons need to have greater kinetic energies than the optical phonon energies of $\approx 65\,\text{meV}$, is also clearly observed above each singularity point.

We now take particular note of the total scattering rates around the conduction band minima, which are governed by intrasubband acoustic phonon scattering and determine the electron mobility. A closer look reveals that the scattering rates exhibit similar values among the three orientations, except at the Van Hove singularity points, which is true for both diameters. The rates are approximately $1\times10^{13}\,\text{s}^{-1}$ for $D=4.5\,\text{nm}$, and approximately $2\times10^{13}\,\text{s}^{-1}$ for $D=3.0\,\text{nm}$. As a result, the electron mobilities decrease with decreasing diameter, as shown in Figure 6.9. It should be noted here that the scattering rate has two contributions from the transition matrix element $|M^{n,n'}(k,k')|^2$ and the density-of-states, as represented in Equation (6.14). We next examine the density-of-states (DOS) for the electrons plotted in Figures 6.10 and 6.11 in more detail.

Looking at the curves for the DOS in Figures 6.10 and 6.11 carefully, we find that they vary in accordance with the orientation and diameter dependences of the effective mass shown in Figure 6.4. More specifically, the density-of-states around the conduction band minima increases with decreasing diameter for the <100>- and <111>-orientations, because of the increased effective mass, while it decreases for the <110>-orientation because of the decreased effective mass. The density-of-states for the <110>-orientation is also found to be smaller than that of the other two orientations, because of its smaller effective mass for both diameters.

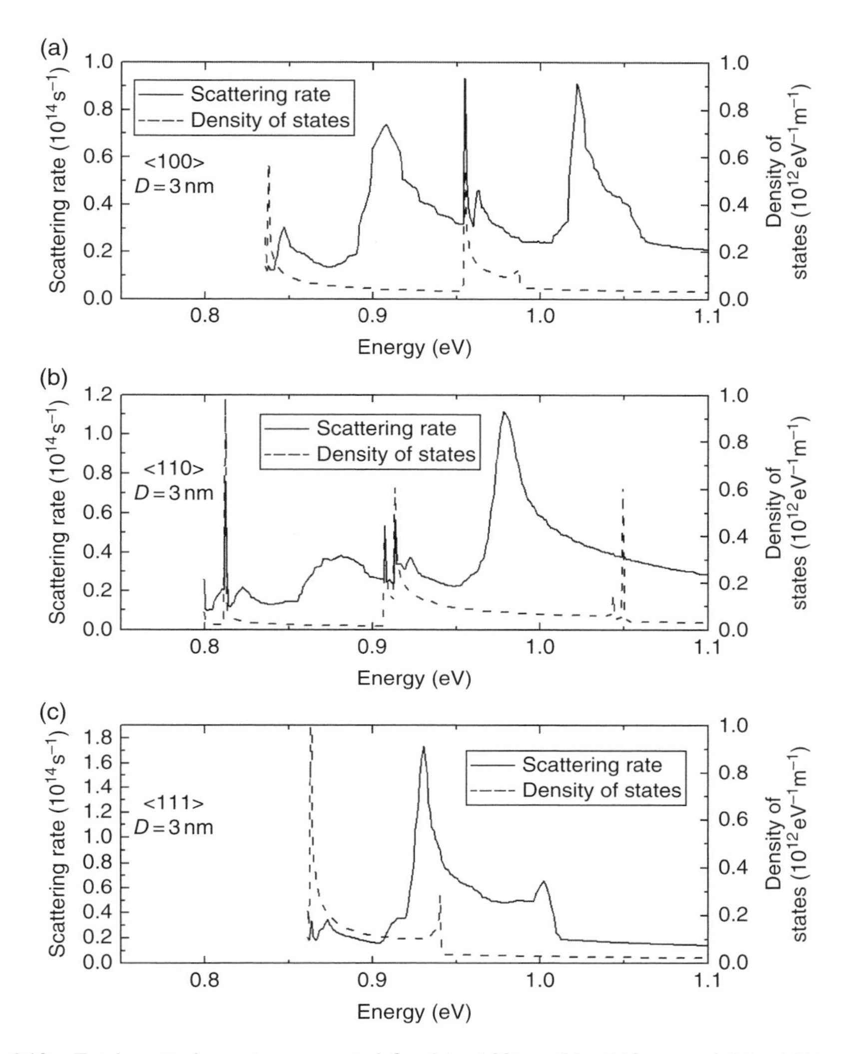

Figure 6.10 Total scattering rates computed for (a) <100>-, (b) <110>-, and (c) <111>-oriented SiNWs with $D = 3$ nm. The solid lines represent the total scattering rates, and the dashed lines represent the DOS for the electrons.

Nevertheless, the total scattering rates are found to be similar among the three orientations, which means that the transition matrix element is larger for the <110>-orientation, compared with the others. This is consistent with the results from elastic simulation for strained-Si – that is, variation of the conduction band minimum because of uniaxial strain, which corresponds to the magnitude of the deformation potential, is larger along the <110>-orientation than along the <100>-orientation [2.14, 2.18]. To summarize the above discussion, the total scattering rates responsible for the mobility are nearly constant among the three orientations, as a consequence of a fortuitous balance of the influences of the density-of-states for electrons and the transition matrix element.

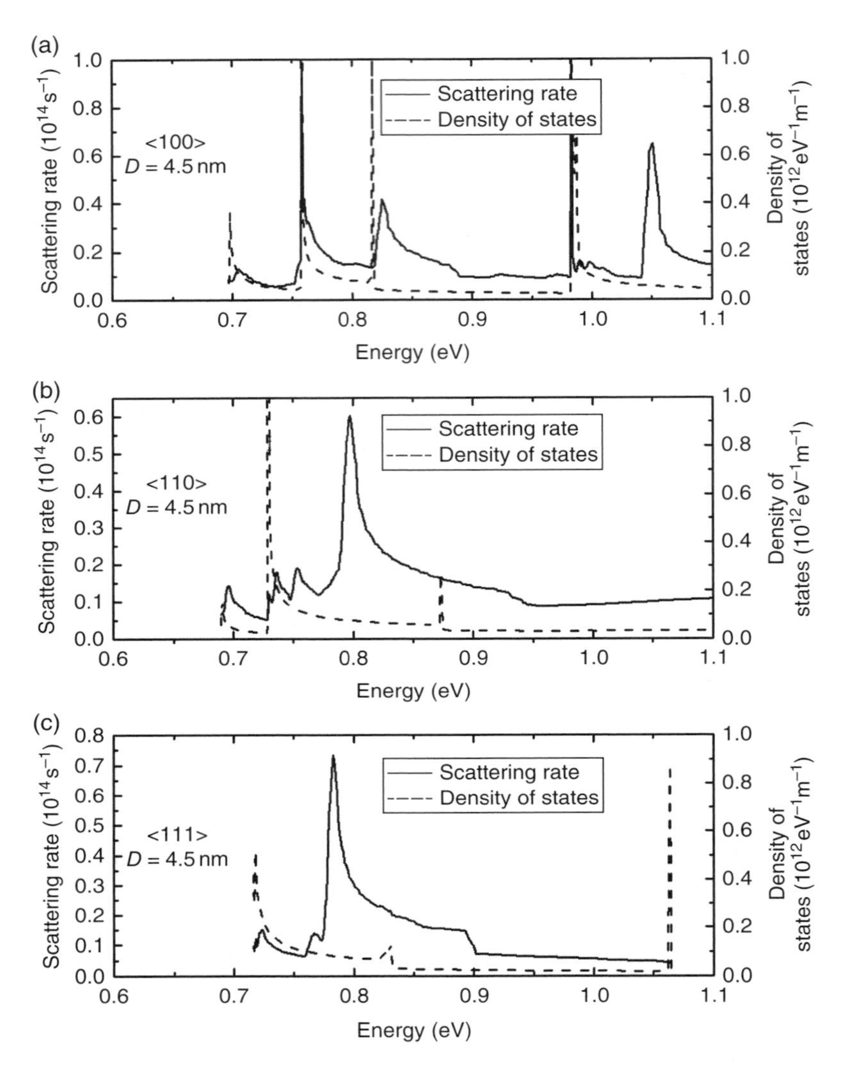

Figure 6.11 Total scattering rates computed for (a) <100>-, (b) <110>-, and (c) <111>-oriented SiNWs with $D = 4.5$ nm. The solid lines represent the total scattering rates, and the dashed lines represent the DOS for the electrons.

Because we considered the diameter-dependent and anisotropic electron-phonon interaction, the mobility for the <110>-orientation in our results decreases monotonically with decreasing diameter. However, in previous simulations [6.6] of atomistic electron band structures computed using the $sp^3d^5s^*$ TB approach, and using diameter-independent bulk phonons, the <110>-orientation exhibits a constant mobility at around 700 cm^2/(V·s) for diameters less than 5 nm. We consider such a discrepancy to be because of the difference in the theoretical treatment of the phonons. However, the phonon-limited mobilities reported in [6.6] indicate a similar magnitude relationship to our results.

In Figure 6.9, the <110>-oriented SiNWs show the highest electron mobility in the present range of diameters, while the <111>-oriented NWs show the lowest. This trend is the same as that in [6.6] and, therefore, the wire-orientation dependence of the electron mobility is primarily governed by the difference between the electron effective masses, as shown in Figure 6.4. Specifically, the <110>-orientation exhibits the smallest effective mass, and the <111>-orientation has the largest, which corresponds to the relative magnitude relationship in the electron mobilities. This is also supported by the fact that the electron mobilities of the <100>- and <110>-orientations approach each other as the diameter increases to 5 nm, which agrees well with the variation in the effective mass. Accordingly, the effective mass of the electrons plays a primary role in determining the atomistic electron mobility of the SiNWs. From another point of view, the qualitative trend in the electron mobility can be understood by considering the atomistic band structures of the electrons. The present results also suggest that the <110>-orientation is promising for high-performance SiNW MOSFETs with diameters of less than 5 nm.

Finally, we compare our atomistic electron mobilities with those calculated by an approach based on an analytical effective mass and a simple isotropic deformation potential. Figure 6.12 shows the electron mobilities as a function of diameter, calculated using the formulation of the effective mass in [6.17] and diameter-independent isotropic deformation potentials [3.62, 6.3], where the value of the deformation potential for acoustic phonons was given as 12 eV according to [3.62], and the parameter set for intervalley scattering was taken from [3.12]. Electronic subband structures were calculated by solving the 2-D Schrödinger and Poisson equations self-consistently [3.62].

The degradation behavior of the electron mobility with decreasing diameter is similar to that in Figure 6.9, but the magnitude relationship between the <100>- and <110>-orientations is reversed, which is mainly because the electron band structures (i.e., the electron effective masses) are different in each approach. These results again suggest that the modification of the electron band structure caused by the 2-D geometric confinement plays an important role in determining the electron mobility.

In conclusion, we have investigated the electron mobility of SiNWs with three crystalline orientations by considering the atomistic electron-phonon interactions. We have calculated the

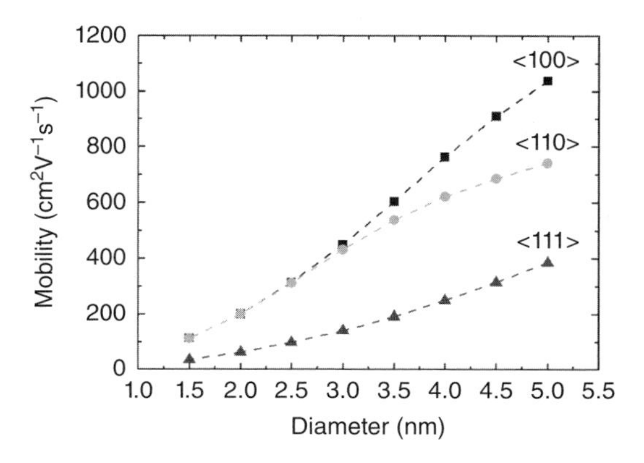

Figure 6.12 Computed electron mobilities of SiNWs using diameter-independent isotropic deformation potentials and self-consistent Schrödinger-Poisson solutions.

electron band structures based on the semi-empirical $sp^3d^5s^*$ TB approach and the phonon band structures, based on the Keating potential model. By combining the electron and phonon eigenstates, based on Fermi's golden rule, and solving the Boltzmann transport equation, we have evaluated the electron mobility of SiNWs. As a result, the electron and phonon eigenstates in SiNWs were found to be significantly dependent on the crystalline orientation and diameter.

As expected, NW confinement modifies the atomic vibration mode and sound velocity, and therefore phonons in SiNWs, behave quite differently from those in bulk Si. However, the electron mobility of SiNWs was found to be primarily governed by the variation in the electron effective mass, rather than that in the phonon eigenstates. Accordingly, the <110>-oriented SiNWs showed the highest electron mobility, because they have the smallest electron effective mass among the three orientations. The results here also suggest that the isotropic deformation potential, using the bulk natures of phonons, still works for projection of a qualitative trend in electron transport in SiNWs.

6.2 *Comparison of Phonon-limited Electron Mobilities between Si and Ge nanowires*

In this section, we apply the atomistic approaches explained in Section 6.1 to the electron mobility calculations for GeNW, and compare its size and orientation dependences with those for SiNW. For the wire orientation, we also considered <112>-direction, because it is expected to have an effective mass smaller than those of <100>- and <111>-directions in Ge.

Figure 6.13 shows the electron band structures calculated for <100>-, <110>-, <111>-, and <112>-oriented GeNWs with a cross-sectional size (W) of 3 nm, where the horizontal axis denotes the wave number normalized by the Brillouin zone width.

First, it is found that all the GeNWs have multiple valleys at the Γ point and at the edge of the Brillouin zone, which can be interpreted by the projection of bulk L, Δ and Γ valleys to

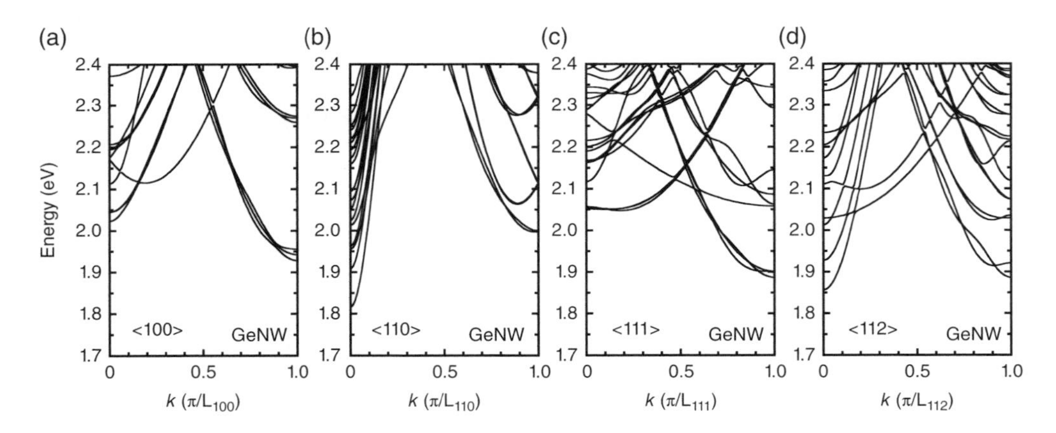

Figure 6.13 Conduction band structures calculated for (a) <100>-, (b) <110>-, (c) <111>-, and (d) <112>-oriented GeNWs with $W \approx 3$ nm, where the horizontal axis denotes wave number normalized by the Brillouin zone width. The origin of the vertical axis corresponds to the Fermi level.

each direction [6.25]. For instance, in <100>-oriented GeNW, all the quantized bulk L valleys are projected to the Brillouin zone edge, whereas the quantized bulk Δ valleys are projected to both Γ and off-Γ valleys. The energy of the valley at the Brillouin zone edge is sufficiently lower than that of the Γ valley, and thus it forms the lowest subband, as shown in Figure 6.13(a).

Next, in <110>-oriented GeNW, valleys both at the Γ point and at the Brillouin zone edge are composed of the bulk L valleys, and quantization with a larger confinement mass leads to form the lowest subband at the Γ point, as shown in Figure 6.13(b). In <111>- and <112>-oriented GeNWs, similar projection of the bulk three valleys produces the lowest subband at the Brillouin zone edge and at the Γ point, respectively, although the projection is quite complicated compared with the previous <100>- and <110>-orientations. For <112>-oriented GeNW, the Γ valley has a small effective mass, which is comparable to that in <110>-orientation, but the energy difference from the higher subband appearing at the Brillouin zone edge is very small. Therefore, scattering should increase, which might be disadvantageous to obtain high electron mobility, as against <110>-oriented GeNWs. We will discuss this point later. Here, we have to add that the band structures in Figure 6.13 are essentially identical to those reported in [6.25].

Figure 6.14 shows the transport effective masses at the conduction band minimum of both GeNWs and SiNWs as a function of W for the four orientations. Note that the horizontal dashed line represents the bulk transverse mass (m_t) of L valleys in Ge.

It is found that, except for <110>-orientation, the transport masses of GeNWs increase as W decreases. This is due to the nonparabolicity of L valleys. The transport mass of <110>-orientation in GeNWs remains almost constant below 0.1 m_0. On the other hand, for SiNWs, the transport mass of <110>-orientation decreases from the bulk m_t of Δ valleys, as W decreases

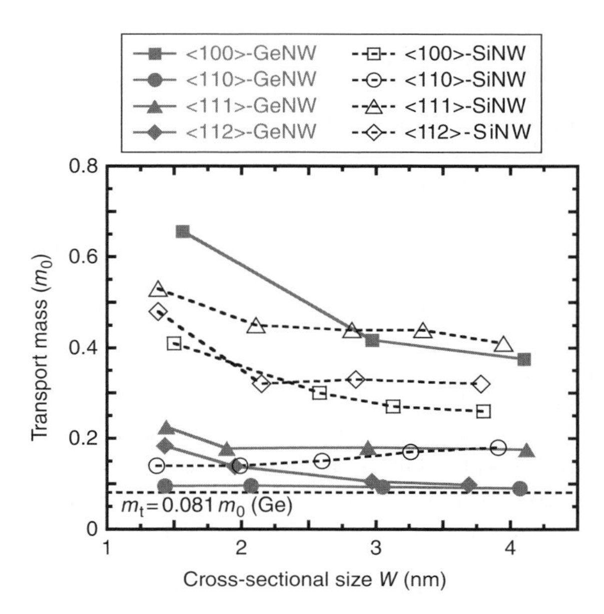

Figure 6.14 Calculated transport effective masses of GeNWs and SiNWs at conduction band minimum as a function of W for four orientations. The horizontal dashed line represents bulk transverse mass (m_t) of L valleys in Ge.

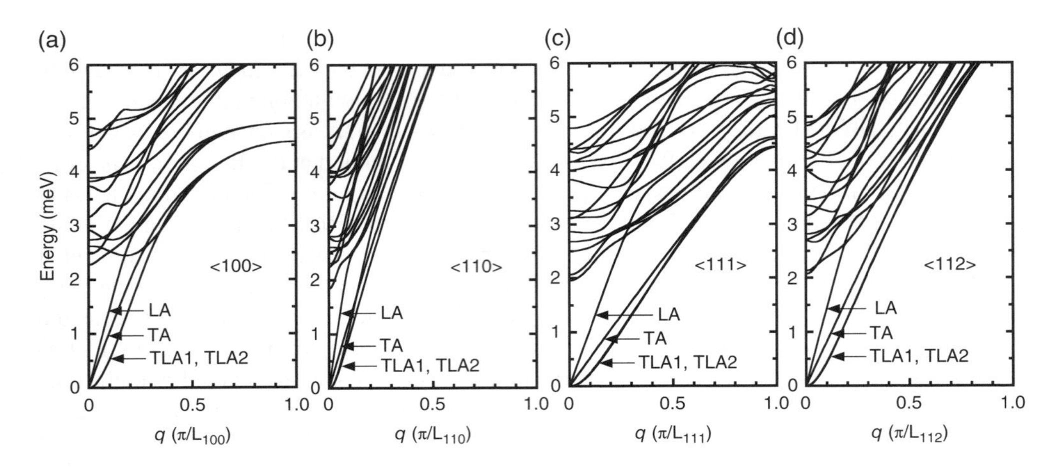

Figure 6.15 Phonon band structures calculated for four orientations in a low-energy regime, where $W \approx 3$ nm and the horizontal axis denotes normalized phonon wave number.

smaller than about 3 nm, which was also shown in Figure 6.4. Consequently, the <110>-oriented SiNW and the <110>-oriented GeNW exhibit a similar effective mass at $W \approx 1.5$ nm, as seen in Figure 6.14. The above variations in the transport masses will result in a drastic change in the electron mobilities, as presented later.

Figure 6.15 shows the phonon band structures calculated for GeNWs with the four orientations in a low-energy regime, where $W \approx 3$ nm and the horizontal axis denotes the normalized phonon wave number.

As discussed in Section 6.1.1, four acoustic modes – TLA1, TLA2, TA, and LA – exist in a long wavelength regime ($q \approx 0$). We notice that TLA1 and TLA2 are degenerate also in the <112>-orientation, because it has rotational symmetries of π. Figure 6.16 shows the phonon DOS calculated for GeNWs corresponding to Figure 6.15.

It is found that the DOS are sharply peaked at around 36 meV for all the orientations, which corresponds to the optical phonon modes. Interestingly, this energy is very close to the optical phonon energy of bulk Ge [3.12], and the DOS are hardly different among the NW orientations. On the other hand, sound velocity exhibits a clear wire orientational dependence, as shown in Figure 6.17. The horizontal dashed lines indicate the theoretical sound velocities for the TA and LA modes in bulk Ge [6.20].

It can be confirmed that the sound velocities of GeNWs decrease almost by half the ones for SiNWs shown in Figure 6.8, which is mainly because Ge crystal has a larger atomic density. The other properties are the same as those in SiNWs, as shown in Figure 6.8.

Figure 6.18 shows the calculated electron mobilities of GeNWs and SiNWs with the four orientations as a function of *W*. It can be seen that the electron mobilities decrease with reducing *W* for all orientations, since the electron-phonon interaction becomes stronger in thinner NWs [6.8].

For $W > 2$ nm, the <110>-oriented GeNWs exhibit the highest electron mobility, whereas the <100>-ones exhibit the lowest. As is the case with SiNWs, the magnitude relation of the electron mobilities in GeNWs is governed by that of the electron effective masses shown in Figure 6.14. In addition, as we have previously pointed out, the <112>-oriented GeNWs

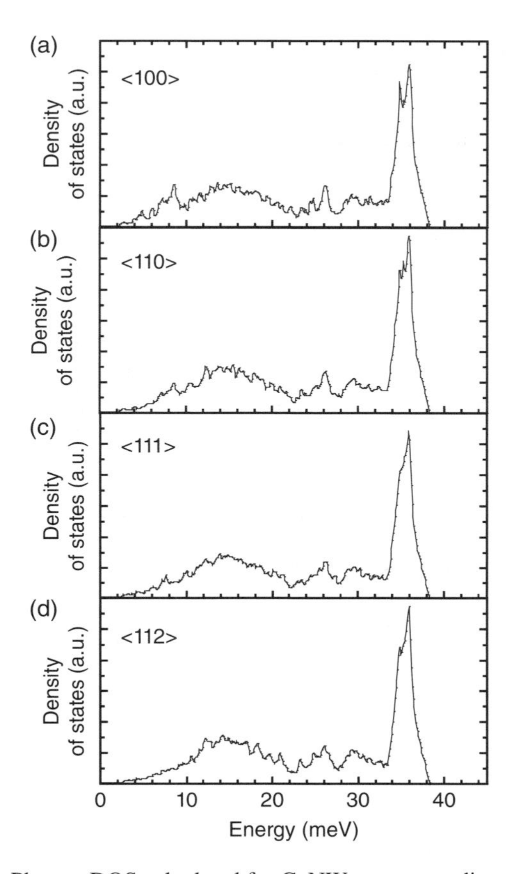

Figure 6.16 Phonon DOS calculated for GeNWs corresponding to Figure 6.15.

indicate smaller electron mobilities than the <110>-oriented GeNWs, in spite of the similar effective masses for $W=3$–$4\,\text{nm}$, which is because they have a small energy difference between the lowest subband at the Γ point and the higher subband at the Brillouin zone edge, resulting in an increased scattering.

In Figure 6.18, we also plotted the electron mobilities calculated for SiNWs. It is found that <110>- and <112>-oriented GeNWs show higher electron mobilities than SiNWs for $W>2\,\text{nm}$, and thus they maintain the advantage over SiNWs. However, it is interesting to note that when W becomes smaller than 2 nm, the electron mobilities of GeNWs with the three orientations, <110>, <112>, and <111>, converge on a similar value. Furthermore, the electron mobilities of <110>-oriented GeNWs coincide with those of <110>-oriented SiNWs. This is considered to be mainly because their electron effective masses, or their conduction band structures, change to be similar when NWs become narrower than 2 nm. In particular, as shown in Figure 6.19, the conduction band structures of <110>-oriented GeNW and SiNW with $W\approx 2\,\text{nm}$ are found to be very similar, including the values of the lowest-subband electron effective masses. The band structure in the <110>-oriented SiNW also has valleys at the Γ, and around the Brillouin zone edge, due to the projection of the bulk Δ valleys [2.4, 2.20, 2.24].

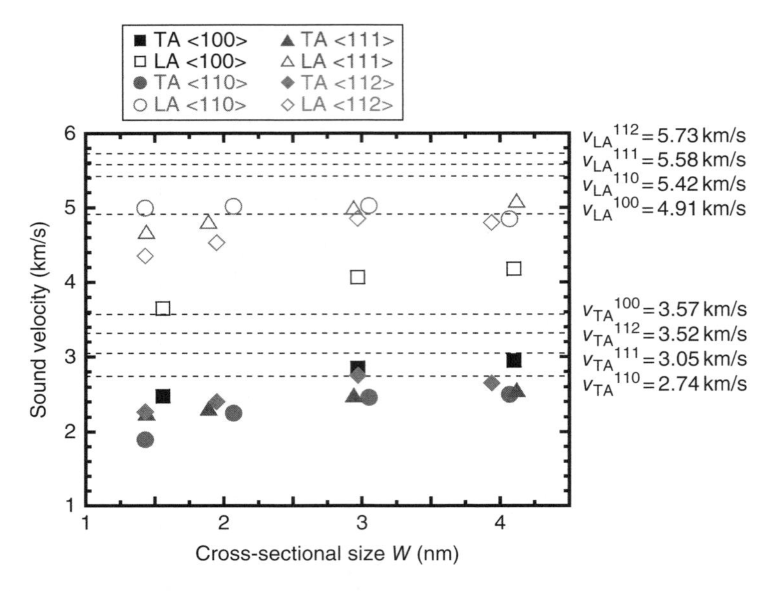

Figure 6.17 Sound velocities of TA and LA phonon modes in GeNWs calculated as a function of *W*. The horizontal dashed lines indicate theoretical sound velocities for TA and LA modes in bulk Ge.

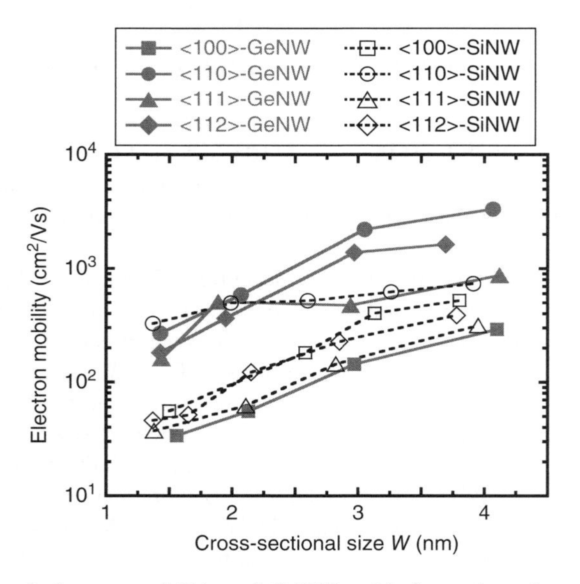

Figure 6.18 Calculated electron mobilities of GeNWs with four orientations as a function of *W*. Electron mobilities calculated for SiNWs are also plotted for comparison.

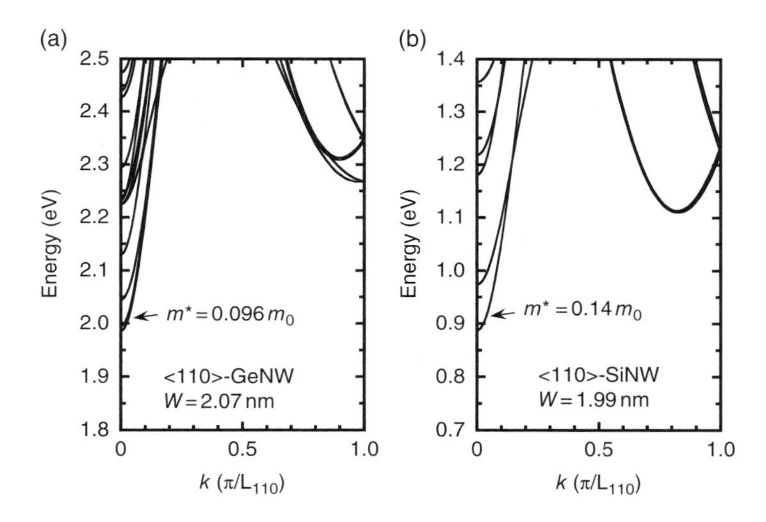

Figure 6.19 Conduction band structures of <110>-oriented (a) GeNW and (b) SiNW with $W \approx 2\,\text{nm}$. In (b), note that <110>-oriented SiNW also has valleys at the Γ and around the Brillouin zone edge, similar to (a).

The above results for $W < 2\,\text{nm}$ remind us of Fischetti and Laux's observation revealing the presence of "universal" speed performance in ultra-small planar MOSFETs, owing to the band structure similarity among technologically significant semiconductors with diamond and zincblende structures at the energy scale of interest [5.10]. Again, getting back to the performance projection, GeNWs are considered to have advantage over SiNWs, since the practical cross-sectional size of NWs will be larger than at least 2 nm, to avoid surface roughness scattering at the gate oxide interface.

In conclusion, it was found that GeNWs have advantage over SiNWs for cross-sectional sizes larger than 2 nm. However, as the cross-sectional size becomes smaller than 2 nm, <110>-oriented GeNWs and SiNWs were shown to exhibit almost identical electron mobilities, because their conduction band structures change to be similar.

6.3 Ballistic Performances of Si and InAs Nanowire MOSFETs

To realize III-VNW MOSFETs with GAA structure, there are many technological issues, such as high-quality and uniform III-VNW formation on Si-platform with superior interface quality, low-resistivity source/drain formation, and so on. In addition to these technological issues, there exists an essential problem associated with III-V compound semiconductors having light effective mass – namely, small DOS for electrons, yielding reduction in the gate capacitance, and thus decrease in the sheet density and the on-current I_{on} [5.7, 5,11, 6.26, 6.27].

Another I_{on} degradation factor in several III-V semiconductors is the transfer of electrons into the L valleys with heavier effective mass [5.4, 5.7, 5.10, 5.16]. Since the 2-D quantum confinement in NW structures may alter these factors through band structure modulation, an atomistic simulation technique is indispensable to estimate device performance of III-VNW MOSFETs. In this section, we consider InAsNW as a III-VNW material, and compute the band structures with three crystal orientations of <100>, <110>and <111>, based on the $sp^3d^5s^*$ TB approach. We then evaluate performance potentials of InAsNW MOSFETs under the ballistic transport, and compare them with those of SiNW MOSFETs [5.66].

6.3.1 Band Structures

We simulated square-shaped NWs grown in three different crystal orientations of <100>, <110>and <111>, as shown in the top of Figure 6.20. Surface atoms are passivated using the sp^3 hybridization scheme [6.12]. Here, note that only the channel atoms enter the atomistic calculation in the Hamiltonian construction.

The gate oxide is not included in the Hamiltonian, but is only treated in the top-of-the-barrier FET model in Section 6.3.2 as a continuum medium. As we have addressed in Section 2.1.1 first-principles calculations of the band structures for Si (001) ultrathin films sandwiched by two kinds of crystalline SiO_2 layers [2.4], the spatial confinement due to the SiO_2 layers has little influence on the effective mass of electron. Therefore, the present scheme should be acceptable also for band structure calculation of NWs. The cross-section is chosen to be about $3\times3\,nm^2$ for all NWs.

We performed $sp^3d^5s^*$ TB band structure calculations, employing Jancu's [6.10] and Boykin's [6.13, 6.14, 6.28] parameters for the InAsNWs and SiNWs, respectively. Figures 6.20 (a)-(c) and (d)-(f) show the computed conduction band structures for the InAsNWs and the SiNWs, respectively. Here, the present InAsNWs have band gaps of ≈ 1 eV or more, which allows us to neglect band-to-band tunneling leakage current at an off-state bias, by choosing a drain bias voltage that is sufficiently small. Effective masses are indicated for the lowest, and some higher, subbands in SiNWs. As described in Section 6.1.1, the <100>- and <110>-oriented SiNWs have conduction band minimum at the Γ point, whereas the <111>-orientation at off-Γ point. The <110>-oriented SiNWs have transport mass smaller than the bulk m_t (=0.19m_0), while the other two orientations have larger transport masses than the bulk m_t, as already shown in Figure 6.4.

On the other hand, the effective masses of the InAsNWs are found to be larger than the InAs bulk value, 0.023m_0 [5.66, 6.26, 6.29–6.32]. This is due to the fact that the quantum confinement cuts through energy contour lines with a smaller curvature caused by highly-non-parabolic dispersion curve of the Γ valley [6.29]. We also added non-parabolicity coefficients along each wire orientation α for the lowest-subband, which was evaluated by fitting the dispersion curves with a relationship of $E(1+\alpha E)=\hbar^2k^2/2m^*$. The <111>-oriented SiNW (Figure 6.20(f)) exhibits $\alpha\approx0$, since the effective mass is quite large and the parabolic relation persists up to several hundreds meV. For the InAsNWs, the non-parabolicity is nearly constant among the three orientations, that is, $\alpha\approx2.0$, regardless of the orientation. On the contrary, the non-parabolicity greatly depends on the wire orientation in SiNWs – that is, the <110>-SiNW indicates twice the value of the <100>-SiNW. Such a high band non-parabolicity in the <110>-SiNW may undermine the advantage of its smaller effective mass, as we will discuss later.

6.3.2 Top-of-the-barrier Model

To evaluate the performance potentials of SiNW and InAsNW MOSFETs, we employ a top-of-the-barrier (ToB) model [6.33]. This model assumes ballistic transport, and then assesses an upper limit performance of atomic-scale MOSFETs. Quantum tunneling is not considered

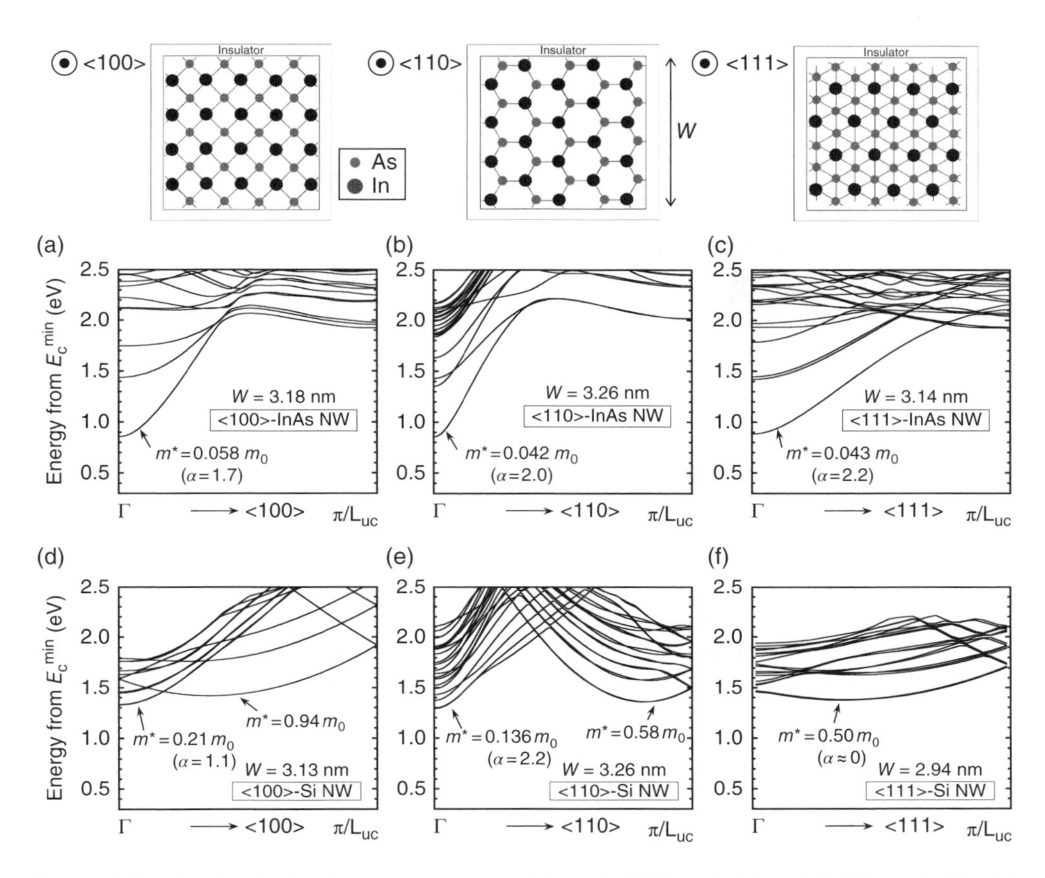

Figure 6.20 Conduction band structures computed for InAs NWs ((a)-(c)) and Si NWs ((d)-(f)) with <100>-, <110>- and <111>-orientations, where the cross-section is about 3×3 nm^2 for all NWs. The origin of the vertical axis (i.e., $E=0$) corresponds to the Fermi level. Effective masses and non-parabolicity coefficients are also indicated.

in this model, so the model was shown to be valid for the Si-MOSFET structure if the channel length is equal to or larger than 10 nm [6.34]. Since the model provides significant insight into the importance of atomistic band structures, with greatly reduced computational time compared with a full self-consistent quantum simulation, the model is suitable for a systematic study comparing the performance limits of various atomistic transistors. In this section, we discuss the electrical characteristics of SiNW and InAsNW MOSFETs, using the ToB model. Figure 6.21(a, b) shows the schematic device structure assumed in the NW MOSFET simulation, and the ToB model represented using three capacitances, C_S, C_G and C_D, respectively. We describe a brief summary of the ToB model below.

When we assume ballistic transport, the drain current of NWs is represented by the Landauer-Büttiker formular as follows.

$$I_D = -q\sum_n \frac{1}{\pi}\int_0^\infty \frac{1}{\hbar}\nabla_k E_n(k)\left[f\left(E_n(k)+U_{scf}-E_{F1}\right)-f\left(E_n(k)+U_{scf}-E_{F2}\right)\right]dk \quad (6.23)$$

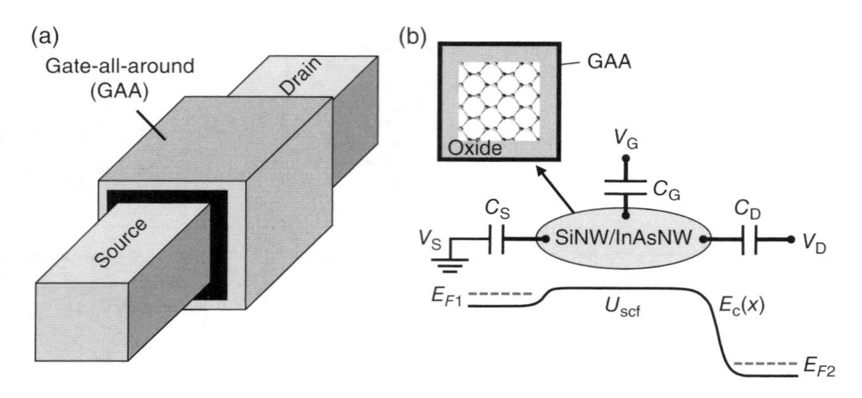

Figure 6.21 (a) Schematic device structure assumed in NW MOSFET simulation; and (b) ToB model represented using three capacitances, C_S, C_G and C_D.

In Equation (6.23), $f(E_n(k)+U_{\text{scf}}-E_{F1})$ and $f(E_n(k)+U_{\text{scf}}-E_{F2})$ are the electron distribution functions of source and drain, respectively, and they are usually given by Fermi-Dirac distribution functions with the Fermi energies E_{F1} and E_{F2}, assuming thermal distribution of electrons in the electrodes. $E_n(k)$ represents the dispersion relation of carriers, which is calculated by the semi-empirical TB approach described in Section 6.1.1. Consequently, we need to define the channel potential U_{scf}, to calculate the drain current using Equation (6.23). Although we usually solve the 2-D or 3-D Poisson's equation and transport equations self-consistently to obtain U_{scf}, the ToB model, expressed using the three capacitances shown in Figure 6.21(b), provides U_{scf} by self-consistently solving the following equations:

$$U_{\text{scf}} = -q\left(\alpha_G V_G + \alpha_D V_D + \alpha_S V_S\right) + \frac{q^2}{C_G + C_D + C_S}\Delta N \tag{6.24}$$

and

$$\Delta N = \left(N_1 + N_2\right) - N_0 \tag{6.25}$$

Here, N_0 is the channel electron density at thermal equilibrium without any biases, and:

$$\alpha_i = \frac{C_i}{C_G + C_D + C_S}, \quad \left(i = G,\ D,\ S\right) \tag{6.26}$$

N_1 and N_2 in Equation (6.25) are the channel electron densities injected from the source and drain under a biased situation, and given by the following equations, respectively:

$$N_1 = \sum_n \frac{1}{\pi}\int_0^\infty f\left(E_n\left(k\right) + U_{\text{scf}} - E_{F1}\right)\, dk \tag{6.27}$$

$$N_2 = \sum_n \frac{1}{\pi}\int_{-\infty}^0 f\left(E_n\left(k\right) + U_{scf} - E_{F2}\right)\, dk \tag{6.28}$$

Note the difference in the integration ranges of Equations (6.27) and (6.28). As mentioned above, the self-consistent calculation is needed to obtain the channel electron density, because the electron density depends on U_{scf}.

6.3.3 $I_D - V_G$ Characteristics

The gate oxide is assumed to be SiO_2, and its thickness – that is, the equivalent oxide thickness (EOT) – is given as 3, 0.5 and 0.25 nm, to examine the influence of quantum capacitance. To clarify the band structure effects, we assumed perfect electrostatic gate control over the channel, i.e., zero capacitive couplings between the source-drain and the channel (C_D/C_G, $C_S/C_G=0$), which gives ideal subthreshold swing. The drain voltage V_D is set at 0.4 V, which is sufficiently small so that band-to-band tunneling leakage current at the off-state bias can be neglected. The source and drain donor concentration is taken as $1.0\times10^{20}\,\text{cm}^{-3}$ for SiNW MOSFETs and $2.0\times10^{19}\,\text{cm}^{-3}$ for InAsNW MOSFETs, considering the maximum doping concentration limited by the solid solubility of donors, as described in Section 5.1.1. On the other hand, the channel is assumed to be undoped.

Figure 6.22 shows the $I_D - V_G$ characteristics at 300 K computed for EOT = 3 nm, 0.5 nm and 0.25 nm. First, we notice that the performance dependence on the wire-orientation is not significant in InAsNW MOSFETs [6.30], compared with SiNW MOSFETs. This is due to the fact that the InAsNWs have effective masses and non-parabolicity coefficients nearly independent of the wire-orientation, as indicated in Figure 6.20, which is owing to the isotropic nature of the Γ valley. This may give us flexibility in the layout design of integrated circuits.

Incidentally, once higher subbands with heavier effective masses start to be occupied by increasing the bias voltages and the cross-sectional dimensions, an obvious wire-orientational dependence appears. Next, it is found that the InAsNW MOSFETs exhibit higher drain current than the SiNW MOSFETs for EOT = 3 nm, as expected. However, as EOT reduces to less than

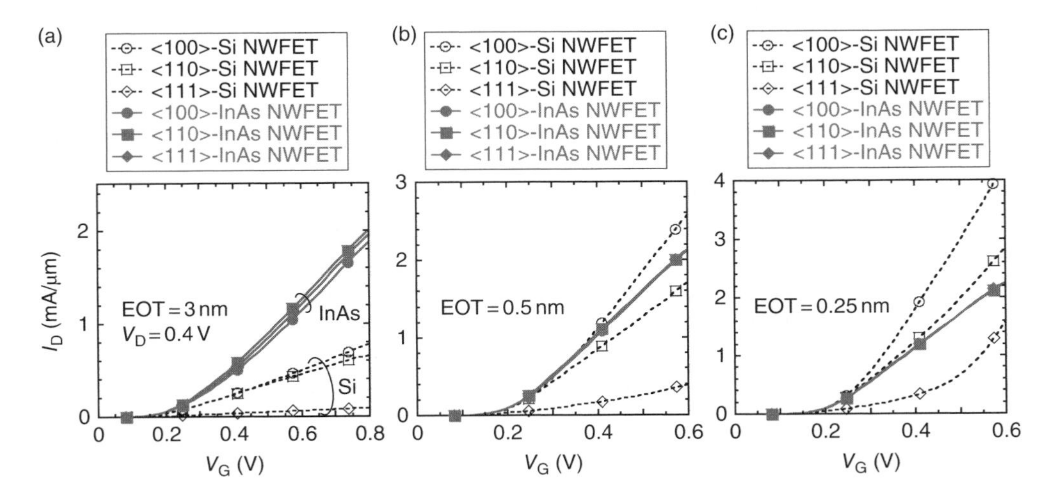

Figure 6.22 $I_D - V_G$ characteristics computed for SiNW and InAsNW MOSFETs with EOT = (a) 3 nm; (b) 0.5 nm; and (c) 0.25 nm. The vertical axis denotes drain current density normalized by the perimeter of NWs. $V_D=0.4$ V. $I_{OFF}=0.01\,\mu$A/μm.

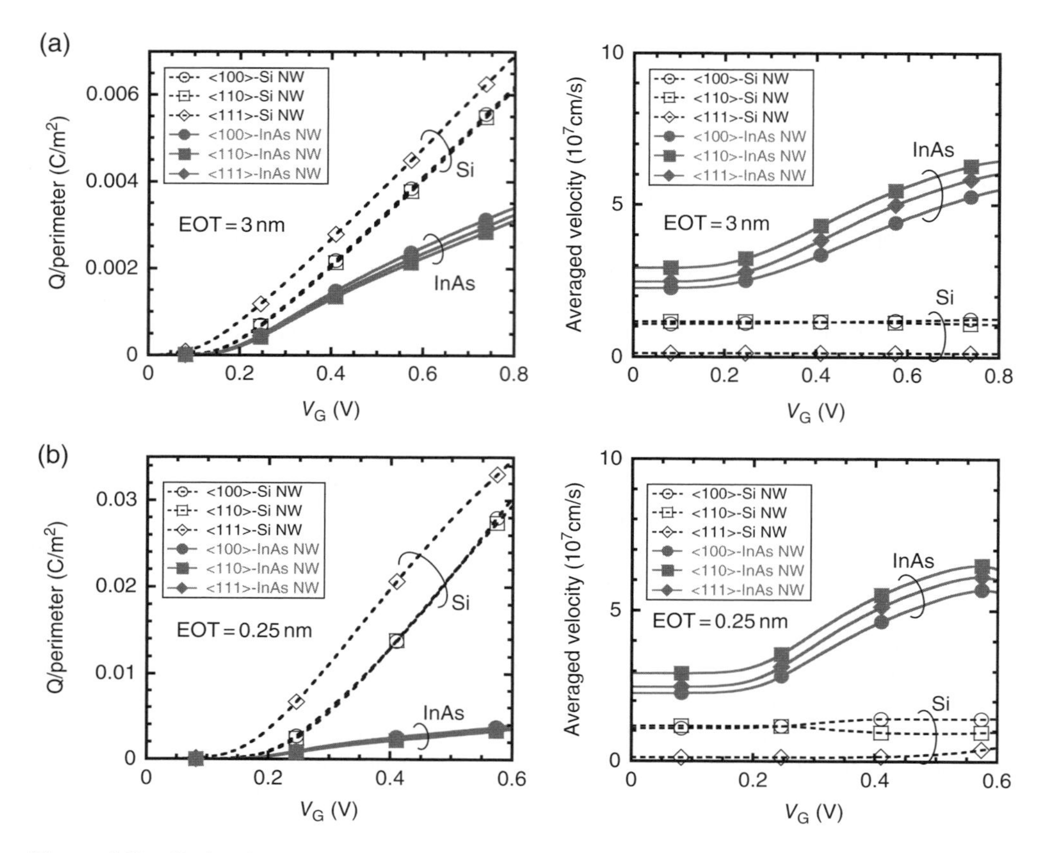

Figure 6.23 (Left column) Computed channel charge densities normalized by the perimeter of NWs and (right column) averaged electron velocities as a function of gate voltage, where EOT = (a) 3 nm; and (b) 0.25 nm.

0.5 nm, the drain current of the InAsNW MOSFETs becomes comparable to, or even smaller than, the SiNW MOSFETs with the <100>- and <110>-orientations.

To explore the cause, we plotted the channel charge densities and the averaged electron velocities as a function of gate voltage as shown in Figure 6.23, where EOT = 3 nm and 0.25 nm. We can confirm that the averaged velocities are identically larger in the InAsNW MOSFETs for both EOTs. On the other hand, the charge densities are smaller in the InAsNW MOSFETs for both EOTs, but they strikingly fall below the SiNW MOSFETs in the case of EOT = 0.25 nm. This is due to a smaller quantum capacitance (C_Q) of InAsNW channels, caused by the smaller DOS for electrons [5.66, 6.26, 6.27], as presented in the next section.

6.3.4 *Quantum Capacitances*

Figure 6.24 shows the computed C_Qs as a function of gate voltage, which indicates that the InAsNW MOSFETs with EOT = 0.25 nm have significantly smaller C_Qs than oxide capacitance C_{ox} ($\equiv \varepsilon_{SiO2}$/EOT), which is displayed by the horizontal dashed line.

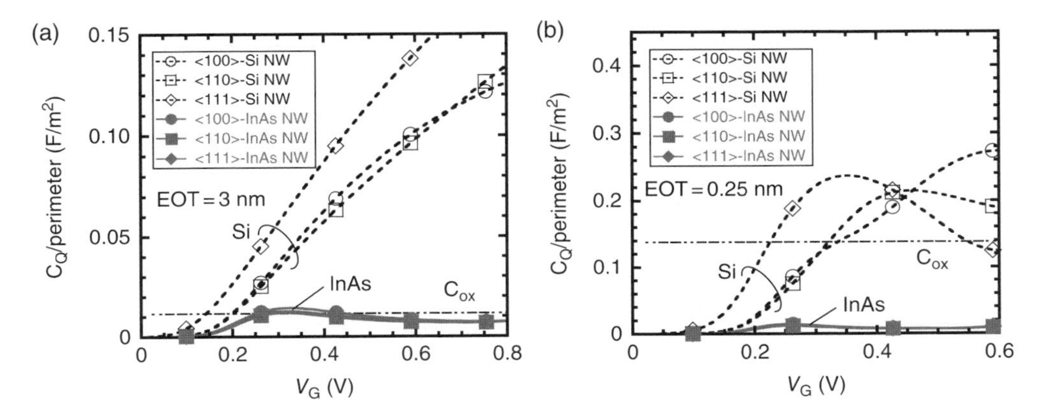

Figure 6.24 Computed quantum capacitance values as a function of gate voltage corresponding to the $I_D - V_G$ curves in Figures 6.22 (a) and (c). The vertical axis is normalized by the perimeter of NWs. The horizontal dashed lines represent the oxide capacitance values, C_{ox}.

In this study, we calculated the quantum capacitance by using $C_Q = \partial(qN)/\partial(-U_{scf}/q)$, where qN and U_{scf} are the charge density per unit length along the channel, and the potential energy at the top of the barrier, respectively [6.30, 6.31, 6.33]. As a result of the QC operation for EOT = 0.25 nm, the total gate capacitance becomes approximately equal to C_Q, and the channel charge drastically decreases.

The above results mean that the lower drain current in the InAsNW MOSFETs with EOT < 0.5 nm is primarily due to the gate capacitance reduction, owing to the smaller C_Q. These results are a general extension of the previous report for <110>-oriented NWFETs [6.26], to the other orientations of <100> and <111>.

In the meantime, as shown in Figure 6.23, the <111>-oriented SiNW MOSFETs exhibit extremely smaller averaged velocity, due to the heavier transport mass, and thus their drain currents are always the lowest, as shown in Figure 6.22. For the <110>-oriented SiNW MOSFETs, the drain current becomes less than that of the <100>-orientation with decreasing EOT, as shown in Figure 6.22. This is because the <110>-oriented SiNW exhibits the higher band non-parabolicity α, which is about twice as large as the <100>-oriented SiNW, as indicated in Figures 6.20(d) and (e), and it should reduce the averaged velocity, as shown in Figure 6.23(b).

6.3.5 Power-delay-product

As found in Figure 6.23, the InAsNW MOSFETs require a much smaller charge density than their Si counterparts for switching a device, especially under the QC limit. Therefore, a lower power switching is still expected using InAsNW MOSFETs, when comparing them under the same I_{on}. We estimated the power-delay-product (PDP) as a function of I_{on}, as shown in Figure 6.25, where EOT = 3 nm and 0.25 nm, and V_D = 0.4 V. The PDP represents the energy required for switching a device, and we calculated it by using the equation of PDP $= I_{ON}V_{DD} \times (Q_{ON} - Q_{OFF})/I_{ON} = (Q_{ON} - Q_{OFF})V_{DD}$ [6.35]. Here, Q_{ON} and Q_{OFF} indicate the total charge densities in the channel at on-state and off-state, respectively. Since the drain voltage hardly affects the

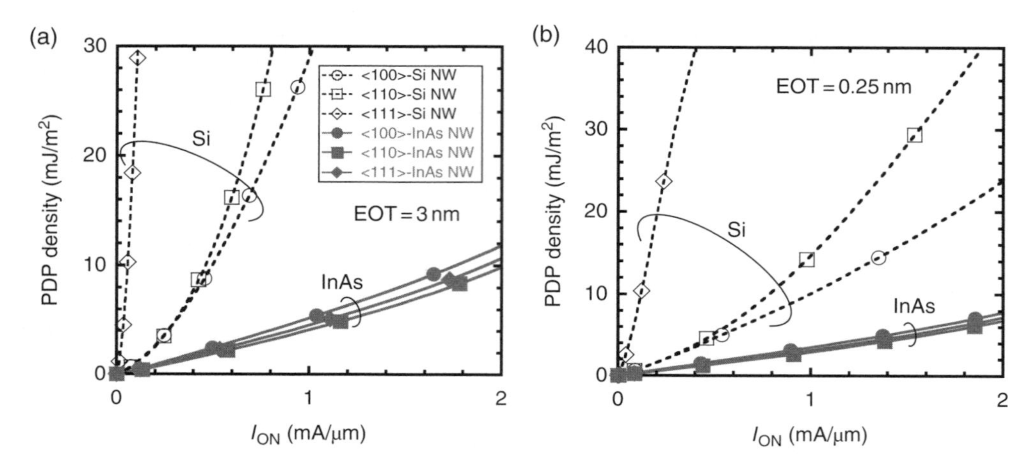

Figure 6.25 Computed PDP densities as a function of ON-current for EOT = (a) 3 nm; and (b) 0.25 nm, where $V_D = 0.4$ V. The vertical axis is divided by the in-plane device area, and $L_{ch} = 10$ nm.

on-state and off-state in the ballistic limit, and we assumed perfect electrostatic gate control over the channel, V_{DD} was substituted with the gate voltage V_G in the present estimation. The vertical axis of Figure 6.25 denotes the PDP density divided by the in-plane device area, and the channel length is set as 10 nm.

We can confirm that the InAsNW MOSFETs have smaller PDP densities than the SiNW MOSFETs for both EOTs, comparing them at the same I_{on}. This is owing to the smaller charge density needed for a switching in the InAsNW MOSFETs, and therefore a lower power switching is, indeed, anticipated in InAsNW MOSFETs, even under the QC limit (EOT < 0.5 nm). However, note that, due to the decreased current drivability in the QC limit, its power saving efficiency is reduced in Figure 6.25(b). We should also add that, among the SiNW MOSFETs, the <100>- and <110>-orientations are considered preferable for a FET channel.

Table 6.1 (a) and (b) show the PDP densities estimated for each orientation and ratio for the best orientations, computed at $I_{ON} = 1.0$ mA/μm for EOT = 1.5 nm and 0.25 nm, respectively. It is found that the reduction ratio in the PDP density increases by about 10 % by reducing EOT from 1.5 nm to 0.25 nm. This is due to the decreased current drivability of the InAsNW MOSFETs in the QC limit as mentioned above, and it represents that the advantage in the lower power operation with InAsNW MOSFETs reduces when the device operates in the QCL [5.66, 6.26]. However, it should be emphasized that, even in the QCL operation, the power can be substantially saved by using InAsNW channel.

In Table 6.1 (c), we further made a similar comparison at $I_{ON} = 2.0$ mA/μm, where EOT is given as 0.25 nm. The reduction ratio is hardly changed by increasing the on-current in this case, since the higher subbands with larger effective masses are not involved in the present bias condition.

In conclusion, it was shown that the wire orientational dependence of the device performance is not significant in InAsNW MOSFETs, owing to the isotropic nature of the Γ valley in InAsNWs. This may remove constraints to choose the channel orientation of NW MOSFETs when actually designing integrated circuits. Next, the InAsNW MOSFETs indicated lower current drivability than <100>- and <110>-oriented Si NWFETs, as the gate oxide thickness

Table 6.1 PDP densities estimated for each orientation and ratio for the best orientations with (a) EOT = 1.5 nm and (b) 0.25 nm, computed at I_{ON} = 1.0 mA/μm. In (c), PDP densities computed at I_{ON} = 2.0 mA/μm are shown, where EOT is given as 0.25 nm.

(a)

EOT = 1.5 nm		I_{ON} = 1.0 (mA/μm)		
–	–	<100>	<110>	<111>
Si NWFET	PDP	18.68	29.41	N/A
InAs NWFET	(mJ/m^2)	4.25	3.45	3.76
Ratio for the best orientations	$\frac{\langle 110\rangle\ \text{InAs}}{\langle 100\rangle\ \text{Si}}$		0.18	

(b)

EOT = 0.25 nm		I_{ON} = 1.0 (mA/μm)		
–	–	<100>	<110>	<111>
Si NWFET	PDP	10.14	14.72	N/A
InAs NWFET	(mJ/m^2)	3.50	2.89	3.14
Ratio for the best orientations	$\frac{\langle 110\rangle\ \text{InAs}}{\langle 100\rangle\ \text{Si}}$		0.29	

(c)

EOT = 0.25 nm		I_{ON} = 2.0 (mA/μm)		
–	–	<100>	<110>	<111>
Si NWFET	PDP	23.78	43.75	N/A
InAs NWFET	(mJ/m^2)	7.85	6.88	7.26
Ratio for the best orientations	$\frac{\langle 110\rangle\ \text{InAs}}{\langle 100\rangle\ \text{Si}}$		0.29	

was scaled below 0.5 nm. The origin was discussed in terms of the gate capacitance reduction caused by the QC operation. However, at the same time, the QC operation requires a much smaller charge density for switching a device and, hence, a lower power switching was found to be still expected in InAsNW MOSFETs for all the three orientations, even under the QC limit. The present results suggest that InAsNW MOSFETs have the advantage over their Si counterparts in terms of lower power operation and flexibility in layout design of practical integrated circuits.

6.4 Ballistic Performances of InSb, InAs, and GaSb Nanowire MOSFETs

As demonstrated in Section 6.3, InAsNW MOSFETs outperform SiNW MOSFETs in terms of low-power operation and flexibility in layout design of integrated circuits. On the other hand, they also showed that, as gate oxide thickness decreases to the QC limit (QCL), InAsNW MOSFETs exhibit drive currents lower than their Si counterparts, due to the degraded DOS for electrons, which results in a decreased charge density compensating a high carrier velocity. This is also called "DOS bottleneck." Hence, to improve the DOS while maintaining a sufficiently high velocity, the use of L-valleys with an anisotropic band structure has been proposed for thin-body planar channel MOSFETs [6.36–6.38].

In these circumstances, we perform a further theoretical study on III-VNW MOSFETs with the other promising III-V semiconductors in this section. In particular, we focus on the significance of the improved DOS on the ballistic device performance of III-VNW MOSFETs, by comparing InSb, InAs, and GaSb NW channel MOSFETs [5.67]. The simulation methods are the same as those used in Section 6.3.

6.4.1 Band Structures

Figure 6.26 shows the computed conduction band structures for InSbNWs, InAsNWs and GaSbNWs, where the wire cross-section is chosen to be about $3 \times 3\,\text{nm}^2$ for all NWs. The origin of the vertical axis (i.e., $E=0$) corresponds to the Fermi level. As described in Section 6.3.1, the lowest subbands at the zone center ($k=0$), which come from the bulk Γ-valley except for the <110>-oriented GaSb NW (the reason will be described later), are found to have larger effective masses than those of the Γ-valley in each bulk material; they are $0.0135\,m_0$ (InSb), $0.026\,m_0$ (InAs), and $0.039\,m_0$ (GaSb) [6.39].

It is worth noting that the InSbNWs and InAsNWs (the In-based NWs) have very similar conduction band structures for all the three orientations – that is, a single subband from the Γ-valley appears at the zone center, and higher subbands are sufficiently separated in energy. Furthermore, the effective masses of those single subbands indicate very close values between the two In-based NWs. Therefore, these subbands give a high electron velocity but a low DOS, which means that the In-based NWs will suffer from the DOS bottleneck.

On the other hand, the GaSbNWs are characterized by the presence of multiple energy subbands at the zone edge ($k=\pi/a$), and also at the zone center for the <110>orientation. Additionally, the subbands at the zone edge, which come from the L-valleys, now appear in energy very close to the lowest subbands at the zone center. This is because the Γ-L separation of GaSb is much smaller than those of InSb and InAs [6.37, 6.39]. The subbands at the zone edge give a small electron velocity, but they significantly improve the DOS, due to the larger effective mass and valley degeneracy [6.36–6.38].

Here, it should be noted that in the <110>-oriented GaSbNW, most of the subbands at the zone center are projected from the L-valleys, and the lowest subband consists of one of the projected L-valleys [6.36–6.38]. Since they originate from the transverse mass of $m_t^*(L) \approx 0.10\,m_0$ [6.39], the effective mass becomes small enough as $0.074\,m_0$. Also, the subbands at the zone edge with a significantly heavier effective mass ($1.21\,m_0$) lie above by $\approx 0.2\,\text{eV}$, which will not be involved with the transport under low gate bias condition. Thus, the <110>-oriented GaSbNW has the potential to improve the DOS while maintaining a high velocity. Detailed discussion on this issue will be given later.

6.4.2 $I_D - V_G$ Characteristics

Next, we computed the ballistic $I_D - V_G$ characteristics at 300 K using the ToB FET model. We employed the GAA structure and assumed perfect electrostatics (i.e., zero capacitive couplings between the source-drain and the channel). The EOT of the gate oxide is given as 3 nm and 0.5 nm. $V_D = 0.4\,\text{V}$, and thus the band-to-band tunneling leakage current at the off-state bias

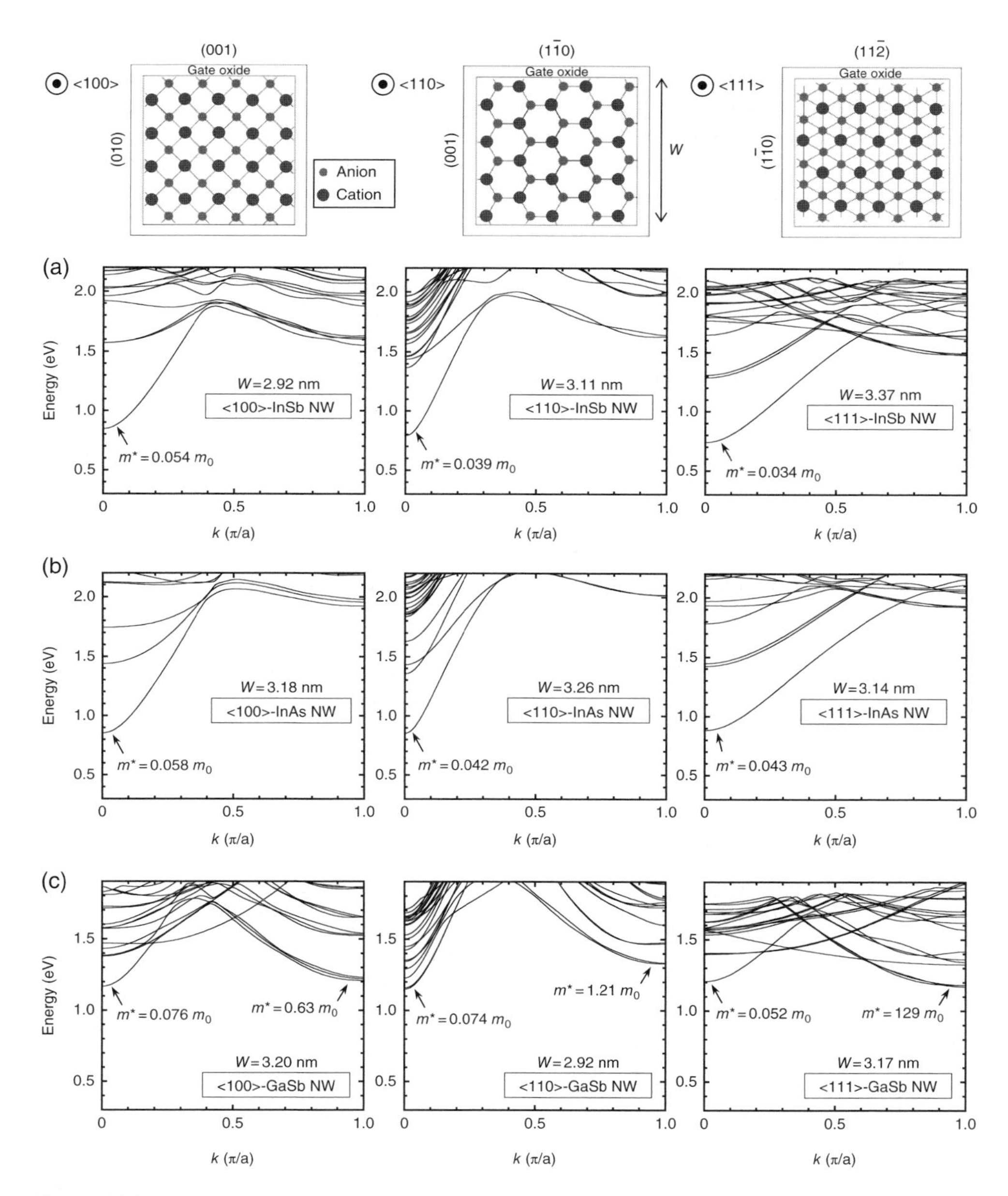

Figure 6.26 Conduction band structures computed for: (a) InSbNWs; (b) InAsNWs; and (c) GaSbNWs, with <100>, <110>, and <111> orientations, where the wire cross-section is about $3 \times 3\,\text{nm}^2$ for all NWs, and a in the horizontal axis represents a length of unit cell. The top part depicts the corresponding atomic models used in the calculations. The origin of the vertical axis ($E=0$) corresponds to the Fermi level. In InSbNWs and InAsNWs, a single subband from the Γ-valley appears at the zone center ($k=0$) with a similar effective mass, while GaSbNWs are characterized by the presence of multiple energy subbands at the zone edge ($k=\pi/a$), and also at the zone center for the <110> orientation, most of them originating from the L-valleys.

was neglected. Figure 6.27 shows the $I_D - V_G$ characteristics computed for EOT = 3 and 0.5 nm, where the vertical axis denotes the drain current density normalized by the perimeter of nanowires to make a comparison under the same gate electrostatics. First of all, in the In-based NW MOSFETs, the performance is nearly independent of the wire orientation [5.66, 5.67, 6.30], and they indicate very similar drain currents for all gate voltages [6.32]. This is due to the fact that the In-based NWs have the similar single subband coming from the Γ-valley at the zone center, with the nearly independent effective masses of the wire orientation, as shown in Figure 6.26.

On the other hand, the GaSbNW MOSFETs exhibit a clear orientation dependence, indicating that the <110>orientation provides the highest drain current for both EOTs. The origin will be discussed in detail later, using channel charge density and averaged electron velocity. Here, note that, as EOT reduces to 0.5 nm, the drain currents of the GaSbNW MOSFETs become larger than those of the In-based ones. The same reversal behavior in drain current has been presented between SiNW and InAsNW MOSFETs in Section 6.3.3, and the reason has been explained in terms of the reduction in the gate capacitance caused by the QC operation.

We then plot the channel charge densities and the averaged electron velocities as a function of gate voltage, as shown in Figure 6.28, where EOT = 3 nm and 0.5 nm. It is confirmed that the In-based NW MOSFETs exhibit smaller charge densities than the GaSbNW MOSFETs, which indicates that the In-based NWs actually suffer from the DOS bottleneck – in particular, in a QCL (EOT = 0.5 nm), as pointed out before. In contrast, the averaged velocities are larger in the In-based NW MOSFETs, owing to smaller effective masses of the single subbands at the zone center. Nevertheless, as a result of the substantially increased charge densities due to the improved DOS, the GaSbNW MOSFETs deliver higher drain currents than the In-based NW MOSFETs in the QCL, as shown in Figure 6.27(b).

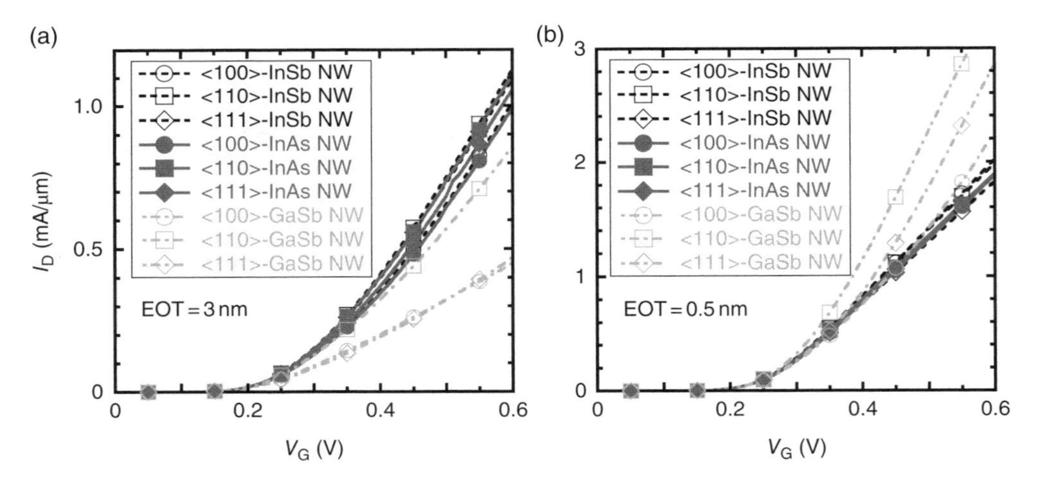

Figure 6.27 Comparison of ballistic $I_D - V_G$ characteristics across materials computed for EOT = (a) 3 nm; and (b) 0.5 nm. The vertical axis denotes the drain current density normalized by the perimeter of NWs. We employed a GAA structure and the gate oxide is assumed to be SiO_2. $V_D = 0.4$ V. $I_{OFF} = 0.01$ μA/μm. $T = 300$ K. The donor density in the source and drain is taken as 2.0×10^{19} cm^{-3}, considering the maximum doping concentration in III-V semiconductors.

Here, we focus on the wire orientation dependence of GaSbNW MOSFETs. First, for <100>- and <111>-oriented GaSbNWs, the multiple subbands at the zone edge are involved with the transport, since they are very close in energy to the subbands at the zone center. Accordingly, their charge densities become the highest and their averaged velocities the lowest, as shown in Figure 6.28.

For the <110>-oriented GaSbNW, multiple subbands appear at the zone center, and the effective mass of $0.074 m_0$ is slightly larger than those of the In-based NWs. As a result, the DOS improves, even though there is no contribution of the subbands at the zone edge. Besides, the effective mass, $0.074 m_0$, is small enough for the transport and, thus, a high electron velocity is obtained, compared with the other two orientations, as shown in Figure 6.28. After all, due to both the modestly improved DOS and the relatively high electron velocity, the <110> orientation exhibits the highest drain current in the GaSbNW MOSFETs, as shown in Figure 6.27.

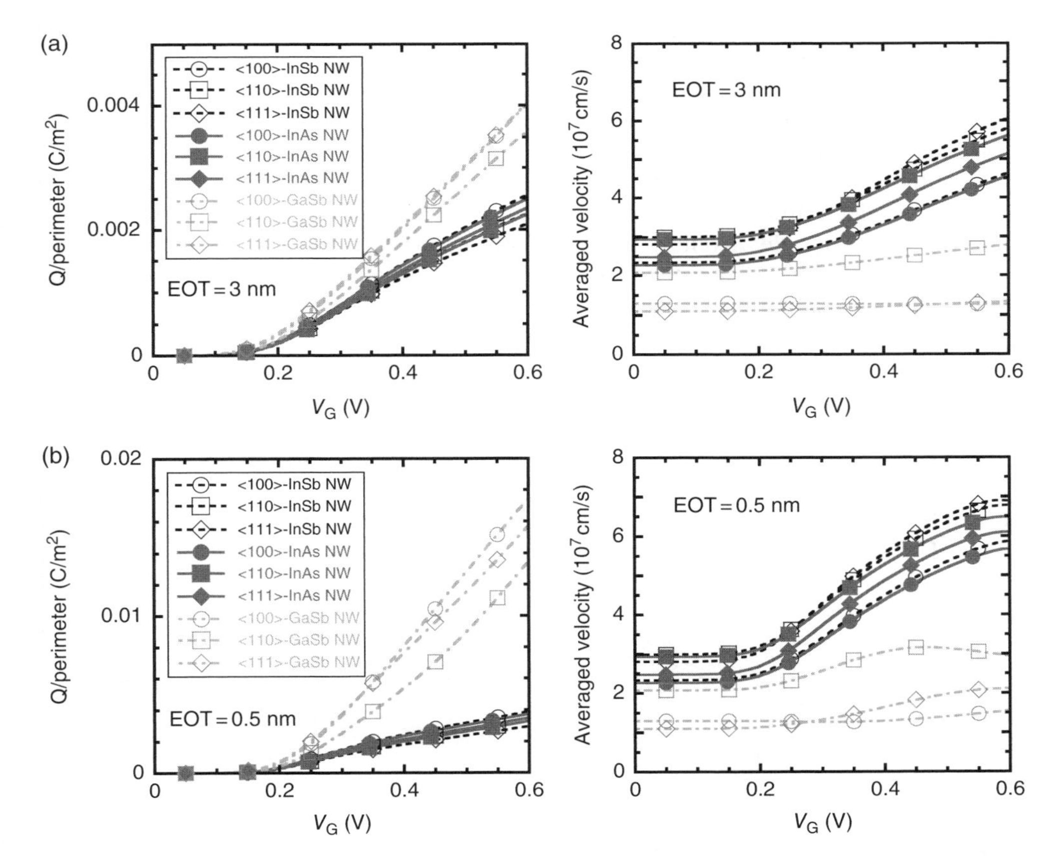

Figure 6.28 (Left column) Channel charge densities normalized by the perimeter of NWs. (Right column) Averaged electron velocities as a function of gate voltage, where EOT = (a) 3 nm; and (b) 0.5 nm.

6.4.3 Power-delay-product

Finally, we compare the PDP computed as a function of I_{on} across materials, as shown in Figure 6.29. The vertical axis denotes the PDP density divided by the in-plane device area. Due to the smaller charge densities needed for a switching in the In-based NW MOSFETs, they exhibit smaller PDP densities than the GaSbNW MOSFETs for both EOTs, comparing at the same I_{on}. This shows that a lower power switching is expected in In-based NW MOSFETs, even under the QCL (EOT = 0.5 nm). However, due to the decreased drain current in the QCL, its power saving efficiency is drastically reduced in Figure 6.29(b).

The above results mean that the formation of multiple subbands at the conduction band minimum (CBM) using the *L*-valleys can improve device performances, such as high drive current and low power consumption of n-channel III-VNW MOSFETs. In particular, <110>-oriented GaSbNW MOSFETs could be a strong competitor to In-based NW MOSFETs in extremely scaled technology nodes.

In conclusion, we have found that InAsNW and InSbNW MOSFETs with nanoscale cross-sections exhibit similar drive currents and similar switching power operations, due to the band structure modulation caused by 2-D quantum confinement. Furthermore, these In-based NW MOSFETs suffer from the DOS bottleneck in the QCL, because of the single energy subband with a small effective mass at the CBM. Nevertheless, a lower power switching is still expected in In-based NW MOSFETs than in their GaSb counterparts.

As for the GaSbNWs, they have multiple energy subbands at the CBM, as a result of the projection of *L*-valleys. Those subbands give a small electron velocity, but they significantly improve the DOS. In particular, the <110>-oriented GaSbNW MOSFET has an improved DOS and a high electron velocity simultaneously, because some of the subbands projected at the zone center from the *L*-valleys lie in the lowest subbands, and they have a sufficiently small effective mass. Consequently, the <110>-oriented GaSbNW MOSFET has the potential

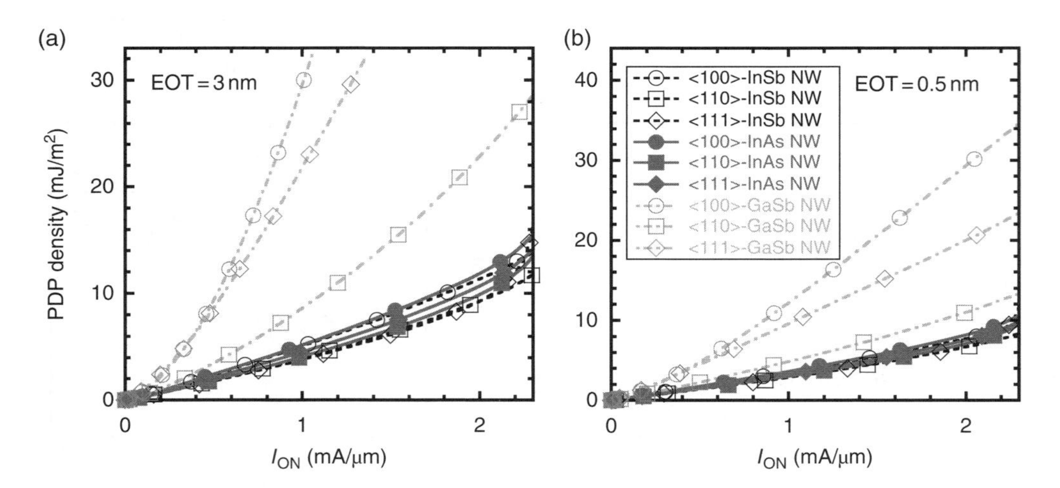

Figure 6.29 Comparison of PDP densities across materials computed as a function of ON current for EOT = (a) 3 nm; and (b) 0.5 nm. V_D is fixed at 0.4 V. The vertical axis denotes the PDP density divided by the in-plane device area, and the channel length is set as 10 nm.

ability to provide high drive current and low-power switching operation in the QCL, provided that the ballistic transport is realized. However, when electron-phonon scattering is taken into account, the presence of multiple energy subbands in GaSbNWs increases the backscattering probability for electrons. Hence, the drastic performance improvement, as obtained in the ballistic limit, might be diminished [6.38].

Appendix A: Atomistic Poisson Equation

The 3-D Poisson equation is solved using an atomistic mesh to define the electrostatic potential at all atomic positions in the unit cell, and the resulting electrostatic potential is substituted into the diagonal on-site matrix element of the TB Hamiltonian in Equation (6.3). In this appendix, we describe a finite volume discretization scheme adopted to solve the atomistic Poisson equation.

First, a 3-D Voronoi tessellation is provided for a given set of atomic positions in the SiNW, and then the Voronoi polyhedron is decomposed into tetrahedra with four triangular pyramids attached at every four equilateral triangles. Therefore, the Voronoi polyhedron has 16 faces, with four regular hexagons and 12 triangles. Here, we define the atomic distance between the first (second) nearest-neighbor atoms as d_1 (d_2), and d_1 and d_2 are represented using the Si lattice constant a, as shown:

$$d_1 = \frac{\sqrt{3}}{4}a, \quad d_2 = \frac{1}{\sqrt{2}}a \tag{A.1}$$

Next, the Voronoi surface between the first (second) nearest-neighbor atoms S_1 (S_2) and the Voronoi volume Δ are given by:

$$S_1 = \sqrt{3}\left(d_1\right)^2, \quad S_2 = \frac{\sqrt{2}}{12}\left(d_1\right)^2, \quad \Delta = \frac{8}{9}\sqrt{3}\left(d_1\right)^3 \tag{A.2}$$

By using these variables, the atomistic 3-D Poisson equation is expressed in the finite volume method using the following equation:

$$\sum_{\eta',\kappa' \in 1NN(\eta,\kappa)} \frac{\varepsilon_{\eta\kappa,\eta'\kappa'}}{d_1} S_1\left(\phi_{\kappa'} - \phi_{\kappa}\right) + \sum_{\eta',\kappa' \in 2NN(\eta,\kappa)} \frac{\varepsilon_{\eta\kappa,\eta'\kappa'}}{d_2} S_2\left(\phi_{\kappa'} - \phi_{\kappa}\right) = en_{\kappa} \tag{A.3}$$

Here, $\varepsilon_{\eta\kappa,\eta'\kappa'}$ is the permittivity, and $1NN\ (\eta, \kappa)$ and $2NN\ (\eta, \kappa)$ represent the first and second nearest-neighbor atoms, respectively. Also, the atomic charge at the (η, κ)-th atom in the right-hand side of Equation (A.3) is calculated by:

$$n_{\kappa} = \frac{4L}{2\pi} \sum_{n} \sum_{\gamma} \int_{0}^{\pi/L} dk \left|c_{\kappa,\gamma}^{n}(k)\right|^2 f_0\left(E_n(k)\right) \tag{A.4}$$

The gate dielectric was considered by giving the permittivity of SiO_2 on the atomistic meshes in the region surrounding the NW, and the gate voltage was provided using the Dirichlet boundary condition at the gate/SiO_2 interfaces.

Appendix B: Analytical Expressions of Electron-phonon Interaction Hamiltonian Matrices

In this study, we assumed the onsite energy to be unchanged by atomic vibration, and hence we describe here the theoretical derivations of the analytical spatial derivatives of the coupling matrices between the *s*-, *p*- and *d*-orbitals.

By using the fact that the first spatial derivatives of the TB Hamiltonian matrix elements are expressed as a function of the directional cosines and the two-center integrals, their analytical expressions can be derived using the Slater-Koster parameters. First, we define the relative position vector from the nearest-neighboring *i*-th to *j*-th atoms $\mathbf{R}_{ij}$ by using its magnitude R_{ij} and unit vectors $\mathbf{e}_x$, $\mathbf{e}_y$ and $\mathbf{e}_z$ as follows:

$$\mathbf{R}_{ij} = \mathbf{R}_j - \mathbf{R}_i = R_{ij}\left(l_{ij}\mathbf{e}_x + m_{ij}\mathbf{e}_y + n_{ij}\mathbf{e}_z\right) \tag{B.1}$$

where l_{ij}, m_{ij} and n_{ij} are integers, and $R_{ij} = \sqrt{(x_j - x_i)^2 + (y_j - y_i)^2 + (z_j - z_i)^2}$ gives the distance between the nearest-neighbor atoms. As mentioned above, we need spatial derivatives of the matrix elements, so we introduce the following relationship between the first derivative of the bond length with respect to $\mathbf{R}_{ij}$ and the directional cosines (l_{ij}, m_{ij}, n_{ij}).

$$\frac{\partial R_{ij}}{\partial \mathbf{R}_{ij}} = \begin{bmatrix} l_{ij} \\ m_{ij} \\ n_{ij} \end{bmatrix} \tag{B.2}$$

Then, the spatial derivative of the two-center integrals is expressed as follows:

$$\frac{\partial V_{i,j}^{\zeta}\left(R_{ij}\right)}{\partial \mathbf{R}_{ij}} = -\frac{n_{\zeta} V_{i,j}^{\zeta}\left(R_{ij}^{0}\right)}{R_{ij}}\left(\frac{R_{ij}^{0}}{R_{ij}}\right)^{n_{\zeta}} \begin{bmatrix} l_{ij} \\ m_{ij} \\ n_{ij} \end{bmatrix} \tag{B.3}$$

Here, n_{ζ} is a generalized Harrison scaling parameter of the ζ bond, which was taken from Reference [6.13] in the calculations.

Also, the first derivative of the directional cosines with respect to $\mathbf{R}_{ij}$ is expressed by:

$$\frac{\partial}{\partial \mathbf{R}_{ij}}\begin{bmatrix} l_{ij} & m_{ij} & n_{ij} \end{bmatrix} = \frac{1}{R_{ij}}\begin{bmatrix} 1-l_{ij}^2 & -l_{ij}m_{ij} & -l_{ij}n_{ij} \\ -m_{ij}l_{ij} & 1-m_{ij}^2 & -m_{ij}n_{ij} \\ -n_{ij}l_{ij} & -n_{ij}m_{ij} & 1-n_{ij}^2 \end{bmatrix} \tag{B.4}$$

In this study, the first derivatives of the Slater-Koster parameters are calculated using Equations (B.3) and (B.4). As an example, for the *s*-orbital and *p*-orbital coupling matrices, the following expressions were obtained:

$$\begin{aligned}
\partial_{\mathbf{R}_{ij}} U_{i,j}^{s,s} &= \partial_{\mathbf{R}_{ij}} V_{i,j}^{ss\sigma}\left(R_{ij}\right) \\
\partial_{\mathbf{R}_{ij}} U_{i,j}^{s,p_x} &= \left(\partial_{\mathbf{R}_{ij}} l_{ij}\right) V_{i,j}^{sp\sigma}\left(R_{ij}\right) + l_{ij}\partial_{\mathbf{R}_{ij}} V_{i,j}^{sp\sigma}\left(R_{ij}\right) \\
\partial_{\mathbf{R}_{ij}} U_{i,j}^{s,p_y} &= \left(\partial_{\mathbf{R}_{ij}} m_{ij}\right) V_{i,j}^{sp\sigma}\left(R_{ij}\right) + m_{ij}\partial_{\mathbf{R}_{ij}} V_{i,j}^{sp\sigma}\left(R_{ij}\right) \\
\partial_{\mathbf{R}_{ij}} U_{i,j}^{s,p_z} &= \left(\partial_{\mathbf{R}_{ij}} n_{ij}\right) V_{i,j}^{sp\sigma}\left(R_{ij}\right) + n_{ij}\partial_{\mathbf{R}_{ij}} V_{i,j}^{sp\sigma}\left(R_{ij}\right)
\end{aligned} \tag{B.5}$$

The first spatial derivatives for all of the other matrix elements can be derived in the same manner. In fact, we derived one hundred expressions for first spatial derivatives of the nearest-neighbor coupling matrices to calculate the electron-phonon interaction Hamiltonian in Section 6.1.2.

References

[6.1] C. Jungemann, A. Emunds and W. L. Engl (Mar. 1993). Simulation of linear and nonlinear electron transport in homogeneous silicon inversion layers. *Solid-State Electronics* **36**(11) 1529–1540.

[6.2] E. B. Ramayya, D. Vasileska, S. M. Goodnick and I. Knezevic (Sep. 2008). Electron transport in silicon nanowires: The role of acoustic phonon confinement and surface roughness scattering. *Journal of Applied Physics* **104**(6), p. 063711.

[6.3] S. Jin, M. V. Fischetti and T.-W. Tang (Oct. 2007). Modeling of electron mobility in gated silicon nanowire at room temperature: Surface roughness scattering, dielectric screening and band nonparabolicity. *Journal of Applied Physics* **102**(8), p. 083715.

[6.4] A. K. Buin, A. Verma and M. P. Anantram (Sep. 2008). Carrier-phonon interaction in small cross-sectional silicon nanowires. *Journal of Applied Physics* **104**(5), p. 053716.

[6.5] F. Murphy-Armando, G. Fagas and J. C. Greer (Feb. 2010). Deformation potentials and electron-phonon coupling in silicon nanowires. *Nano Letters* **10**, 869-873.

[6.6] N. Neophytou and H. Kosina (Aug. 2011). Atomistic simulations of low-field mobility in Si nanowires: Influence of confinement and orientation. *Physical Review B* **84**(8), p. 085313.

[6.7] W. Zhang, C. Delerue, Y.-M. Niquet, G. Allan and E. Wang (Sep. 2010). Atomistic modeling of electron-phonon coupling and transport properties in n-type [110] silicon nanowires. *Physical Review B* **82**(11), p. 115319.

[6.8] Y. Yamada, H. Tsuchiya and M. Ogawa (Mar. 2012). Atomistic modeling of electron-phonon interaction and electron mobility in Si nanowires. *Journal of Applied Physics* **111**(6), p. 063720.

[6.9] J. C. Slater and G. F. Koster (June 1954). Simplified LCAO method for the periodic potential problem. *Physical Review* **94**(6), 1498–1524.

[6.10] J.-M. Jancu, R. Scholz, F. Beltram and F. Bassani (Mar. 1998). Empirical $spds^*$ tight-binding calculation for cubic semiconductors: General method and material parameters. *Physical Review B* **57**(11), 6493–6507.

[6.11] P. N. Keating (May 1966). Effect of invariance requirements on the elastic strain energy of crystals with application to the diamond structure. *Physical Review* **145**(2), 637–645.

[6.12] S. Lee, F. Oyafuso, P. v. Allmen and G. Klimeck (Jan. 2004). Boundary conditions for the electronic structure of finite-extent embedded semiconductor nanostructures. *Physical Review B* **69**(4), p. 045316.

[6.13] T. B. Boykin, G. Klimeck and F. Oyafuso (Mar. 2004). Valence band effective-mass expressions in the $sp^3d^5s^*$ empirical tight-binding model applied to a Si and Ge parametrization. *Physical Review B* **69**(11), p. 115201.

[6.14] T. B. Boykin, N. Kharche and G. Klimeck (July 2007). Brillouin-zone unfolding of perfect supercells having nonequivalent primitive cells illustrated with a Si/Ge tight-binding parameterization. *Physical Review B* **76**(3), p. 035310.

[6.15] D. R. Fokkema, G. L. G. Sleijpen and H. A. Van der Vorst (1998). Jacobi-Davidson style QR and QZ algorithms for the reduction of matrix pencils. *SIAM Journal on Scientific Computing* **20**(1), 94–125.

[6.16] Y. Saad (2003). *Iterative methods for sparse linear systems*, 2nd edition. Society for Industrial Mathematics, Philadelphia.
[6.17] M. Bescond, N. Cavassilas and M. Lannoo (May 2007). Effective-mass approach for n-type semiconductor nanowire MOSFETs arbitrarily oriented. *Nanotechnology* **18**(25), p. 255201.
[6.18] E. Anderson, Z. Bai, C. Bischof, S. Blackford, J. Demmel, J. Dongarra, J. Du Croz, A. Greenbaum, S. Hammarling, A. McKenney and D. Sorensen (1999). *LAPACK Users' Guide*. SIAM, Philadelphia.
[6.19] T. Thonhauser and G. D. Mahan (Feb. 2004). Phonon modes in Si [111] nanowires. *Physical Review B* **69**(7), p. 075213.
[6.20] P. Y. Yu and M. Cardona (2010). *Fundamentals of semiconductors*, 4th edition, Chapter 3, pp. 107–158. Springer-Verlag, Berlin, Heidelberg.
[6.21] M. Luisier and G. Klimeck (Oct. 2009). Atomistic full-band simulations of silicon nanowire transistors: Effects of electron-phonon scattering. *Physical Review B* **80**(15), p. 155430.
[6.22] J. M. Ziman (2001). *Electrons and phonons: the theory of transport phenomena in solids*. Oxford University Press, USA.
[6.23] G. Mahan (1990). *Many-Particle Physics*, 2nd edition. Plenum, New York.
[6.24] O. D. Restrepo, K. Varga and S. T. Pantelides (May 2009). First-principles calculations of electron mobilities in silicon: Phonon and Coulomb scattering. *Applied Physics Letters* **94**(21), p. 212103.
[6.25] S. Mori, N. Morioka, J. Suda and T. Kimoto (Mar. 2013). Orientation and shape effects on ballistic transport properties in gate-all-around rectangular germanium nanowire nFETs. *IEEE Transactions on Electron Devices* **60**(3), 944–950.
[6.26] N. Takiguchi, S. Koba, H. Tsuchiya and M. Ogawa (Jan. 2012). Comparisons of performance potentials of Si and InAs nanowire MOSFETs under ballistic transport. *IEEE Transactions on Electron Devices* **59**(1), 206–211.
[6.27] A. Rahman, G. Klimeck and M. Lundstrom (Dec. 2005). Novel channel materials for ballistic nanoscale MOSFETs - bandstructure effects. *IEDM Technical Digest* 601–604.
[6.28] T. B. Boykin (Sep. 2002). Diagonal parameter shifts due to nearest-neighbor displacements in empirical tight-binding theory. *Physical Review B* **66**(12), p. 125207.
[6.29] Y. Lee, K. Kakushima, K. Natori and H. Iwai (Apr. 2012). Gate capacitance modeling and diameter-dependent performance of nanowire MOSFETs. *IEEE Transactions on Electron Devices* **59**(4), 1037–1045.
[6.30] K. Alam and R. N. Sajjad (Nov. 2010). Electronic properties and orientation-dependent performance of InAs nanowire transistors. *IEEE Transactions on Electron Devices* **57**(11), 2880–2885.
[6.31] E. Lind, M. P. Persson, Y.-M. Niquet and L.-E. Wernersson (Feb. 2009). Band structure effects on the scaling properties of [111] InAs nanowire MOSFETs. *IEEE Transactions on Electron Devices* **56**(2), 201–205.
[6.32] M. A. Khayer and R. K. Lake (Feb. 2008). Performance of n-type InSb and InAs nanowire field-effect transistors. *IEEE Transactions on Electron Devices* **55**(11), 2939–2945.
[6.33] A. Rahman, J. Guo, S. Datta and M.S. Lundstrom (Sep. 2003). Theory of ballistic nanotransistors. *IEEE Transactions on Electron Devices* **50**(9), 1853–1864.
[6.34] A. Paul, S. Mehrotra, G. Klimeck and M. Luisier (May 2009). *On the validity of the top of the barrier quantum transport model for ballistic nanowire MOSFETs*. Proceedings of International Workshop on Computational Electronics (IWCE), Beijing, 173–176.
[6.35] S. O. Koswatta, M. S. Lundstrom and D. E. Nikonov (Mar. 2009). Performance comparison between p-i-n tunneling transistors and conventional MOSFETs. *IEEE Transactions on Electron Devices* **56**(3), 456–465.
[6.36] M. Rodwell, W. Frensley, S. Steiger, E. Chagarov, S. Lee, H. Ryu, Y. Tan, G. Hegde, L. Wang, J. Law, T. Boykin, G. Klimek, P. Asbeck, A. Kummel and J. N. Schulman (June 2010). *III-V FET channel designs for high current densities and thin inversion layers*. Proceedings of 68th Device Research Conference (DRC), 149–152.
[6.37] R. Kim, T. Rakshit, R. Kotlyar, S. Hasa and C. E. Weber (June 2011). Effects of surface orientation on the performance of idealized III-V thin-body ballistic n-MOSFETs. *Electron Device Letters* **32**(6), 746–748.
[6.38] M. Luisier (Dec. 2011). Performance comparison of GaSb, strained-Si and InGaAs double-gate ultrathin-body n-FETs. *Electron Device Letters* **32**(12), 1686–1688.
[6.39] I. Vurgaftman, J. R. Meyer and L. R. Ram-Mohan (June 2001). Band parameters for III-V compound semiconductors and their alloys. *Journal of Applied Physics* **89**(11), 5815–5875.

7

2-D Materials and Devices

To overcome the performance limit of CMOS technologies, new device structures such as ultrathin body MOSFETs and FinFETs have been proposed to enhance the electrostatic gate controllability over the channel. Furthermore, new channel materials beyond silicon, with higher mobilities, have also been extensively studied. Among the various alternative channel materials for CMOS technology, 2-D materials are becoming attractive candidates, because the materials possess certain properties that may provide a path towards the ultimate physical scaling limit beyond what is predicted for conventional semiconductor materials. These properties include:

1. high carrier mobility;
2. presence of a large bandgap suitable for electronic switching and negligible band-to-band tunneling;
3. ultrathin channel structure, providing excellent gate controllability;
4. absence of dangling bonds at the surface;
5. atomically flat surface;
6. relatively high effective mass of carriers, suppressing source-drain direct tunneling.

In this chapter, we discuss the electronic properties of 2-D materials composed of carbon (C), silicon (Si) and germanium (Ge) elements, and further examine their performance potentials as an FET channel.

7.1 2-D Materials

Graphene, a 2-D sheet of carbon atoms in a honeycomb lattice, has attracted significant research interest, due to its unique and excellent electronic properties, such as linear energy dispersion relation, extremely high electron mobility, and so on [7.1]. Experimentally,

Carrier Transport in Nanoscale MOS Transistors, First Edition. Hideaki Tsuchiya and Yoshinari Kamakura.

electron mobility as high as ≈ 120 000 cm^2/Vs at 240 K has been measured in suspended graphene [7.2], and a higher electron mobility, exceeding 10^7 cm^2/Vs at temperatures up to 50 K, has been demonstrated using graphene layers decoupled from bulk graphite [7.3].

The discovery of graphene and the succeeding tremendous advancement in this field of research have promoted researchers to search theoretically and experimentally for similar 2-D materials composed of group IV elements, especially Si and Ge. The Si equivalent of graphene, silicene, was first mentioned in a theoretical study by Takeda and Shiraishi [7.4] in 1994. However, silicene does not seem to exist in nature and, as a consequence, pure 2-D silicene layers cannot be generated by the exfoliation method, as performed initially in the case of graphene [7.1]. More sophisticated methods had to be searched out for the growth or synthesis of silicene.

Recently, Fleurence *et al.* [7.5] have demonstrated that epitaxial 2-D silicene forms spontaneously on (0001)-oriented thin films of zirconium diboride (ZrB_2) that were grown epitaxially on Si (111) wafers. Through a detailed characterization of the structure and electronic properties, they have further shown that silicene is more flexible than graphene, which allows an engineering of the band structure through epitaxy. As will be described later in detail, silicene and germanene exhibit similar linear-band dispersions near the Fermi level, and have a semi-metallic nature without a bandgap. The discovery of the reproducible method for fabricating silicene mentioned above will pave the way for the further study of practical device applications using silicene. In this section, we describe the fundamental features of 2-D materials and advantages of MOSFETs using 2-D material channels.

7.1.1 Fundamental Properties of Graphene, Silicene and Germanene

Graphene is a one-atom-thick carbon sheet arranged in hexagonal structure, as shown in Figure 7.1(a). It has a rhombus-shaped unit cell as depicted with the dashed line, where two C atoms (symbolized by A and B) are included in the unit cell [7.6].

The primitive lattice vectors are written as:

$$\mathbf{a}_1 = \frac{a}{2}\left(3, \sqrt{3}\right), \quad \mathbf{a}_2 = \frac{a}{2}\left(3, -\sqrt{3}\right) \tag{7.1}$$

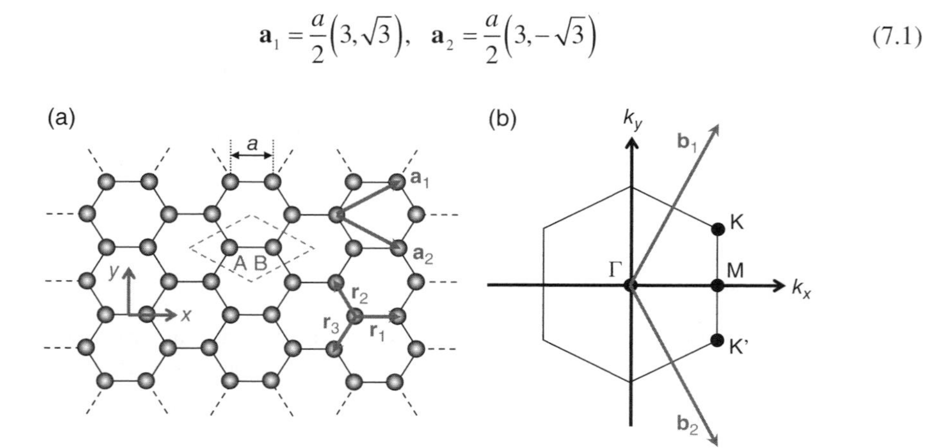

Figure 7.1 (a) Honeycomb lattice; and (b) its first Brillouin zone (BZ). In (a), $\mathbf{a}_1$ and $\mathbf{a}_2$ are the primitive lattice vectors, and $\mathbf{r}_1$, $\mathbf{r}_2$, and $\mathbf{r}_3$ are the nearest-neighbor vectors. In (b), $\mathbf{b}_1$ and $\mathbf{b}_2$ are the primitive reciprocal lattice vectors, and the Dirac cones are located at the *K* and *K'* points.

where $a \approx 1.42\,\mathring{A}$ is the interatomic (C-C) distance. The primitive reciprocal lattice vectors are given by:

$$\mathbf{b}_1 = \frac{2\pi}{3a}\left(1, \sqrt{3}\right), \quad \mathbf{b}_2 = \frac{2\pi}{3a}\left(1, -\sqrt{3}\right) \tag{7.2}$$

Of particular importance for the physics of graphene are the two points K and K' at the corners of the first Brillouin zone (BZ) as shown in Figure 7.1(b). These are named Dirac points, and their positions in momentum space are given by:

$$\mathbf{K} = \left(\frac{2\pi}{3a}, \frac{2\pi}{3\sqrt{3}a}\right), \quad \mathbf{K}' = \left(\frac{2\pi}{3a}, -\frac{2\pi}{3\sqrt{3}a}\right) \tag{7.3}$$

The three nearest-neighbor vectors in real-space are given by:

$$\mathbf{r}_1 = a\left(1,\ 0\right), \quad \mathbf{r}_2 = \frac{a}{2}\left(-1,\ \sqrt{3}\right), \quad \mathbf{r}_3 = \frac{a}{2}\left(-1,\ -\sqrt{3}\right) \tag{7.4}$$

Assuming that electron orbitals are localized at each atom, the wave function Ψ is expressed as a linear combination of the atomic orbitals (LCAO), as follows:

$$\Psi = c_A \psi_A + c_B \psi_B \tag{7.5}$$

Here, we employ a p_z orbital basis set of the π-electron belonging to the atoms A and B as the wave functions ψ_A and ψ_B, throughout this chapter. One p_z orbital per atom is enough for atomistic physical description since s, p_x, and p_y are far from the Fermi level and do not play an important role for electron transport. c_A and c_B in Equation (7.5) are the amplitudes of the wave functions. Substituting Equation (7.5) into the Shrödinger equation, we obtain:

$$H\left(c_A \psi_A + c_B \psi_B\right) = \varepsilon\left(c_A \psi_A + c_B \psi_B\right) \tag{7.6}$$

When we multiply Equation (7.6) by ψ_A^* or ψ_B^* and further integrate them with respect to $\mathbf{r}$, the following two equations are obtained:

$$c_A \langle \psi_A | H | \psi_A \rangle + c_B \langle \psi_A | H | \psi_B \rangle = \varepsilon\left(c_A \langle \psi_A | \psi_A \rangle + c_B \langle \psi_A | \psi_B \rangle \right) \tag{7.7}$$

$$c_A \langle \psi_B | H | \psi_A \rangle + c_B \langle \psi_B | H | \psi_B \rangle = \varepsilon\left(c_A \langle \psi_B | \psi_A \rangle + c_B \langle \psi_B | \psi_B \rangle \right) \tag{7.8}$$

To simplify Equations (7.7) and (7.8), we introduce the Röwdin orthogonal basis [7.7], which has orthonormality. However, since it does not satisfy Bloch's theorem, we express the wave functions ψ_A and ψ_B, using a Bloch summation as follows:

$$\psi_A = \sum_A \exp\left(i\mathbf{k} \cdot \mathbf{r}_A\right) X\left(\mathbf{r} - \mathbf{r}_A\right) \tag{7.9}$$

$$\psi_B = \sum_B \exp\left(i\mathbf{k} \cdot \mathbf{r}_B\right) X\left(\mathbf{r} - \mathbf{r}_B\right) \tag{7.10}$$

where $\mathbf{r}_A$ and $\mathbf{r}_B$ denote the position vectors of the atoms A and B in the unit cell, respectively, and $X(\mathbf{r})$ is the Röwdin orthogonal orbital representing the π-electron p_z orbital. Substituting

Equations (7.9) and (7.10) into Equations (7.7) and (7.8), and further using the orthonormality of the wave functions, we obtain the matrix equation as:

$$\begin{bmatrix} \langle \psi_A | H | \psi_A \rangle & \langle \psi_A | H | \psi_B \rangle \\ \langle \psi_B | H | \psi_A \rangle & \langle \psi_B | H | \psi_B \rangle \end{bmatrix} \begin{bmatrix} c_A \\ c_B \end{bmatrix} = \varepsilon \begin{bmatrix} c_A \\ c_B \end{bmatrix} \tag{7.11}$$

The matrix elements on the left-hand side of Equation (7.11) are called tight-binding (TB) Hamiltonian. When we consider only the nearest-neighbor interaction of the wave functions, each matrix element is given by:

$$\langle \psi_A | H | \psi_A \rangle = \exp[i\mathbf{k}\cdot(\mathbf{r}_A - \mathbf{r}_A)]\varepsilon_{2p} = \varepsilon_{2p} \tag{7.12}$$

$$\langle \psi_A | H | \psi_B \rangle = \sum \exp[i\mathbf{k}\cdot(\mathbf{r}_B - \mathbf{r}_A)]\gamma_0 = \gamma_0[\exp(i\mathbf{k}\cdot\mathbf{r}_1) + \exp(i\mathbf{k}\cdot\mathbf{r}_2) + \exp(i\mathbf{k}\cdot\mathbf{r}_3)] \tag{7.13}$$

$$\langle \psi_B | H | \psi_A \rangle = \sum \exp[i\mathbf{k}\cdot(\mathbf{r}_A - \mathbf{r}_B)]\gamma_0 = \gamma_0[\exp(i\mathbf{k}\cdot\mathbf{r}_1) + \exp(i\mathbf{k}\cdot\mathbf{r}_2) + \exp(i\mathbf{k}\cdot\mathbf{r}_3)]^* \tag{7.14}$$

$$\langle \psi_B | H | \psi_B \rangle = \exp[i\mathbf{k}\cdot(\mathbf{r}_B - \mathbf{r}_B)]\varepsilon_{2p} = \varepsilon_{2p} \tag{7.15}$$

where $\mathbf{r}_1$, $\mathbf{r}_2$ and $\mathbf{r}_3$ are the nearest-neighbor vectors in real-space given by Equation (7.4), and ε_{2p} and γ_0 are parameters representing the integrals of the wave functions, defined as:

$$\varepsilon_{2p} = \langle X(\mathbf{r}-\mathbf{r}_A) | H | X(\mathbf{r}-\mathbf{r}_A) \rangle = \langle X(\mathbf{r}-\mathbf{r}_B) | H | X(\mathbf{r}-\mathbf{r}_B) \rangle \tag{7.16}$$

$$\gamma_0 = \langle X(\mathbf{r}-\mathbf{r}_A) | H | X(\mathbf{r}-\mathbf{r}_B) \rangle = \langle X(\mathbf{r}-\mathbf{r}_B) | H | X(\mathbf{r}-\mathbf{r}_A) \rangle \tag{7.17}$$

ε_{2p} and γ_0 are called onsite energy and hopping energy, respectively, and are given as two-center integrals using the same wave functions and the different wave functions, respectively. Substituting Equations (7.12) to (7.15) into Equation (7.11), we finally obtain the following matrix equation:

$$\begin{bmatrix} \varepsilon_{2p} & \gamma_0 f \\ \gamma_0 f^* & \varepsilon_{2p} \end{bmatrix} \begin{bmatrix} c_A \\ c_B \end{bmatrix} = \varepsilon \begin{bmatrix} c_A \\ c_B \end{bmatrix} \tag{7.18}$$

Where:

$$f = \exp(i\mathbf{k}\cdot\mathbf{r}_1) + \exp(i\mathbf{k}\cdot\mathbf{r}_2) + \exp(i\mathbf{k}\cdot\mathbf{r}_3) \tag{7.19}$$

By solving Equation (7.18) for wave vectors along the special points in the first BZ, the band structure of graphene is obtained as shown in Figure 7.2 (a), where we used the TB parameters as $\varepsilon_{2p} = 0\,\text{eV}$ and $\gamma_0 = 2.6\,\text{eV}$ [7.8] and, thus, the Fermi energy is located at $E = 0$. It is found that the dispersion relation is symmetric around zero energy in this first nearest-neighbor approximation. If the second nearest-neighbor hopping energy is considered, the electron-hole symmetry is broken and the π and π^* bands become asymmetric [7.6]. In addition, the two dispersion curves cross at the K point, indicating that graphene has no bandgap.

In Figure 7.2 (b), we show a zoom in of the band structure close to the K point. The most striking difference between this result and the usual semiconductor case is that the linear

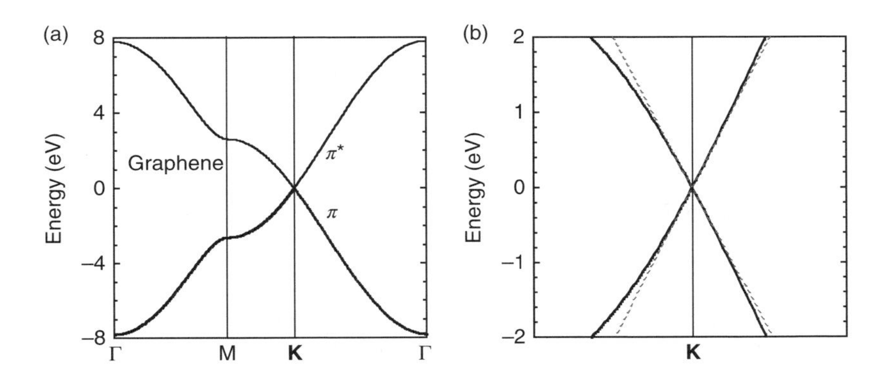

Figure 7.2 (a) Electronic band structure of graphene computed along the special points in the first BZ. (b) Zoom-in of the band structure close to the *K* point, where the dashed lines represent tangent lines at the *K* point for reference, to indicate how the dispersion relation actually holds its linearity. $E=0$ corresponds to the Fermi energy.

dispersion relation appears in graphene. From Figure 7.2 (b), we can see that the linear dispersion relation holds until about 0.7 eV in the energy region.

The analytical dispersion relation has been derived from the TB Hamiltonian, as given by [7.6, 7.9]:

$$E_{\pm}(\mathbf{k}) = \pm\ \gamma_0\sqrt{3+g(\mathbf{k})} \tag{7.20}$$

$$g(\mathbf{k}) = 2\cos\left(\sqrt{3}k_y a\right) + 4\cos\left(\frac{\sqrt{3}}{2}k_y a\right)\cos\left(\frac{3}{2}k_x a\right) \tag{7.21}$$

where the sign applies to the upper (π) and the minus sign the lower (π*) band. Equation (7.20) can be approximated by expanding it close to the **K** (or **K**′) vector, Equation (7.3), as $\mathbf{k}=\mathbf{K}+\mathbf{q}$ with $|\mathbf{q}| << |\mathbf{K}|$, as follows [7.9]:

$$E_{\pm}(\mathbf{k}) \approx \pm\ v_F\,|\mathbf{q}| + O\left[(q/K)^2\right] \tag{7.22}$$

Where **q** is the momentum measured relatively to the Dirac points and v_F is the Fermi velocity, given by $v_F=3\gamma_0 a/2$, with a value of $v_F \approx 1\times 10^6$ m/s [7.9]. Equation (7.22) indicates that the Fermi velocity does not depend on the energy or momentum whereas, in the usual semiconductor materials, we have $v = k/m = \sqrt{2E/m}$ using the electron mass m and, hence, the velocity changes substantially with energy E.

Figure 7.3 shows the atomic structure of silicene and germanene consisting of one honeycomb lattice. Germanene is the Ge equivalent of graphene. As shown in Figure 7.1 and described above, all atoms line up in the same plane in the case of graphene, whereas the structures of silicene and germanene are slightly buckled, as shown in Figure 7.3. Compared to graphene, the interatomic distance is larger in Si and Ge (*cf.* Table 7.1) and, hence, the diminished $\pi-\pi$ overlaps can no longer maintain the planar stability. Consequently, the sp^2

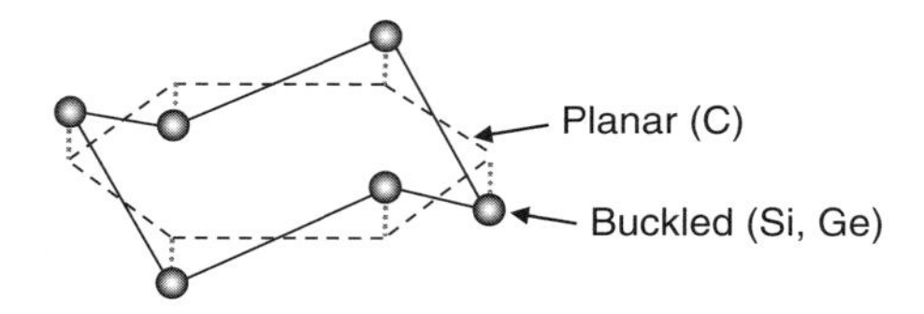

Figure 7.3 Low buckled geometry in silicene and germanene.

Table 7.1 Parameters used for TB band structure calculations. a and γ_0 are the bond lengths and the hopping energies, respectively. The evaluated Fermi velocities are also indicated.

	a (Å)	γ_0 (eV)	v_F (10^6 m/s)
graphene	1.42	2.60	0.85
silicene	2.25	1.03	0.57
germanene	2.37	1.05	0.53

hybrid orbitals are slightly dehybridized to form sp^3–like orbitals, and the stable structure for silicene and germanene results in a low buckled geometry, as shown in Figure 7.3.

Theoretical studies [7.4, 7.10, 7.11] have predicted that silicene and germanene exhibit similar linear-band dispersions near the Fermi level, and have a semi-metallic nature without a bandgap. Therefore, experimental work on the growth of silicene nanoribbon (SiNR) was performed to induce a bandgap [7.12].

The electronic and atomic structures of SiNR and germanene nanoribbon (GeNR) were also investigated, based on first-principle simulation [7.13], which has shown that the bond-length distribution is nearly uniform, except for a sudden decrease at the edges, which was the same as in armchair-edged graphene nanoribbons (A-GNRs) [7.14]. Therefore, the electronic structures of hydrogen-terminated armchair-edged SiNRs (A-SiNRs) and armchair-edged GeNRs (A-GeNRs) have been successfully reproduced by a first nearest-neighbor TB approach with different TB parameters at the edges [7.13]. Accordingly, we employ the first nearest-neighbor TB approach mentioned above to the band structure calculations of silicene and germanene in this section, and also of A-SiNRs and A-GeNRs in Section 7.4.

Table 7.1 shows the bond lengths (a) and the hopping energies (γ_0) obtained by fitting to the first-principles results [7.13], where the parameters for graphene [7.14] are also included. Figure 7.4(a) shows the band structures of graphene, silicene and germanene computed along the special points in the first BZ using the first-nearest TB approach. As expected, the dispersion curves cross at the K point (i.e., silicene and germanene have no bandgap, similar to graphene).

Similar results have been reported using first-principle simulations in [7.4, 7.10, 7.11]. The Fermi velocities are calculated from the results in Figure 7.4, as shown in Table 7.1, which correspond well to the equation $v_F = 3\gamma_0 a/2$ derived from the analytical dispersion relation in Equation (7.20). In Figure 7.4(b), we show a zoom in of the band structures close to the K point, which indicates that the linear dispersion relation in silicene and germanene is broken at a lower energy than it is in graphene.

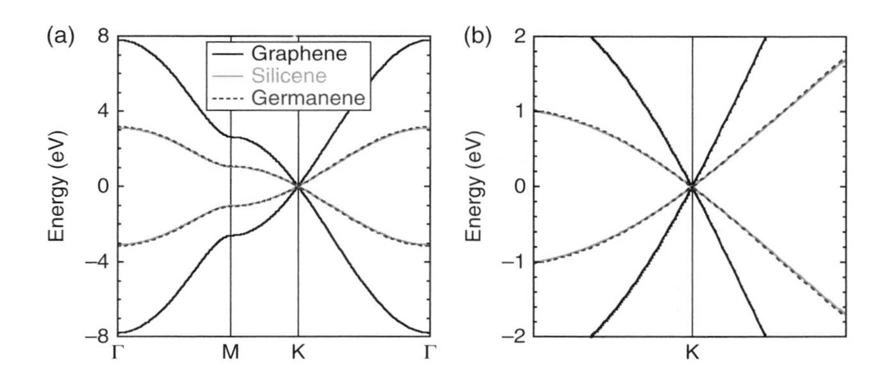

Figure 7.4 (a) Electronic band structure of graphene, silicene and germanene, computed along the special points in the first BZ. (b) Zoom-in of the band structures close to the K point. It is found that the linear dispersion relationship in silicene and germanene is broken at a lower energy than in graphene.

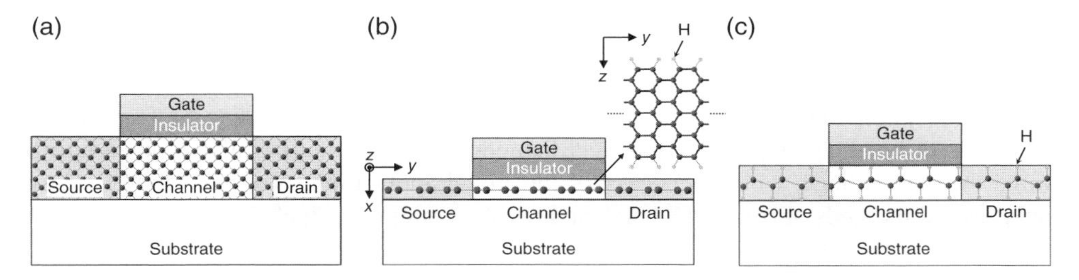

Figure 7.5 Schematics of (a) conventional Si MOSFET; (b) graphene nanoribbon FET; and (c) germanane FET [7.15]. The generation of dangling bonds at the surfaces is one of the major drawback in a conventional Si MOSFET.

7.1.2 Features of 2-D Materials as an FET Channel

Figure 7.5 shows the schematics of conventional Si MOSFET and two examples of 2-D material MOSFETs – that is, graphene nanoribbon FET and germanane FET [7.15]. In conventional Si MOSFETs, it is very difficult to scale down bulk silicon thickness to a few nanometers, since there exist dangling bonds at the surfaces of the silicon channel, as shown in Figure 7.5(a). As a result, the realization of atomically flat channel surfaces is impractical.

In other words, channel thickness fluctuation is inevitable in Si MOSFETs, and it causes a new type of surface roughness scattering in a few nanometer thick silicon channels [3.24]. On the other hand, 2-D materials such as graphene [7.1], silicene [7.5] and germanene are atomically flat and have no dangling bonds at the surfaces, owing to the robust sp^2 hybridization, as shown in Figure 7.5(b). However, such atomically flat materials have no bandgap themselves, as shown in Figure 7.4 and, hence, a nanoscale processing technique, such as a nanoribbon structure [7.12, 7.16, 7.17], shown in Figure 7.5(b), is needed to open a bandgap for switching devices.

In addition, alternative 2-D materials such as silicane and germanane [7.18] are hydrogen-terminated Si and Ge monolayers, as shown in Figure 7.5(c), and have an analogous geometry to a Si / Ge (111) surface in which every Si / Ge atom is terminated with hydrogen (H) atoms above or below the layer. Thus, no dangling bonds are generated at the surfaces. The atomic structure of germanane and its electronic properties will be discussed in Section 7.7.

Up until now, a lot of theoretical studies, in parallel with experimental efforts, have been performed to the understanding of the material properties of 2-D materials, but the device performance evaluation of 2-D material FETs is not yet sufficient. In view of practical use, 2-D materials from group IV elements may have better integration and process or materials compatibility with the present Si-platform, although transition metal dichalcogenide 2-D materials, typified by MoS_2 [7.19–7.21] and black phosphorus [7.22, 7.23] have also been actively researched, since they has a large bandgap of about 1.6 eV. Therefore, in this chapter, we intend to analyze atomic monolayer semiconductor FETs composed of Si, Ge and C elements, and to examine their upper limit performance.

7.2 Graphene Nanostructures with a Bandgap

As described in Section 7.1, graphene, silicene and germanene are expected to be candidate materials for use in the channels of next-generation FETs, because of their extremely high mobility and Fermi velocity, excellent gate controllability, absence of dangling bonds at the surface, and so on. However, an important drawback of graphene regarding FET application is the lack of bandgap. Therefore, the electrical conduction cannot be fully switched off by tuning the gate voltage, which is necessary for digital applications. To overcome this drawback, several methods of opening a bandgap have been proposed, including:

(i) patterning monolayer graphene into nanoribbons (graphene nanoribbons: GNRs) [7.16, 7.17]; and
(ii) utilizing bilayer graphene (BLG) under a vertical electric field [7.24] or inducing symmetry breaking between two carbon layers via the interaction between a graphene layer and its substrate [7.25, 7.26].

Previous theoretical calculations have shown that armchair-edged GNRs (A-GNRs) exhibit larger bandgaps than those of BLGs applied by a vertical electric field and, more interestingly, they have quite different dispersion relations in their low-energy spectra. For instance, in BLGs, a Mexican hat structure with a negative effective mass appears even under a relatively small vertical electric field [7.8, 7.27–7.29]. On the other hand, A-GNRs still sustain a linear dispersion relation until the opened bandgap energy increases up to several hundred millielectronvolts by decreasing the ribbon width [7.14, 7.29–7.33]. Therefore, superior carrier transport in A-GNRs has been predicted theoretically [7.29]. An alternative GNR – that is, zigzag-edged GNR (Z-GNR) – has also been investigated theoretically. In Z-GNRs, however, localized edge states appear in the bandgap [7.34–7.39]. Thus, they may not be suitable for FET application [7.30].

As mentioned above, the electronic states of A-GNRs and BLGs highly depend on geometrical configurations and operating conditions; thus, systematic investigation is needed to understand their relative advantages for use in FET channels. With the recent progress in atomically precise fabrication techniques both for GNRs [7.40] and BLGs [7.41], it becomes important to clarify the intrinsic effects of the bandgap opening on electron transport in

A-GNR-FETs and BLG-FETs. In this section, we address the subject and investigate systematically the ultimate device performances of both FETs, based on atomistic and ballistic simulation approaches.

7.2.1 Armchair-edged Graphene Nanoribbons (A-GNRs)

Figure 7.6 shows the atomic structure of hydrogen-terminated A-GNR, where N represents the number of atoms in the width direction. To take the influence of the bandgap opening on the electron transport into account, we first calculate the electronic band structures of A-GNRs, by applying the nearest-neighbor TB approach described in Section 7.1 to the unit cell of A-GNR shown in Figure 7.6.

First, Figure 7.7 shows the detail of the unit cell and the first BZ of A-GNRs [7.15]. Since A-GNR is a 1-D crystal, its primitive lattice vector is defined by **a** as shown in Figure 7.7 (a), and it is given by:

$$\mathbf{a} = (3a, 0, 0) \tag{7.23}$$

The corresponding primitive reciprocal lattice vector **b** is obtained using the relationship of $\mathbf{a}\cdot\mathbf{b}=2\pi$ as:

$$\mathbf{b} = 2\pi\left(\frac{1}{3a}, 0, 0\right) \tag{7.24}$$

Accordingly, the first BZ of A-GNRs also becomes 1-D, as shown in Figure 7.7(b). Here, the point M is a midpoint of the **b** vector, whose coordinates in the k-space are $M : \pi(1/3a, 0, 0)$.

The atom pair in the unit cell, A and B, is labeled using $C1$, $C2$, ... C_N for the carbon atoms, and using H_l and H_u for the hydrogen atoms, as shown in Figure 7.7(a). Furthermore, the atoms A and B in the C_i group are expressed as C_{iA} and C_{iB}, which are separated from the nearest-neighbor atoms by a spatial vector, $\mathbf{r}_1$, $\mathbf{r}_2$, $\mathbf{r}_3$ or $-\mathbf{r}_1$, $-\mathbf{r}_2$, $-\mathbf{r}_3$. Based on the above notations, the TB Hamiltonian for A-GNRs is formulated below.

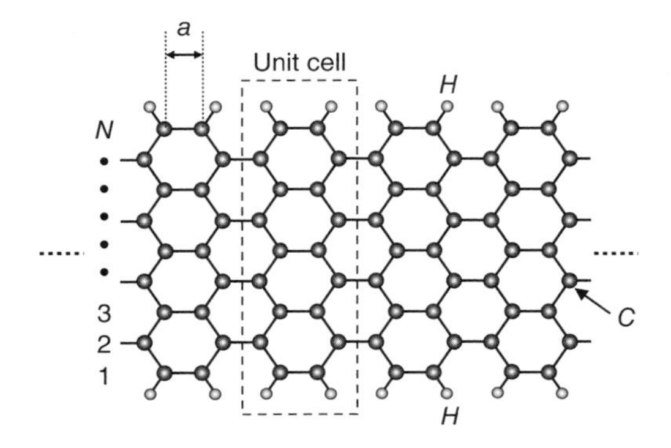

Figure 7.6 Atomic structure of hydrogen-terminated A-GNRs, where hydrogen atoms inactivate dangling bonds generated at the edges. N represents the number of atoms in the width direction, and a the interatomic distance. A unit cell is depicted by the dashed lines.

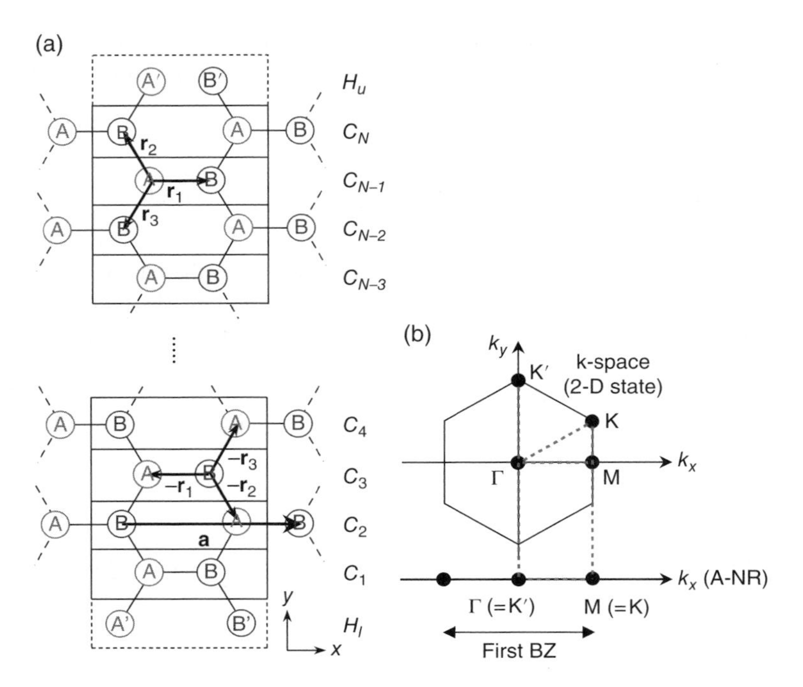

Figure 7.7 (a) Unit cell of A-GNRs. The symbols A and B represent carbon atoms, whereas A' and B' at the top and bottom groups represent hydrogen atoms. **a** is a primitive lattice vector. (b) first BZ of A-GNRs.

When the number of atoms in the width direction is N, the electron wave function inside the unit cell is written as a linear superposition of the π-electron p_z orbitals, $\psi_{iX}(i=1,2,\cdots,N;X=A, B)$, coming from all atoms as follows:

$$\Psi = c_{1A}\psi_{1A} + c_{1B}\psi_{1B} + c_{2A}\psi_{2A} + c_{2B}\psi_{2B} + \cdots + c_{NA}\psi_{NA} + c_{NB}\psi_{NB} \tag{7.25}$$

where $c_{iX}(i=1,2,\cdots,N;X=A, B)$ is the amplitude of each orbital. In this case, the Schrödinger equation and the TB Hamiltonian H' are, respectively, given by:

$$H'\begin{bmatrix} c_{1A} \\ c_{1B} \\ c_{2A} \\ c_{2B} \\ \vdots \\ c_{N-1A} \\ c_{N-1B} \\ c_{NA} \\ c_{NB} \end{bmatrix} = \varepsilon \begin{bmatrix} c_{1A} \\ c_{1B} \\ c_{2A} \\ c_{2B} \\ \vdots \\ c_{N-1A} \\ c_{N-1B} \\ c_{NA} \\ c_{NB} \end{bmatrix} \tag{7.26}$$

and

$$
H' = \begin{array}{c} \\ \langle 1A| \\ \langle 1B| \\ \langle 2A| \\ \langle 2B| \\ \vdots \\ \langle N-1A| \\ \langle N-1B| \\ \langle NA| \\ \langle NB| \end{array}
\begin{array}{ccccc} |1A\rangle & |1B\rangle & |2A\rangle & |2B\rangle & \cdots \\
H_{1A,1A} & H_{1A,1B} & H_{1A,2A} & H_{1A,2B} & \cdots \\
H_{1B,1A} & H_{1B,1B} & H_{1B,2A} & H_{1B,2B} & \cdots \\
H_{2A,1A} & H_{2A,1B} & H_{2A,2A} & H_{2A,2B} & \cdots \\
H_{2B,1A} & H_{2B,1B} & H_{2B,2A} & H_{2B,2B} & \cdots \\
\vdots & \vdots & \vdots & \vdots & \ddots \\
H_{N-1A,1A} & H_{N-1A,1B} & H_{N-1A,2A} & H_{N-1A,2B} & \cdots \\
H_{N-1B,1A} & H_{N-1B,1B} & H_{N-1B,2A} & H_{N-1B,2B} & \cdots \\
H_{NA,1A} & H_{NA,1B} & H_{NA,2A} & H_{NA,2B} & \cdots \\
H_{NB,1A} & H_{NB,1B} & H_{NB,2A} & H_{NB,2B} & \cdots
\end{array}
$$

$$
\begin{array}{c} \\ \langle 1A| \\ \langle 1B| \\ \langle 2A| \\ \langle 2B| \\ \vdots \\ \langle N-1A| \\ \langle N-1B| \\ \langle NA| \\ \langle NB| \end{array}
\begin{array}{ccccc} \cdots & |N-1A\rangle & |N-1B\rangle & |NA\rangle & |NB\rangle \\
\cdots & H_{1A,N-1A} & H_{1A,N-1B} & H_{1A,NA} & H_{1A,NB} \\
\cdots & H_{1B,N-1A} & H_{1B,N-1B} & H_{1B,NA} & H_{1A,NB} \\
\cdots & H_{2A,N-1A} & H_{2A,N-1B} & H_{2A,NA} & H_{1A,NB} \\
\cdots & H_{2B,N-1A} & H_{2B,N-1B} & H_{2B,NA} & H_{1A,NB} \\
\ddots & \vdots & \vdots & \vdots & \vdots \\
\cdots & H_{N-1A,1B} & H_{N-1A,N-1B} & H_{N-1A,NA} & H_{N-1A,NB} \\
\cdots & H_{N-1B,N-1A} & H_{N-1B,N-1B} & H_{N-1B,NA} & H_{N-1B,NB} \\
\cdots & H_{NA,N-1A} & H_{NA,N-1B} & H_{NA,NA} & H_{NA,NB} \\
\cdots & H_{NB,N-1A} & H_{NB,N-1B} & H_{NB,NA} & H_{NB,NB}
\end{array} \tag{7.27}
$$

The TB matrix element $H_{iX,\,jY}(i=1,2,\cdots,N;X=A,\,B)$ represents $<\psi_{iX} \mid H \mid \psi_{jY}>$, which will be expressed using the onsite energy ε_{2p} and the hopping energy γ_0, defined in Equations (7.16) and (7.17), below.

First, we consider the two atoms, A and B, belonging to the bottom carbon group C_1 in Figure 7.7 (a). Since the two-center integrals using the same wave functions, $<\psi_{iX} \mid H \mid \psi_{iX}>$, gives the onsite energy identical to Equation (7.16), we have the relation:

$$\langle \psi_{iX} | H | \psi_{iX} \rangle = \varepsilon_{2p} \; (i=1,2,\cdots,N; \;\; X = A,\,B) \tag{7.28}$$

In other words, Equation (7.28) is common to all carbon atoms. On the other hand, focusing on the C_{1A} atom, it has the hopping energy elements with the C_{1B} and C_{2B} atoms, while none with other atoms within the present nearest-neighbor approximation. Hence, we can write the hopping energy elements with respect to the C_{1A} atom as follows:

$$\langle \psi_{1A} | H | \psi_{1B} \rangle = \gamma_0 \exp(i\mathbf{k} \cdot \mathbf{r}_1) \tag{7.29}$$

$$\langle \psi_{1A} | H | \psi_{2B} \rangle = \gamma_0 \exp(i\mathbf{k} \cdot \mathbf{r}_2) \tag{7.30}$$

$$\langle \psi_{1A} | H | \psi_{iX} \rangle = 0 \quad (iX \neq 1A, 1B, 2B) \tag{7.31}$$

Similarly, the C_{1B} atom has the hopping energy elements only with the C_{1A} and C_{2A} atoms, as follows.

$$\langle \psi_{1B} | H | \psi_{1A} \rangle = \gamma_0 \exp(i\mathbf{k} \cdot \mathbf{r}_1)^* \tag{7.32}$$

$$\langle \psi_{1B} | H | \psi_{2A} \rangle = \gamma_0 \exp(i\mathbf{k} \cdot \mathbf{r}_3)^* \tag{7.33}$$

$$\langle \psi_{1B} | H | \psi_{iX} \rangle = 0 \quad (iX \neq 1A, 1B, 2A) \tag{7.34}$$

Next, we consider the two atoms, A and B, belonging to the top carbon group C_N. Since the nearest-neighbor atoms from the C_{NA} atom are C_{N-1B} and C_{NB}, we have the relations as:

$$\langle \psi_{NA} | H | \psi_{N-1B} \rangle = \gamma_0 \exp(i\mathbf{k} \cdot \mathbf{r}_3) \tag{7.35}$$

$$\langle \psi_{NA} | H | \psi_{NB} \rangle = \gamma_0 \exp(i\mathbf{k} \cdot \mathbf{r}_1) \tag{7.36}$$

$$\langle \psi_{NA} | H | \psi_{iX} \rangle = 0 \quad (iX \neq N-1B, NA, NB) \tag{7.37}$$

Similarly, the C_{NB} atom has the hopping energy elements with the C_{N-1A} and C_{NA} atoms, as follows:

$$\langle \psi_{NB} | H | \psi_{N-1A} \rangle = \gamma_0 \exp(i\mathbf{k} \cdot \mathbf{r}_2)^* \tag{7.38}$$

$$\langle \psi_{NB} | H | \psi_{NA} \rangle = \gamma_0 \exp(i\mathbf{k} \cdot \mathbf{r}_1)^* \tag{7.39}$$

$$\langle \psi_{NB} | H | \psi_{iX} \rangle = 0 \quad (iX \neq N-1A, NA, NB) \tag{7.40}$$

Finally, we describe the hopping energy elements for carbon atoms belonging to the inner group $c_i (1<i<N)$. The C_{iA} and C_{iB} atoms have three nearest-neighbor atoms, C_{i-1B}, C_{iB}, C_{i+1B}, and C_{i-1A}, C_{iA}, C_{i+1A}, respectively, and thus they have the following hopping energy elements:

$$\langle \psi_{iA} | H | \psi_{i-1B} \rangle = \gamma_0 \exp(i\mathbf{k} \cdot \mathbf{r}_3) \tag{7.41}$$

$$\langle \psi_{iA} | H | \psi_{iB} \rangle = \gamma_0 \exp(i\mathbf{k} \cdot \mathbf{r}_1) \tag{7.42}$$

$$\langle \psi_{iA} | H | \psi_{i+1B} \rangle = \gamma_0 \exp(i\mathbf{k} \cdot \mathbf{r}_2) \tag{7.43}$$

$$\langle \psi_{iA} | H | \psi_{jX} \rangle = 0 \quad (jX \neq i-1B, iA, iB, i+1B) \tag{7.44}$$

and

$$\langle \psi_{iB} | H | \psi_{i-1A} \rangle = \gamma_0 \exp(i\mathbf{k} \cdot \mathbf{r}_2)^* \quad (7.45)$$

$$\langle \psi_{iB} | H | \psi_{iA} \rangle = \gamma_0 \exp(i\mathbf{k} \cdot \mathbf{r}_1)^* \quad (7.46)$$

$$\langle \psi_{iB} | H | \psi_{i+1A} \rangle = \gamma_0 \exp(i\mathbf{k} \cdot \mathbf{r}_3)^* \quad (7.47)$$

$$\langle \psi_{iB} | H | \psi_{jX} \rangle = 0 \quad (jX \neq i-1A, iA, iB, i+1A) \quad (7.48)$$

Substituting Equations (7.28) to (7.48) into Equation (7.27) and expressing the phase factor as $p_i = \exp(i\mathbf{k} \cdot \mathbf{r}_i)$ $(i = 1, 2, 3)$ for simplicity, the TB Hamiltonian H' is rewritten by:

$$H' = \begin{array}{c|ccccccccc} & |1A\rangle & |1B\rangle & |2A\rangle & |2B\rangle & \cdots & |N-1A\rangle & |N-1B\rangle & |NA\rangle & |NB\rangle \\ \hline \langle 1A| & \varepsilon_{2p} & \gamma_0 p_1 & 0 & \gamma_0 p_2 & \cdots & 0 & 0 & 0 & 0 \\ \langle 1B| & \gamma_0 p_1^* & \varepsilon_{2p} & \gamma_0 p_3^* & 0 & \cdots & 0 & 0 & 0 & 0 \\ \langle 2A| & 0 & \gamma_0 p_3 & \varepsilon_{2p} & \gamma_0 p_1 & \cdots & 0 & 0 & 0 & 0 \\ \langle 2B| & \gamma_0 p_2^* & 0 & \gamma_0 p_1^* & \varepsilon_{2p} & \cdots & 0 & 0 & 0 & 0 \\ \vdots & \vdots & \vdots & \vdots & \vdots & \ddots & \vdots & \vdots & \vdots & \vdots \\ \langle N-1A| & 0 & 0 & 0 & 0 & \cdots & \varepsilon_{2p} & \gamma_0 p_1 & 0 & \gamma_0 p_2 \\ \langle N-1B| & 0 & 0 & 0 & 0 & \cdots & \gamma_0 p_1^* & \varepsilon_{2p} & \gamma_0 p_3^* & 0 \\ \langle NA| & 0 & 0 & 0 & 0 & \cdots & 0 & \gamma_0 p_3 & \varepsilon_{2p} & \gamma_0 p_1 \\ \langle NB| & 0 & 0 & 0 & 0 & \cdots & \gamma_0 p_2^* & 0 & \gamma_0 p_1^* & \varepsilon_{2p} \end{array} \quad (7.49)$$

7.2.2 *Relaxation Effects of Edge Atoms*

According to the structure relaxation analysis based on first-principles calculation, the interatomic distance a is decreased to a_{edge} at the edges by the hydrogen termination, as shown in Figure 7.8 and, correspondingly, the hopping energy is increased to γ_0^{edge}, to reflect that edge bonds are stronger than those in the inner bonds [7.13, 7.14].

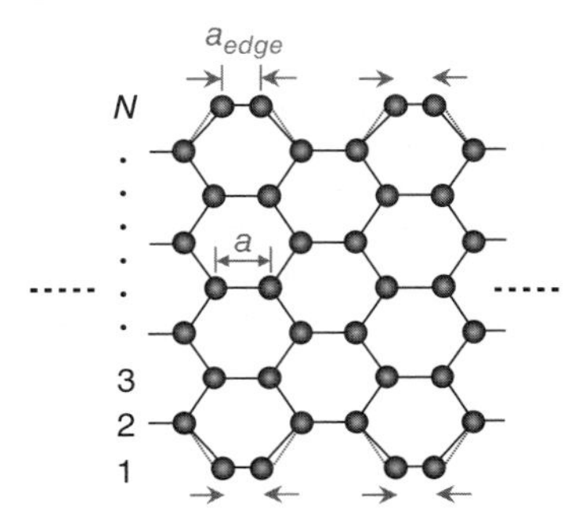

Figure 7.8 Edge relaxation effects in A-NRs due to hydrogen termination of edge atoms.

For instance, carbon-carbon bonds at the edges of A-GNRs, which are also bonded to hydrogen atoms to terminate dangling bonds, are about 3.5% shorter than other carbon-carbon bonds inside the ribbon [7.14]. Therefore, these bond lengths have been determined to be 0.137 nm. Furthermore, to fit the first-principles band structure results, a different TB parameter of $\gamma_0^{edge} = 1.12\,\gamma_0$ is used for the edge bonds [7.14, 7.42].

Such relaxation effects of the edge atoms can be taken into account in the TB approach by introducing γ_0^{edge} and $p_1' = exp(i\mathbf{k}\cdot\mathbf{r}_1')$ with replacing $\mathbf{r}_1 \rightarrow \mathbf{r}_1'$ in the x-direction. As a result, the TB Hamiltonian considering the edge effects is given by:

$$
H' = \begin{array}{c} \\ \langle 1A| \\ \langle 1B| \\ \langle 2A| \\ \langle 2B| \\ \vdots \\ \langle N-1A| \\ \langle N-1B| \\ \langle NA| \\ \langle NB| \end{array}
\begin{array}{c} |1A\rangle \quad |1B\rangle \quad |2A\rangle \quad |2B\rangle \quad \cdots \quad |N-1A\rangle \quad |N-1B\rangle \quad |NA\rangle \quad |NB\rangle \\
\left[\begin{array}{ccccccccc}
\varepsilon_{2p} & \gamma_0^{edge} p_1' & 0 & \gamma_0 p_2 & \cdots & 0 & 0 & 0 & 0 \\
\gamma_0^{edge} p_1'^{*} & \varepsilon_{2p} & \gamma_0 p_3^{*} & 0 & \cdots & 0 & 0 & 0 & 0 \\
0 & \gamma_0 p_3 & \varepsilon_{2p} & \gamma_0 p_1 & \cdots & 0 & 0 & 0 & 0 \\
\gamma_0 p_2^{*} & 0 & \gamma_0 p_1^{*} & \varepsilon_{2p} & \cdots & 0 & 0 & 0 & 0 \\
\vdots & \vdots & \vdots & \vdots & \ddots & \vdots & \vdots & \vdots & \vdots \\
0 & 0 & 0 & 0 & \cdots & \varepsilon_{2p} & \gamma_0 p_1 & 0 & \gamma_0 p_2 \\
0 & 0 & 0 & 0 & \cdots & \gamma_0 p_1^{*} & \varepsilon_{2p} & \gamma_0 p_3^{*} & 0 \\
0 & 0 & 0 & 0 & \cdots & 0 & \gamma_0 p_3 & \varepsilon_{2p} & \gamma_0^{edge} p_1' \\
0 & 0 & 0 & 0 & \cdots & \gamma_0 p_2^{*} & 0 & \gamma_0^{edge} p_1'^{*} & \varepsilon_{2p}
\end{array}\right]
\end{array}
\tag{7.50}
$$

The TB parameters used in the present study are summarized in Table 7.2, which were obtained by fitting to the first-principles results [7.13, 7.14].

Figure 7.9 shows the band structures of A-GNRs with N=12, 13, and 14, computed by the TB approach considering the edge effects and first-principles method [7.14]. The TB results exhibit very good agreement with the first-principles method, especially in the low-energy region. This is due to the first-nearest approximation that we employed.

To improve the accuracy in high-energy region, it is well known that a third-nearest TB approach is necessary [7.43, 7.44]. However, as found from Figure 7.9, the first-nearest TB approach is sufficient to discuss the electron transport under the small bias condition considered in this study. Incidentally, the bandgap underestimation problem, well known for first-principles simulations based on the density-functional theory (DFT) (see Section 2.1.1),

Table 7.2 TB parameters used in the present study. a (γ_0) and a_{edge} (γ_0^{edge}) are the interatomic distances (hopping energies) in the inner region and at the edges, respectively.

	a (Å)	a_{edge} (Å)	γ_0 (eV)	γ_0^{edge} (eV)
A-GNR	1.42	1.37	2.60	2.91
A-SiNR	2.25	2.20	1.03	1.15
A-GeNR	2.37	2.32	1.05	1.13

is much less serious for GNRs, because the bandgap of GNRs is caused by quantum confinement of the gapless p_z orbital bands of graphene. This is well described by the DFT, unlike the bandgap originating from lattice potential as in Si [7.33, 7.45].

7.2.3 Electrical Properties of A-GNR-FETs Under Ballistic Transport

We first examine the effect of edge bond relaxation on the bandgap and the electron effective mass [7.33]. Figures 7.10 and 7.11 show the bandgap energy and the electron effective mass as a function of ribbon width, respectively. The effective mass was calculated by performing

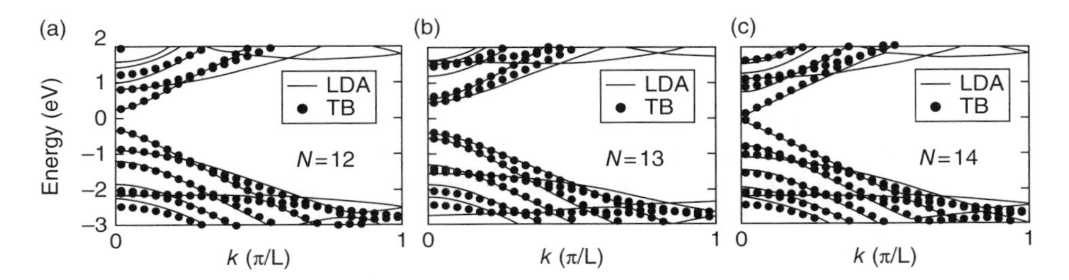

Figure 7.9 Band structures computed for A-GNRs with N=(a) 12, (b) 13, and (c) 14, using TB approach with edge effects and first-principles method within local density approximation (LDA) [7.14].

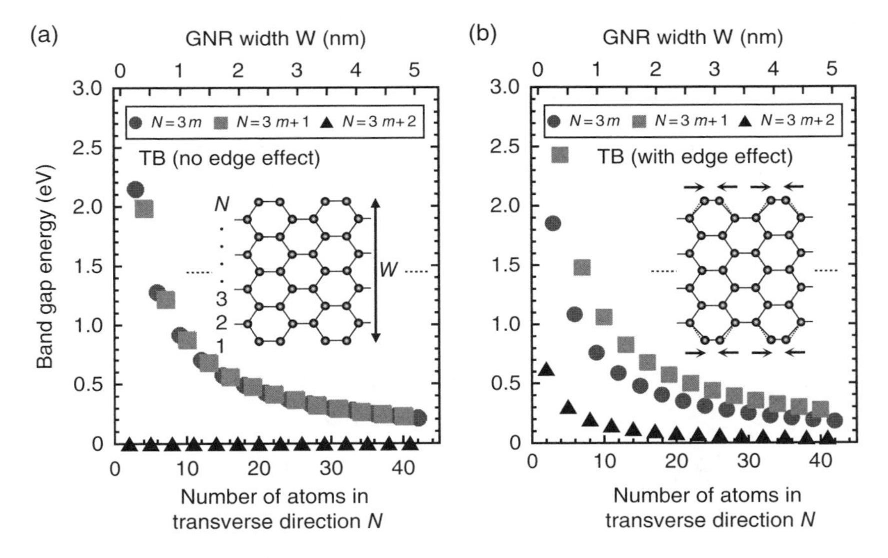

Figure 7.10 Bandgap energy computed as a function of ribbon width, where (a) and (b) correspond to the results without and with the presence of edge bond relaxation, respectively, and m in the legend represents a positive integer. The lower and upper horizontal axes denote number of carbon atoms in transverse direction N and actual ribbon width W in units of nm, respectively. The insets depict the atomic models of A-GNRs used in the simulation. Note that in (b), the carbon-carbon bond lengths at the edges are shortened by about 3.5%, compared with other carbon-carbon bonds inside the ribbon.

a second order differential of the *E*-*k* dispersion relation for the first conduction subband minimum with respect to wave number. Here, (a) and (b) correspond to the results without and with the presence of edge bond relaxation, respectively. The insets in Figure 7.10 depict the atomic models of A-GNRs used in the simulation.

Note that in Figure 7.10(b), the carbon-carbon bond lengths at the edges are shortened by about 3.5%, compared to other carbon-carbon bonds inside the ribbon. As reported in [7.14], the edge bond relaxation has a significant influence on the bandgap as shown in Figure 7.10. In fact, without the edge bond relaxation, A-GNRs are metallic if $N=3m+2$ (where m is a positive integer) – or, otherwise, semiconducting, as shown in Figure 7.10(a). However, when the edge bond relaxation is considered, there are no metallic nanoribbons and, furthermore, the bandgaps are well separated into three different groups, as shown in Figure 7.10(b). Such a change in the bandgap hierarchy mentioned above has been attributed to the 12 % increase of the hopping integrals between carbon atoms at the edges – that is, $\gamma_0^{edge} = 1.12\,\gamma_0$ for the edge bonds [7.14, 7.33, 7.42].

Next, as shown in Figure 7.11 the edge bond relaxation has also a significant influence on the electron effective mass. The results for $N=3m+2$ are omitted in Figure 7.11, because effective mass can not be defined in that case due to their linear dispersion relation. Due to the edge bond relaxation, the effective mass decreases in $3m$ A-GNRs and, by contrast, increases in $(3m+1)$ A-GNRs, as shown in Figure 7.11(b). Consequently, $3m$ A-GNRs seem to be favorable for use in channels of FETs. However, the bandgap energy also decreases (increases) in $3m$ A-GNRs [$(3m+1)$ A-GNRs], as shown in Figure 7.10(b), so a comparison of the transport properties under the same bandgap energy is important in terms of the FET application.

To this end, we focus on two pairs of A-GNRs with $N=6$ and 10, which have $E_G \approx 1.1$ eV, and A-GNRs with $N=12$ and 19, which have $E_G \approx 0.6$ eV. Figure 7.12 shows the band

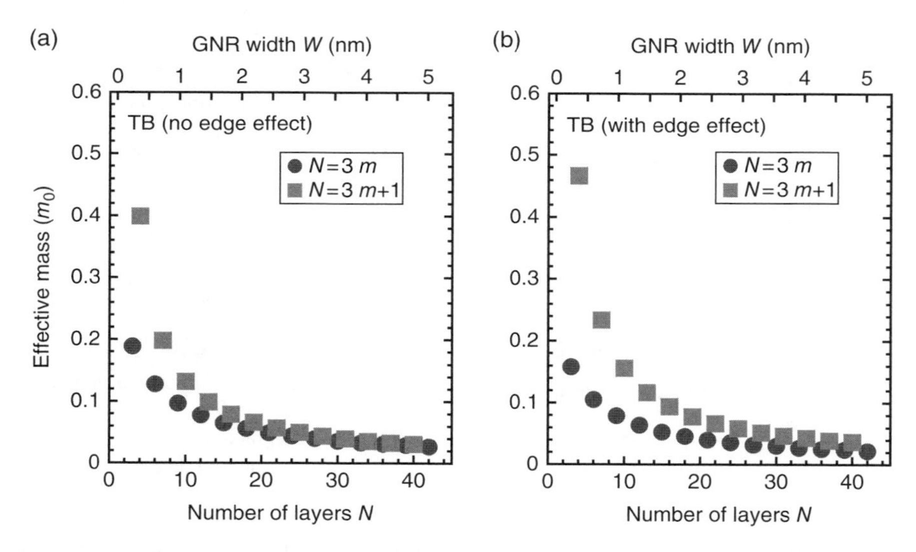

Figure 7.11 Effective mass at conduction band minimum computed as a function of ribbon width, where (a) and (b) correspond to the results without and with the presence of edge bond relaxation, respectively. Note that the results for $N=3m+2$ are omitted, because effective mass can not be defined in that case, due to their linear dispersion relation.

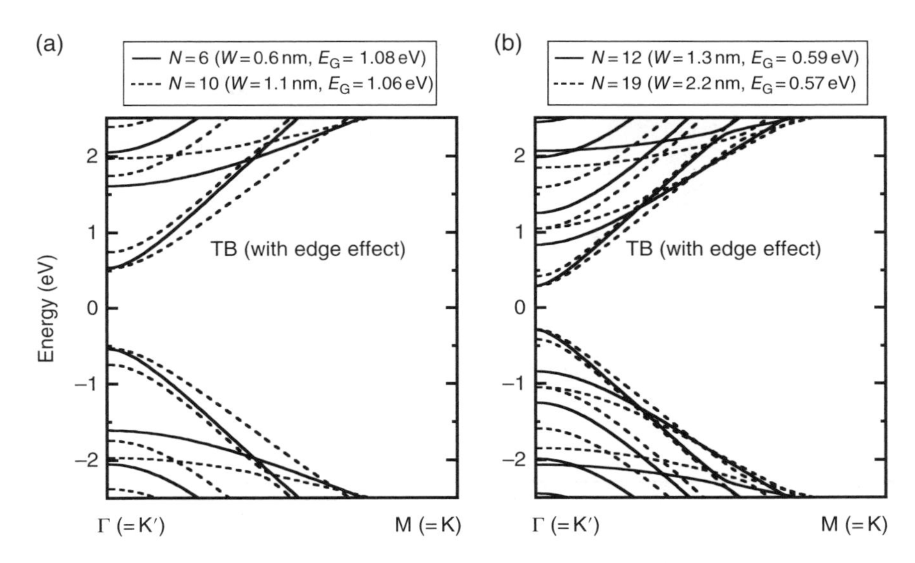

Figure 7.12 Band structures of A-GNRs with (a) $N=6$ and 10 ($E_G \approx 1.1\,\text{eV}$); and (b) $N=12$ and 19 ($E_G \approx 0.6\,\text{eV}$).

structures computed for the two pairs of A-GNRs, where the edge bond relaxation is considered. It is found that the A-GNRs with $N=3m$ family have a steeper slope than those with $N=3m+1$ family, though almost the same bandgap energies are created. As a result, smaller effective masses are obtained for the $N=3m$ family, even under the condition of the same bandgap energy. They are $m=0.105\,m_0$ ($0.156\,m_0$) for $N=6$ ($N=10$), and $0.064\,m_0$ ($0.078\,m_0$) for $N=12$ ($N=19$). Then, we will next discuss the electrical characteristics of A-GNR-FETs by fully considering the atomistic band structures.

To directly examine the influences of the atomistic band structures on the device performances, we compute the electrical characteristics under ballistic transport, based on the top-of-the-barrier (ToB) model introduced in Section 6.3.2 [6.33]. This model has been proven to be suitable for a systematic study comparing the performance limits of atomistic transistors.

Figure 7.13 shows the schematic diagram of the simulated A-GNR-FETs, where the source and drain are assumed to be heavily doped A-GNR contacts, while the channel is intrinsic. Figure 7.14 shows the drain current versus gate voltage ($I_D - V_G$) characteristics and the intrinsic device delays versus the ON-OFF current ratio, where the intrinsic device delay is calculated as $\tau=(Q_{ON}-Q_{OFF})/I_{ON}$, where Q_{ON} and Q_{OFF} are the total charge in the channel at on-state and off-state, respectively, and I_{ON} is the on-current. The gate insulator (SiO_2) thickness is $T_{ox}=0.5\,\text{nm}$ and temperature is 300 K. The drain voltage is set sufficiently small ($V_D=0.4\,\text{V}$), so band-to-band tunneling is ignored in this study.

To make a reasonable comparison between different device architectures, we used a technique to consider both the on-state and off-state at the same power supply voltage in Figure 7.14(b), where the curves were obtained by sweeping a constant V_{DD} (= 0.4 V)-bias window along the V_G axis in the $I_D - V_G$ characteristics [7.46]. As expected from the difference in the band structures between the $N=3m$ and $3m+1$ families mentioned before, the drain current (the delay) is always larger (smaller) in the $N=3m$ family, if the same bandgap FETs

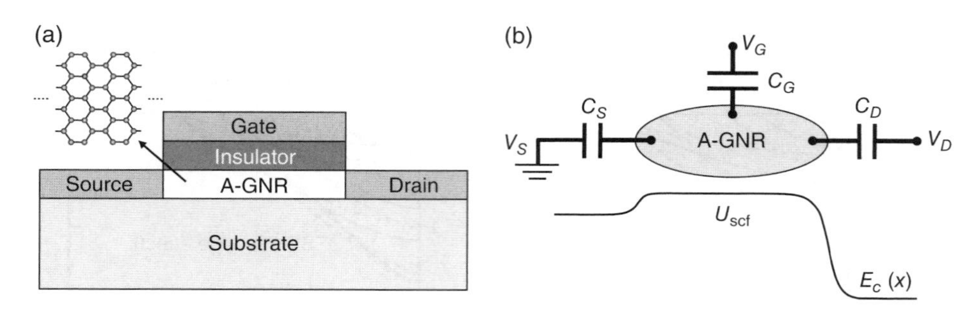

Figure 7.13 (a) Schematic diagram of the simulated A-GNR-FETs, where the source and drain are assumed to be heavily doped A-GNR contacts while the channel is intrinsic. The gate insulator was assumed to be SiO_2. (b) ToB model for A-GNR-FETs represented using three capacitances.

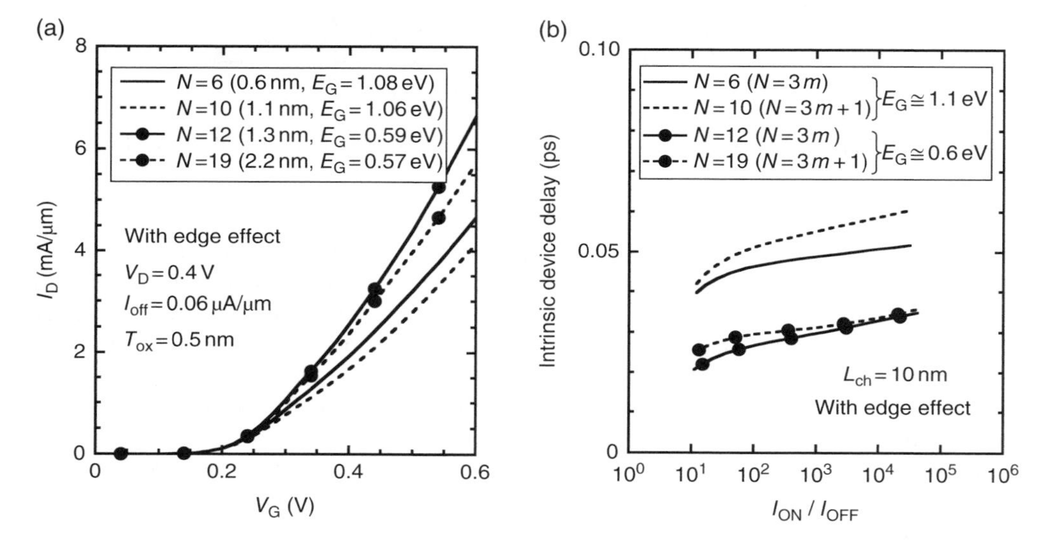

Figure 7.14 (a) $I_D - V_G$ characteristics and (b) intrinsic device delays versus I_{ON}/I_{OFF} ratio for $N=6$, 10, 12 and 19. $V_D=0.4$ V and $T_{ox}=0.5$ nm. In (a), the simulations were performed at the same OFF-current density ($I_{off}=0.06$ μA/μm). In (b), the channel length is assumed to be 10 nm. Temperature is 300 K.

are compared. Similar results have been obtained for a thicker gate insulator with $T_{ox}=1.5$ nm. Therefore, A-GNR-FETs with the $N=3m$ family are preferable as a high-performance digital switch in future logic circuits [7.33].

On the other hand, the subthreshold leakage current due to the source-drain direct tunneling and band-to-band tunneling can increase more rapidly in $3m$ A-GNR-FETs with smaller bandgaps, as the channel length decreases. From that perspective, superior off-state performance may be expected in $(3m+1)$ A-GNR-FETs with larger bandgaps in the case of sub-10 nm channel length [7.47].

In summary, using the TB band structure calculation and the ballistic transport model, it is found that the edge bond relaxation in A-GNRs has a significant influence, not only on the bandgap opening due to the quantum confinement, but also on the determination of electron effective mass. In particular, the edge bond relaxation was found to decrease (increase) the electron effective mass of $3m$ A-GNRs [$(3m+1)$ A-GNRs]. Thus, we further performed a ballistic MOSFET simulation using the atomistic band structures, and demonstrated that $3m$ A-GNRs exhibit superior device performances over $(3m+1)$ A-GNRs, in principle.

7.2.4 Bilayer Graphenes (BLGs)

A bilayer graphene (BLG) consists of two atomic carbon sheets which are spatially separated by a distance d. To understand the geometrical features of BLGs, we briefly summarize several stacking structures of 3-D graphite. As shown in Figure 7.15, a graphite can take several stabilized structures such as AA stacking, AB stacking (or Bernal stacking), ABC stacking (or Rhombohedral stacking), and so on. Among these, the AB stacking is more stable than the other structures and, hence, the bilayer structure with the AB stacking is usually considered.

In this section, we also examine the bilayer structure with the AB stacking. BLGs are interesting because they have a small effective mass, and also a bandgap can open between the conduction and valence bands, as demonstrated below.

Figure 7.16 shows the atomic structure and the unit cell of the BLG with the AB stacking. The right figure shows the in-plane atomic structure as seeing from the top. The BLG has the same periodic structure as that of graphene, and thus possesses the primitive lattice vectors and the first BZ, identical to those in Section 7.1.1. However, unlike graphene, there are two kinds of atomic sheets, vertically separated by the distance d in the unit cell, where $d = 3.354\,\text{Å}$ in BLG with the AB stacking. In other words, it has the A and B sublattice system – the A (B) sublattices are vertically overlapped (non-overlapped). Therefore, the BLG has four hopping energy parameters – γ_0, γ_1, γ_3 and γ_4 [7.6, 7.8] – as shown in Figure 7.17.

As mentioned above, there are four carbon atoms inside the unit cell of the BLG and, hence, the electron wave function inside the unit cell is written as:

$$\Psi = c_A \psi_A + c_B \psi_B + c_{A'} \psi_{A'} + c_{B'} \psi_{B'} \tag{7.51}$$

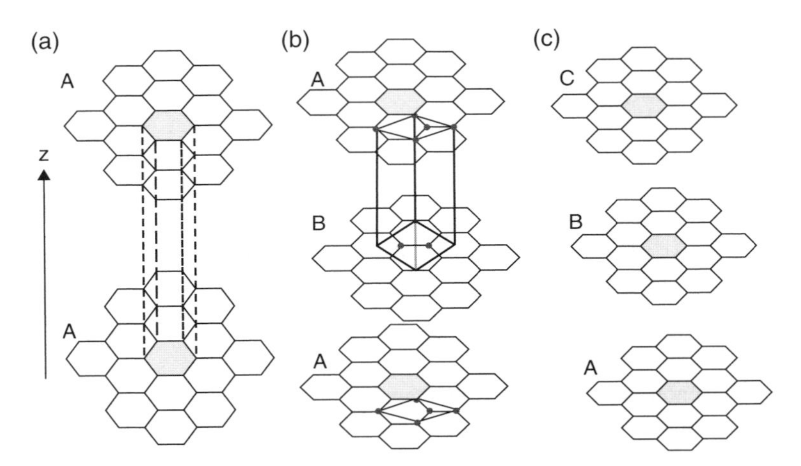

Figure 7.15 Representative stacking structures of 3-D graphite. (a) AA stacking; (b) AB stacking (or Bernal stacking); (c) ABC stacking (or Rhombohedral stacking).

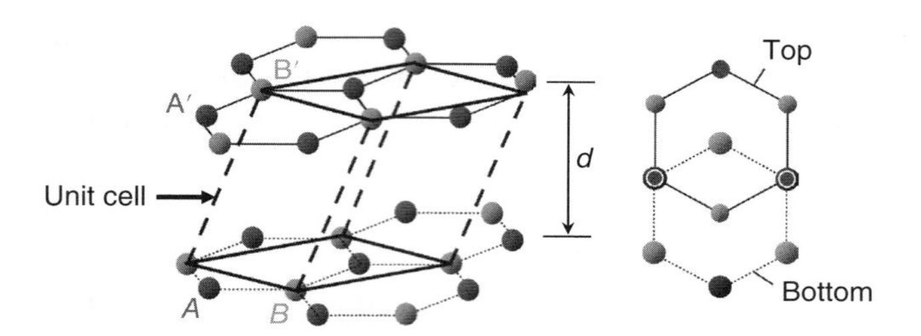

Figure 7.16 Atomic structure and unit cell of BLG with the AB stacking. The right figure shows the in-plane atomic structure as seeing from the top. It has the A and B sublattice system – that is, the A (B) sublattices are vertically overlapped (non-overlapped). $d = 3.354$ Å is the interlayer distance.

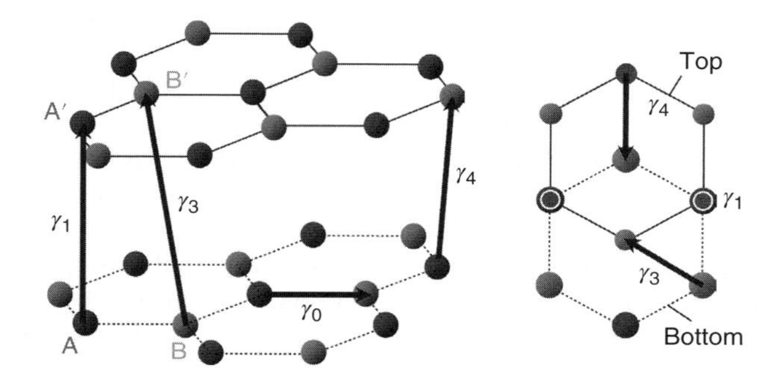

Figure 7.17 Four hopping energy parameters in BLG, γ_0, γ_1, γ_3 and γ_4 [7.6, 7.8].

where c_X ($X = A, B, A', B'$) is the amplitude of each orbital. In this case, the Schrödinger equation and the TB Hamiltonian H' are, respectively, given by:

$$H' \begin{bmatrix} c_A \\ c_B \\ c_{A'} \\ c_{B'} \end{bmatrix} = \varepsilon \begin{bmatrix} c_A \\ c_B \\ c_{A'} \\ c_{B'} \end{bmatrix} \tag{7.52}$$

and

$$H' = \begin{matrix} & |A\rangle & |B\rangle & |A'\rangle & |B'\rangle \\ \begin{matrix} \langle A| \\ \langle B| \\ \langle A'| \\ \langle B'| \end{matrix} & \left[\begin{matrix} \langle \psi_A|H|\psi_A\rangle \\ \langle \psi_B|H|\psi_A\rangle \\ \langle \psi_{A'}|H|\psi_A\rangle \\ \langle \psi_{B'}|H|\psi_A\rangle \end{matrix} \right. & \begin{matrix} \langle \psi_A|H|\psi_B\rangle \\ \langle \psi_B|H|\psi_B\rangle \\ \langle \psi_{A'}|H|\psi_B\rangle \\ \langle \psi_{B'}|H|\psi_B\rangle \end{matrix} & \begin{matrix} \langle \psi_A|H|\psi_{A'}\rangle \\ \langle \psi_B|H|\psi_{A'}\rangle \\ \langle \psi_{A'}|H|\psi_{A'}\rangle \\ \langle \psi_{B'}|H|\psi_{A'}\rangle \end{matrix} & \left. \begin{matrix} \langle \psi_A|H|\psi_{B'}\rangle \\ \langle \psi_B|H|\psi_{B'}\rangle \\ \langle \psi_{A'}|H|\psi_{B'}\rangle \\ \langle \psi_{B'}|H|\psi_{B'}\rangle \end{matrix} \right] \end{matrix} \tag{7.53}$$

First, the two-center integrals using the same wave functions, $<\psi_X \mid H \mid \psi_X>$, gives the onsite energy identical to Equation (7.16), and we have the relation as:

$$\langle \psi_A|H|\psi_A\rangle = \langle \psi_B|H|\psi_B\rangle = \langle \psi_{A'}|H|\psi_{A'}\rangle = \langle \psi_{B'}|H|\psi_{B'}\rangle = \varepsilon_{2p} \tag{7.54}$$

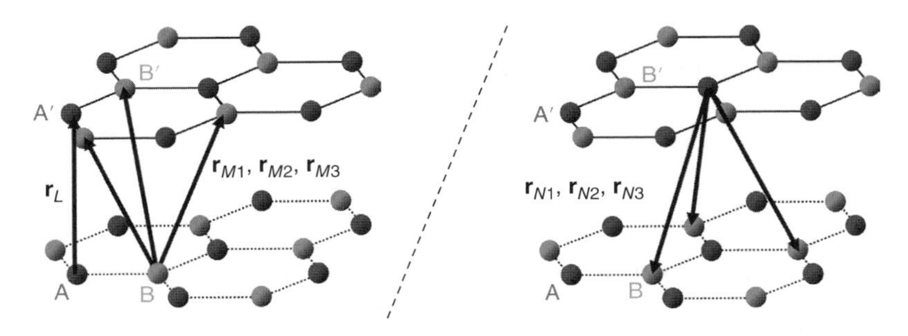

Figure 7.18 (Left) Positional vectors from the atom B to the atom B' group, $\mathbf{r}_{M1}$, $\mathbf{r}_{M2}$, $\mathbf{r}_{M3}$, and (right) those from the atom A' to the atom B group, $\mathbf{r}_{N1}$, $\mathbf{r}_{N2}$, $\mathbf{r}_{N3}$.

Next, the hopping energy elements will be formulated by considering the positional relation shown in Figures 7.17 and 7.18.

First of all, the hopping energy elements between intralayer atoms – that is, $\langle\psi_A|H|\psi_B\rangle$, $\langle\psi_B|H|\psi_A\rangle$, $\langle\psi_{A'}|H|\psi_{B'}\rangle$ and $\langle\psi_{B'}|H|\psi_{A'}\rangle$ – have positional relations identical to graphene and GNRs, and thus the following equations hold:

$$\langle\psi_A|H|\psi_B\rangle = \langle\psi_{A'}|H|\psi_{B'}\rangle = \gamma_0 f \tag{7.55}$$

$$\langle\psi_B|H|\psi_A\rangle = \langle\psi_{B'}|H|\psi_{A'}\rangle = \gamma_0 f^* \tag{7.56}$$

where γ_0 and f are defined by Equations (7.17) and (7.19), respectively.

Secondly, the hopping energy elements in the vertical direction are formulated below, which correspond to the elements, $\langle\psi_A|H|\psi_{A'}\rangle$ and $\langle\psi_{A'}|H|\psi_A\rangle$. These hopping elements are peculiar to BLGs, and we newly formulate them using the positional vectors indicated in Figure 7.18. Since the positional vector from the atom A to the atom A', $\mathbf{r}_L$, is expressed using the interlayer distance d as:

$$\mathbf{r}_L = d(0,0,1) \tag{7.57}$$

and the two-center integrals of the matrix elements in this direction correspond to γ_1 in Figure 7.17, we have the relations as follows:

$$\langle\psi_A|H|\psi_{A'}\rangle = \gamma_1 p_L \tag{7.58}$$

$$\langle\psi_A|H|\psi_{A'}\rangle = \gamma_1 p_L{}^* \tag{7.59}$$

where we defined as $p_L = \exp(i\mathbf{k}\cdot\mathbf{r}_L)$. Thirdly, the hopping energy elements between the non-overlapped sublattices, namely, $\langle\psi_B|H|\psi_{B'}\rangle$ and $\langle\psi_{B'}|H|\psi_B\rangle$, are formulated. In Figure 7.18, the positional vectors from the atom B to the atom B' group, $\mathbf{r}_{M1}$, $\mathbf{r}_{M2}$, $\mathbf{r}_{M3}$, are expressed as

$$\mathbf{r}_{Mi} = \mathbf{r}_L + \mathbf{r}_i\ (i = 1,2,3) \tag{7.60}$$

and the two-center integrals of the matrix elements in this direction correspond to γ_3 in Figure 7.17, we have the relations as follows:

$$\langle \psi_B | H | \psi_{B'} \rangle = \gamma_3 f_M \tag{7.61}$$

$$\langle \psi_{B'} | H | \psi_B \rangle = \gamma_3 f_M{}^* \tag{7.62}$$

where we defined as $f_M = \exp(i\mathbf{k}\cdot\mathbf{r}_{M1}) + \exp(i\mathbf{k}\cdot\mathbf{r}_{M2}) + \exp(i\mathbf{k}\cdot\mathbf{r}_{M3})$. Lastly, remaining hopping energy elements, $\langle \psi_A | H | \psi_{B'} \rangle$, $\langle \psi_B | H | \psi_{A'} \rangle$, $\langle \psi_{A'} | H | \psi_B \rangle$ and $\langle \psi_{B'} | H | \psi_A \rangle$, are formulated. In Figure 7.18, the positional vectors from the atom A' to the atom B group, $\mathbf{r}_{N1}$, $\mathbf{r}_{N2}$, $\mathbf{r}_{N3}$, are expressed as:

$$\mathbf{r}_{Ni} = -\mathbf{r}_L + \mathbf{r}_i (i = 1,2,3) \tag{7.63}$$

and the two-center integrals of the matrix elements in this direction correspond to γ_4 in Figure 7.17, we have the relations as follows:

$$\langle \psi_{A'} | H | \psi_B \rangle = \langle \psi_{B'} | H | \psi_A \rangle = \gamma_4 f_N \tag{7.64}$$

$$\langle \psi_A | H | \psi_{B'} \rangle = \langle \psi_B | H | \psi_{A'} \rangle = \gamma_4 f_N{}^* \tag{7.65}$$

where we defined as $f_N = \exp(i\mathbf{k}\cdot\mathbf{r}_{N1}) + \exp(i\mathbf{k}\cdot\mathbf{r}_{N2}) + \exp(i\mathbf{k}\cdot\mathbf{r}_{N3})$. Substituting the hopping energy elements formulated above into Equation (7.53), the TB Hamiltonian H' is rewritten by:

$$H' = \begin{array}{c} \\ \langle A| \\ \langle B| \\ \langle A'| \\ \langle B'| \end{array} \begin{array}{c} \begin{array}{cccc} |A\rangle & |B\rangle & |A'\rangle & |B'\rangle \end{array} \\ \begin{bmatrix} \varepsilon_{2p} & \gamma_0 f & \gamma_1 p_L & \gamma_4 f_N{}^* \\ \gamma_0 f^* & \varepsilon_{2p} & \gamma_4 f_N{}^* & \gamma_3 f_M \\ \gamma_1 p_L{}^* & \gamma_4 f_N & \varepsilon_{2p} & \gamma_0 f \\ \gamma_4 f_N & \gamma_3 f_M{}^* & \gamma_0 f^* & \varepsilon_{2p} \end{bmatrix} \end{array} \tag{7.66}$$

Figure 7.19 shows the computed band structure of the BLG with the AB stacking, where we used $\varepsilon_{2p} = 0$ eV, $\gamma_0 = 2.6$ eV, $\gamma_1 = 0.34$ eV, $\gamma_3 = 0.3$ eV, $\gamma_4 = 0$ eV and $d = 3.354$ Å [7.8]. The two parabolic bands touch at $E = 0$, and there are two additional bands that start at $E \pm \gamma_1$ [7.6].

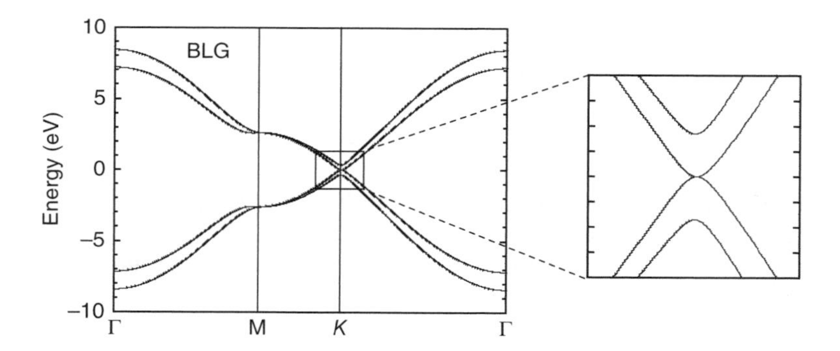

Figure 7.19 Band structure computed for BLG with the AB stacking. The zoom-in of the band structure close to the K point is indicated in the right figure.

The dispersion is electron-hole symmetric. The above results mean that BLG is metallic without bandgap as in graphene, within this approximation. In addition, the calculated effective mass is $0.053\,m_0$, which is small enough for electron device applications.

When we add a potential energy term $V/2$, which is half of the shift in electrochemical potential between the two layers, into the diagonal terms in the TB Hamiltonian (7.66), as follows:

$$H' = \begin{array}{c} \langle A| \\ \langle B| \\ \langle A'| \\ \langle B'| \end{array} \overset{\begin{array}{cccc} \quad |A\rangle & \qquad |B\rangle & \qquad |A'\rangle & \qquad |B'\rangle \end{array}}{\begin{bmatrix} \varepsilon_{2p} + \frac{V}{2} & \gamma_0 f & \gamma_1 p_L & \gamma_4 f_N{}^* \\ \gamma_0 f^* & \varepsilon_{2p} + \frac{V}{2} & \gamma_4 f_N{}^* & \gamma_3 f_M \\ \gamma_1 p_L{}^* & \gamma_4 f_N & \varepsilon_{2p} - \frac{V}{2} & \gamma_0 f \\ \gamma_4 f_N & \gamma_3 f_M{}^* & \gamma_0 f^* & \varepsilon_{2p} - \frac{V}{2} \end{bmatrix}} \tag{7.67}$$

the equivalence of the two layers or, alternatively, the inversion symmetry is broken. This gives rise to the opening of a bandgap close to, but not directly at, the K point [7.6, 7.8], as discussed below.

The potential energy term $V/2$ will appear if a potential bias is applied between the layers, or if the bilayer is placed in an external electric field, or if symmetry breaking between the layers is induced via the interaction between a graphene layer and its substrate, and so on. In the remaining part of this section, we formulate the TB Hamiltonian under the influence of an external electric field.

When an external electric field E_{ext} is applied perpendicular to the graphene planes, the potential energy term is simply expressed using the interlayer distance d as:

$$V = E_{ext} \cdot d \tag{7.68}$$

However, since the external potential is strongly screened by charge transfer between the layers, the interlayer electric field E_L is weakened and is approximately given by [7.8, 7.48]:

$$E_L \approx \frac{E_{ext}}{3} = \frac{1}{3}\frac{V}{d} \tag{7.69}$$

Consequently, the TB Hamiltonian under the influence of a vertical electric field is represented as:

$$H' = \begin{array}{c} \langle A| \\ \langle B| \\ \langle A'| \\ \langle B'| \end{array} \overset{\begin{array}{cccc} \quad |A\rangle & \qquad |B\rangle & \qquad |A'\rangle & \qquad |B'\rangle \end{array}}{\begin{bmatrix} \varepsilon_{2p} + \frac{E_L d}{2} & \gamma_0 f & \gamma_1 p_L & \gamma_4 f_N{}^* \\ \gamma_0 f^* & \varepsilon_{2p} + \frac{E_L d}{2} & \gamma_4 f_N{}^* & \gamma_3 f_M \\ \gamma_1 p_L{}^* & \gamma_4 f_N & \varepsilon_{2p} - \frac{E_L d}{2} & \gamma_0 f \\ \gamma_4 f_N & \gamma_3 f_M{}^* & \gamma_0 f^* & \varepsilon_{2p} - \frac{E_L d}{2} \end{bmatrix}} \tag{7.70}$$

Figure 7.20 shows the band structure computed for BLG under $E_{ext}=1.65$ V/nm, which has a bandgap of $E_G=0.15$ eV [7.48]. As seen in the right figure, a Mexican hat structure with a negative effective mass appears at the K point. Figure 7.21 shows the bandgap of the BLG as a function of the vertical electric field E_{ext}, where the approximate relationship of $E_L \approx E_{ext}/3$ was used to consider the screening effect of the external interlayer potential [7.48]. It is found that the bandgap of the BLG increases with E_{ext}, but saturates at around 0.25–0.26 eV beyond $E_{ext}=5$ V/nm. The electrical characteristics of BLG-FETs and A-GNR-FETs will be compared in detail under the ballistic transport condition in Section 7.3.

7.2.5 Graphene Nanomeshes (GNMs)

As demonstrated in the previous section, a bandgap can be opened up in BLG by applying a vertical electric field. Although BLG can be fabricated using a large-area graphene sheet, stacking only two layers uniformly in a large-area is still difficult, using the current technologies. In addition, the bandgap of BLG is saturated, as shown in Figure 7.21 [7.8, 7.48, 7.49], and has negative effective masses at the K and K' points [7.8, 7.28, 7.48, 7.50, 7.51].

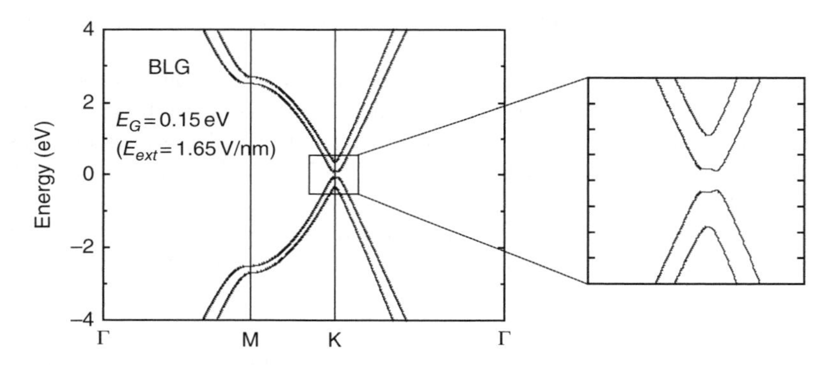

Figure 7.20 Band structure computed for BLG under $E_{ext}=1.65$ V/nm, which has a bandgap of $E_G=0.15$ eV [7.48]. As seen in the zoomed-in figure, a Mexican hat structure with a negative effective mass appears at the K point (Dirac point).

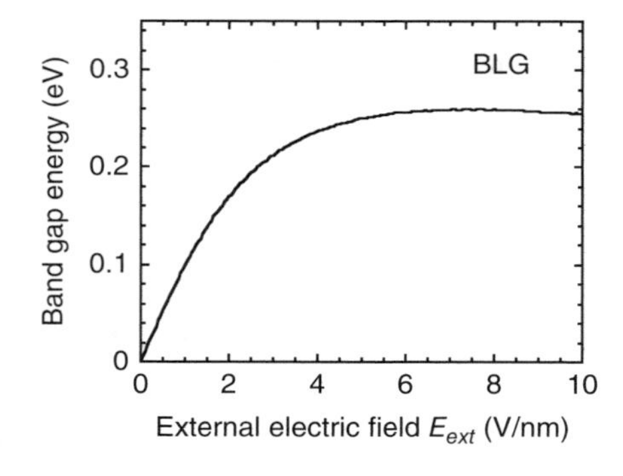

Figure 7.21 Bandgap of BLG computed as a function of vertical electric field E_{ext}, where the approximate relationship of $E_L \approx E_{ext}/3$ was used to consider the screening effect of the external interlayer potential [7.48].

Figure 7.22 An example of graphene nanomesh. In this structure, the horizontal and vertical sides of nanoholes have an armchair-edged and zigzag-edged configuration, respectively [7.59].

As for A-GNRs, the bandgap is generated by quantum confinement effect and increases inversely with the ribbon width, as shown in Figure 7.10. According to Figures 7.10 and 7.21, A-GNRs are found to exhibit bandgaps larger than those of BLG, and also indicate no Mexican-hat structure as shown in Figure 7.12 [2.4, 7.14, 7.30, 7.31, 7.33]. Therefore, A-GNR is attractive for a FET channel material, but its electron transport properties are extremely sensitive to ribbon edge fluctuation or configuration. The fabrication of A-GNR channels with a certain amount of length and the production of dense arrays of ordered nanoribbons still remains a significant challenge, and drastic degradation in the device performance, due to atomic fluctuation of the ribbon width and the edge conformation, has been pointed out [7.52–7.56].

Under the circumstances, a new graphene nanostructure for opening a finite bandgap, which is called graphene nanomesh (GNM), has been reported [7.57, 7.58]. GNM is composed of single-layer graphene, into which a high-density array of nanoscale holes are punched, as shown in Figure 7.22, for example [7.59]. This can open up a bandgap comparable with the values achieved in GNRs in a large sheet of graphene thin film.

Actually, it was demonstrated that GNM-FETs provide driving currents nearly 100 times greater than individual GNR-FETs, and the on-off current ratio – that is, the bandgap opening can be tuned by varying the neck width [7.57, 7.58]. GNM lattices are also considered to be much easier to produce and handle than GNRs. These features indeed make GNM attractive for a FET channel material. However, since the nanomeshes have variable periodicities, neck widths, shapes of nanohole, edge configurations, and so on, their electric properties indicate quite complicated behaviors. Thus, more fundamental theoretical and experimental researches will be needed to understand their characteristics and possibilities.

7.3 Influence of Bandgap Opening on Ballistic Electron Transport in BLG and A-GNR-MOSFETs

In this section, the influence of the bandgap opening on electron transport in BLG and A-GNR-FETs under the ballistic condition is investigated by simulating their ultimate device performances using the ToB model [7.48], where the electronic band structures of BLG and A-GNR were calculated using the TB approach with a p_z orbital basis set, as described in

Section 7.2. Figure 7.23 shows the schematic diagram of the simulated BLG and A-GNR-FETs. BLG is placed in a vertical electric field E_{ext}, and AGNR has the number of atoms in its transverse direction N and the edges are assumed to be terminated by hydrogen atoms. The source and drain are heavily doped BLG/A-GNR contacts, with $N_D = 1 \times 10^{20}\,\text{cm}^{-3}$ ($3.35 \times 10^{12}\,\text{cm}^{-2}$) and the channel is intrinsic. The gate insulator is assumed to be HfO_2 with a dielectric constant of 23.4 ε_0, which contributes to reduce the electric field impressed to the gate insulator.

For BLG-FET, we assumed that the vertical electric field E_{ext} is spatially constant throughout the device, and is independent of both gate and drain voltages. This implies that E_{ext} is solely induced by impurity concentration in the substrate. Therefore, charge density in the channel is controlled by the top gate electrostatics while maintaining a constant bandgap, which means that a gate voltage-induced bandgap modulation and screening of the vertical electric field by positive charges injected in the channel, due to band-to-band tunneling [7.60, 7.61], and also the effect of edge roughness [7.52–7.56] are ignored in this study. We consider that these simplifications are reasonable for ultimate device performance estimation in the on-state. For A-GNR-FET, we adopted 1-D perfect source and drain having the same width as the channel, since metal-induced-gap states are not produced in such 1-D perfect contacts [7.62, 7.63]. Furthermore, we assume that the gate has perfect electrostatic control over the channel.

In terms of the FET application, a comparison of the device performances under the same bandgap energy is important, because an off-state leakage current due to band-to-band tunneling is strongly dependent on the bandgap energy. Thus, the electrical characteristics of BLG and A-GNR-FETs are compared under the two conditions of $E_G = 0.15\,\text{eV}$ and $0.25\,\text{eV}$ in the next section. In addition, we focus on A-GNR-FETs with the $N = 3m$ group, since they are expected to provide superior device performance over the $N = 3m + 1$ and $3m + 2$ groups [7.33, 7.64], as discussed in Section 7.2.3.

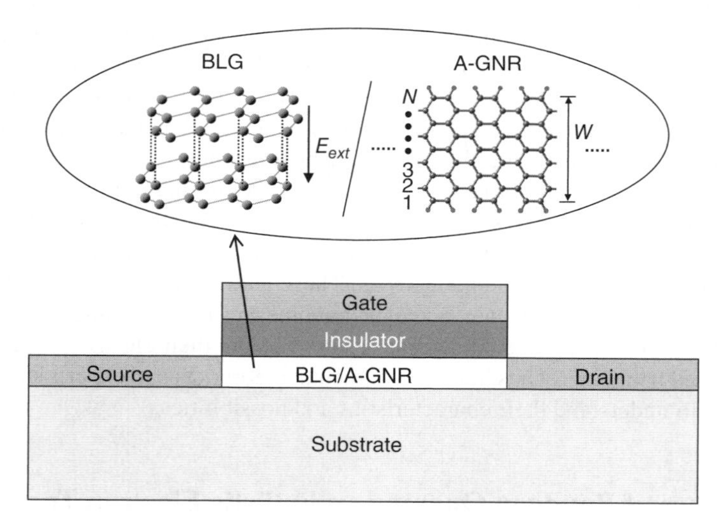

Figure 7.23 Schematic diagram of the simulated BLG and A-GNR-FETs. BLG is placed in a vertical electric field E_{ext}, and AGNR has the number of atoms in its transverse direction N and the edges are assumed to be terminated by hydrogen atoms. The source and drain are heavily doped BLG/A-GNR contacts while the channel is intrinsic. The gate insulator is assumed to be HfO_2 with a dielectric constant of 23.4 ε_0.

7.3.1 Small Bandgap Regime

In this section, we discuss the electrical characteristics under a relatively small bandgap energy of $E_G = 0.15$ eV, but being sufficiently larger than the thermal energy at room temperature. First, Figure 7.24 shows the band structures computed for the BLG under $E_{ext} = 1.65$ V/nm and the A-GNR with $W = 6.1$ nm, both of which have a band-gap energy of $E_G = 0.15$ eV. As shown in the inset of Figure 7.24(a), a Mexican hat structure with a negative effective mass appears in the BLG at the K point. On the other hand, a linear dispersion relationship is still observed in the A-GNR of Figure 7.24(b), where its electron effective mass is estimated to be $0.019\,m_0$, despite the opening of a finite bandgap.

Now, we evaluate the ultimate device performances when the BLG and A-GNR are applied to the FET channels as shown in Figure 7.23. Figure 7.25 shows comparisons in the drain current versus gate voltage ($I_D - V_G$) characteristics and in the averaged electron velocities and the transit times computed for the BLG- and A-GNR-FETs with gate oxide thickness of $T_{ox} = 9$ nm (EOT = 1.5 nm) and 3 nm (EOT = 0.5 nm). In the calculation of the transit times, the channel length is assumed to be 10 nm.

Here, the current drive is expressed in units of drain current per unit width, even for A-GNR-FET, which will be suited to represent the current drive of closely spaced nanoribbon arrays such as nanomesh [7.57]. Since the channel length is set as 10 nm, and the high-k gate oxide is employed, we can expect excellent electrostatic control to prevent source-drain direct tunneling. Furthermore, the drain voltage is set sufficiently small ($V_D = 0.3$ V), and we also ignored band-to-band tunneling in this study. For $T_{ox} = 9$ nm, the A-GNR-FET provides not only larger drain current, but also higher averaged velocity than the BLG-FET. This is the anticipated result, considering the difference between their dispersion relationships in the low energy regime, as shown in Figure 7.24.

Here, we should point out that the averaged electron velocity of the A-GNR-FET is nearly independent of the gate bias voltage, since the A-GNR channel has an almost linear dispersion relationship, as shown in Figure 7.24(b). In other words, at the on-state, the Fermi level is deep inside the conduction band and higher subbands get populated. Though carriers in the higher subbands transport under the influence of a partly parabolic dispersion relationship, a large

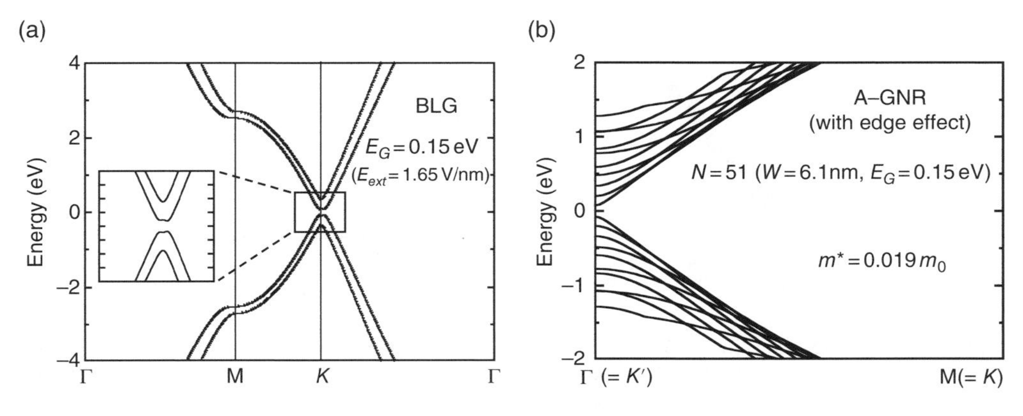

Figure 7.24 Band structures computed for (a) BLG under $E_{ext} = 1.65$ V/nm and (b) A-GNR with $W = 6.1$ nm, both of which have a band-gap energy of $E_G = 0.15$ eV.

part of carriers are still distributed into a linear part of the dispersion curves. Therefore, the average electron velocity hardly changes, even if the gate bias increases.

For the same reason, the averaged electron velocities are almost independent of the gate oxide thickness in the A-GNR-FET. On the other hand, in the BLG-FET with smaller T_{ox}, a number of electrons increasingly occupy a higher-momentum region, which has linear dispersion relationship even in BLGs. Therefore, as shown in Figure 7.25 (b) the average electron velocity increases to improve the $I_D - V_G$ characteristics and the transit times of the BLG-FET by decreasing T_{ox}.

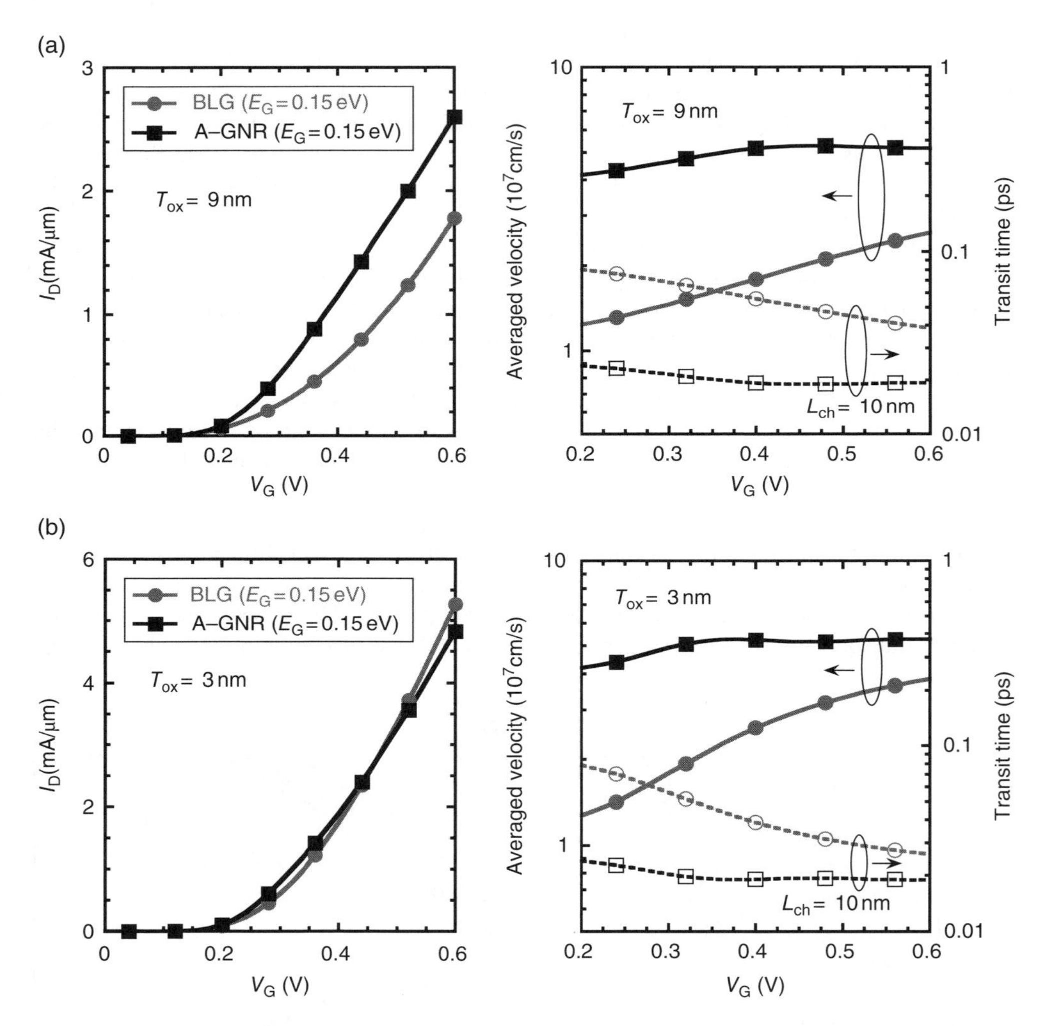

Figure 7.25 Comparisons in $I_D - V_G$ characteristics and in averaged electron velocities and transit times computed for BLG- and A-GNR-FETs with E_G=0.15 eV for (a) T_{ox}=9 nm (EOT=1.5 nm); and (b) 3 nm (EOT=0.5 nm), where the gate oxide was assumed to be HfO_2. In the right column figures, ■ and □ represent A-GNR, and ● and ○ BLG. The current density at V_G=0 V is set as 0.06 μA/μm, and channel length is assumed to be 10 nm in the calculation of transit times. V_D is set at 0.3 V and temperature is 300 K.

It is also noteworthy that for $T_{ox}=3\,\text{nm}$ in Figure 7.25(b), the current drive of A-GNR-FET is almost identical to that of BLG-FET, even though the carrier velocity, which is proportional to the reciprocal of device delay, is undoubtedly higher than the BLG-FET. This is due to the reduction in the total gate capacitance influenced by the quantum capacitance of the channels C_Q, particularly, in the A-GNR-FET, as described in Section 6.3.4. Since the C_Q is determined by the DOS for carriers, A-GNR-FET with a smaller effective mass is more likely to be affected by C_Q under such a very small T_{ox}. We actually computed the quantum capacitances for both FETs by using the ToB model and compare them, as shown in Figure 7.26, where (a) $T_{ox}=9\,\text{nm}$ and (b) 3 nm. The horizontal dashed lines represent the oxide capacitances determined by $C_{ox}=\varepsilon_{ox}/T_{ox}$. ε_{ox} is the permittivity of the gate oxide. The quantum capacitance is calculated using the same method as in Section 6.3.4.

It is found that the C_Q in A-GNR-FET is remarkably smaller than the C_{ox} when T_{ox} is 3 nm. On the other hand, the C_Q in BLG-FET is still comparable to the C_{ox} at the on state for $T_{ox}=3\,\text{nm}$, because the Mexican hat structure provides larger DOS for carriers than those in A-GNR. Consequently, the current drive of A-GNR-FET degrades considerably due to the quantum capacitance under such a very small T_{ox}, as shown in Figure 7.25 (b).

7.3.2 *Large Bandgap Regime*

Next, we study the performances under a larger bandgap energy of $E_G=0.25\,\text{eV}$ – that is, close to the maximum bandgap energy in BLG. Figure 7.27 shows the band structures computed for BLG under $E_{ext}=5.0\,\text{V/nm}$ and A-GNR with $W=3.6\,\text{nm}$, both of which have a bandgap energy of $E_G=0.25\,\text{eV}$.

In Figure 7.27 (a), the result computed for $E_{ext}=8.0\,\text{V/nm}$ is also plotted for further discussion, where the bandgap energy only slightly increases to $E_G=0.26\,\text{eV}$, but its dispersion relationship drastically changes, as shown in the inset. The Mexican hat structure becomes

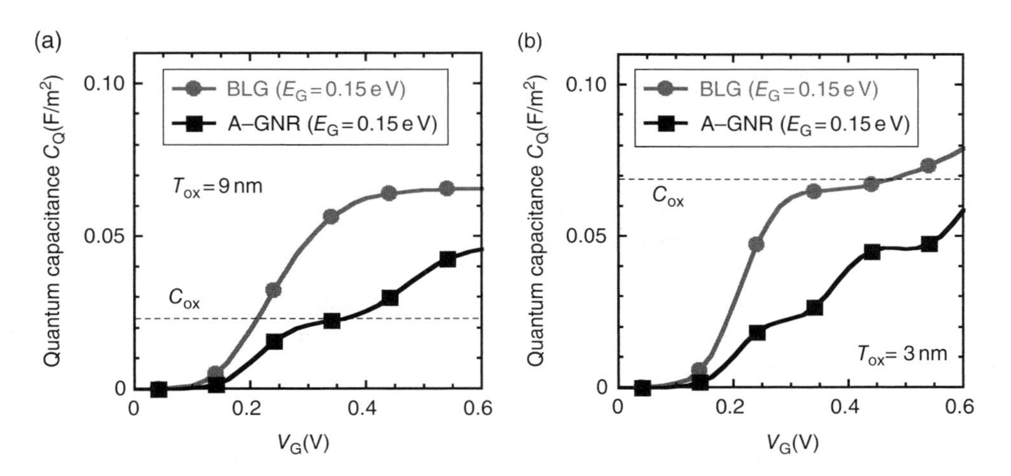

Figure 7.26 Comparisons in quantum capacitances computed for BLG- and A-GNR-FETs with $E_G=0.15\,\text{eV}$, where (a) $T_{ox}=9\,\text{nm}$ and (b) 3 nm. The horizontal dashed lines represent the oxide capacitances determined by $C_{ox}=\varepsilon_{ox}/T_{ox}$.

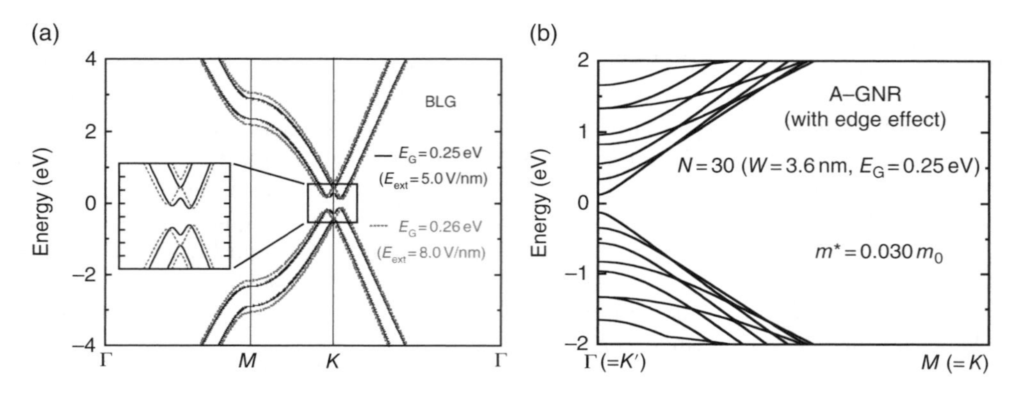

Figure 7.27 Band structures computed for (a) BLG under E_{ext}=5.0V/nm and (b) A-GNR with W=3.6nm, both of which have a bandgap energy of E_G=0.25eV. In (a), the result computed for E_{ext}=8.0V/nm is also plotted, where the bandgap energy only slightly increases to E_G=0.26eV, while its dispersion relationship drastically changes, as shown in the inset.

more warped by increasing the vertical electric field, even though the opened bandgap energy is almost the same. Thus, de-acceleration of electrons due to the negative effective mass becomes more serious, and probably causes further performance degradation. Such consideration is actually confirmed by simulating the electrical characteristics as shown in Figure 7.28, where the calculation conditions were the same as in Figure 7.25. For both of T_{ox}=9nm and 3nm, the BLG-FETs exhibit much inferior performances compared to the A-GNR-FET. In particular, the increasing E_{ext} substantially degrades the electrical characteristics of BLG-FETs, as expected above. This also means that the device performance of BLG-FETs can significantly differ, even if they have nearly equal bandgap energies.

In addition, it is found that the characteristics of the A-GNR-FET are almost unchanged by increasing the bandgap energy from 0.15eV (Figure 7.25) to 0.25eV (Figure 7.28), since the effective mass only slightly increases to 0.030m_0. This behavior is in contrast to that of BLG-FETs. However, as reported in [7.29], A-GNR-FETs with larger bandgap energy than ≈0.4eV exhibit marked degradation of the device performances, since a parabolic dispersion relationship with a heavier effective mass appears in A-GNRs. Consequently, there should be a trade-off relationship between bandgap energy and carrier velocity in A-GNR-FETs, as well as in BLG-FETs. Nonetheless, the present results suggest that ideal A-GNR-FETs outperform, in principle, BLG-FETs consisting of the multilayer graphene architecture [7.65].

In summary, the intrinsic effects of the bandgap opening on electron transport in A-GNR-FETs and BLG-FETs were systematically investigated, based on TB band structure calculation and the ballistic MOSFET model. We found that increasing the vertical electric field in BLG-FET to obtain a larger bandgap energy significantly degrades the electrical characteristics, because a Mexican hat structure becomes more warped, and de-acceleration of electrons due to a negative effective mass causes a serious performance degradation. Therefore, in principle, A-GNR-FETs outperform BLG-FETs. However, we also pointed out that there is a trade-off relationship between bandgap energy and carrier velocity in applying such semiconducting graphene channels into high-speed logic circuits.

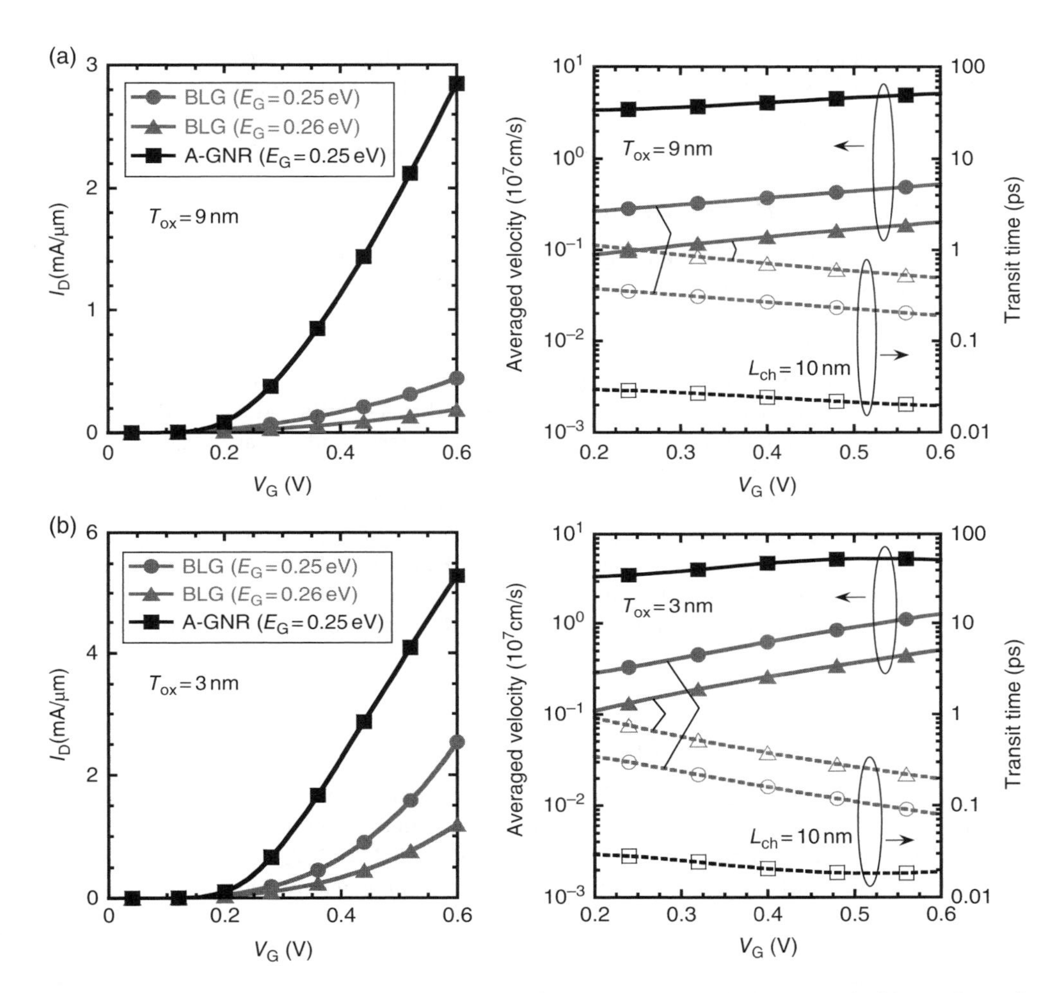

Figure 7.28 Comparisons in $I_D - V_G$ characteristics and in averaged electron velocities and transit times, computed for BLG- and A-GNR-FETs with E_G=0.25 eV. For BLG-FET, the results computed for E_{ext}=8.0 V/nm, providing E_G=0.26 eV, are also plotted. The gate oxide thickness is taken as (a) T_{ox}=9 nm and (b) 3 nm, where the gate oxide was assumed to be HfO_2. In the right column figures, ■ and □ represent A-GNR, and ●, ▲, ○ and Δ represent BLGs. The calculation conditions were the same as in Figure 7.25.

7.4 Silicene, Germanene and Graphene Nanoribbons

Silicene or germanene is a monolayer honeycomb lattice made of Si or Ge, similar to graphene made of carbon. As shown in Section 7.1.1, silicene and germanene are semi-metallic, without a bandgap and, hence, experimental works on growth of silicene nanoribbon (SiNR) and germanene nanoribbon (GeNR) have been performed to induce a bandgap. In this section, we evaluate the performance potentials of FETs with A-SiNR and A-GeNR channels, and compare them with that of A-GNR-FETs. As described in Section 7.1, we employ the first nearest-neighbor TB approach in the band structure calculations of A-SiNRs and A-GeNRs, and evaluate their device performances using the ToB model [7.15, 7.66]. The edges of A-SiNRs and A-GeNRs are also assumed to be terminated by hydrogen atoms, and we used the TB parameters shown in Table 7.2.

7.4.1 Bandgap vs Ribbon Width

Figure 7.29 shows the variation in bandgap energies with the ribbon width given in terms of N, calculated for A-GNRs, A-SiNRs and A-GeNRs. A-NRs having widths from $N=3$ to 20 were investigated here.

One can see three branches with decreasing behaviors for all the three A-NRs considered in this study, which indicates that the bandgap opening originates from the quantum confinement effect. Incidentally, this decreasing behavior represents that A-SiNRs and A-GeNRs with infinite ribbon width – that is, silicene and germanene – have zero bandgap, as already discussed in Section 7.1.1. Here, it is worth noting that comparing at the same number of N, A-SiNRs and A-GeNRs exhibit smaller bandgaps than those of A-GNRs. This is due to the smaller γ_0s in A-SiNRs and A-GeNRs, as shown in Table 7.2. If the bandgap is different among channel materials, off-state leakage current due to band-to-band tunneling can be exponentially changed by the bias condition used. Hence, to make a comparison of their performance potentials as a FET channel, we use three kinds of A-NRs with the same bandgap energy, as discussed below.

7.4.2 Comparison of Band Structures

Hereafter, we focus on three kinds of A-NRs with approximately the same bandgap of ≈ 0.5 eV, which is larger than the energy of the drain bias assumed in the electrical characteristics simulation described later. Specifically, they are A-GNR with $N=15$, A-SiNR with $N=6$, and A-GeNR with $N=6$, where we adopted the $N=3m$ group (m: a positive integer) in this study,

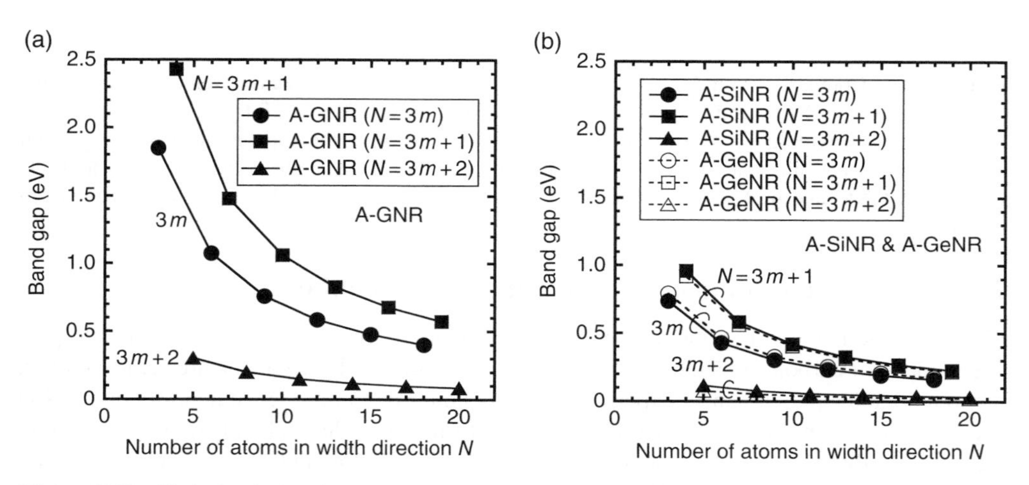

Figure 7.29 Variation in band gap energies with the ribbon width given in terms of N, calculated for (a) A-GNRs and (b) A-SiNRs and A-GeNRs.

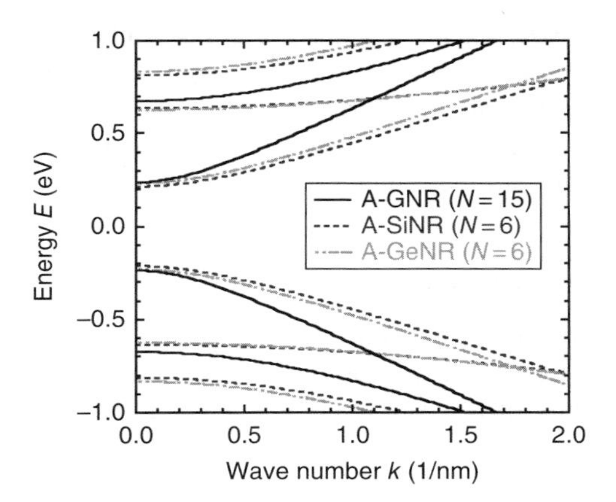

Figure 7.30 Band structures computed for A-GNR, A-SiNR, and A-GeNR with a band gap of about 0.5 eV. Specifically, the A-GNR has $N = 15$, and the A-SiNR and A-GeNR have $N = 6$. A-GNR has the smallest effective mass, $m^*_{GNR} = 0.051\,m_0$, while A-SiNR and A-GeNR have almost the same and larger masses, $m^*_{SiNR} = 0.106\,m_0$ and $m^*_{GeNR} = 0.100\,m_0$.

since it was reported to be more suitable for high-performance use, due to its smaller effective masses than those of the $N = 3m + 1$ group [7.33, 7.64].

Figure 7.30 shows the band structures computed for those A-NRs. It can be seen that, when the band structures are compared under the same bandgap of ≈ 0.5 eV, the A-GNR has the smallest effective mass, $m^*_{GNR} = 0.051\,m_0$, while the A-SiNR and A-GeNR have almost the same and larger ones, $m^*_{SiNR} = 0.106\,m_0$ and $m^*_{GeNR} = 0.100\,m_0$. This is because the effective masses are qualitatively related to $(\gamma_0 \times d)^{-1}$, which gives smaller effective masses for A-GNRs. Accordingly, A-GNR-FET is expected to provide better device performance than A-SiNR and A-GeNR FETs under the ballistic transport, which will be discussed in the next section.

7.5 Ballistic MOSFETs with Silicene, Germanene and Graphene nanoribbons

We compare the performance potentials of A-SiNR, A-GeNR and A-GNR-FETs under the ballistic transport using the ToB model. The device structure and the calculation conditions are set at the same as in Section 7.3, except the gate insulator, which is assumed to be SiO_2 with thicknesses (T_{ox}) of 3 and 0.5 nm.

7.5.1 $I_D - V_G$ *Characteristics*

Figure 7.31 shows the $I_D - V_G$ characteristics computed for $T_{ox} = 3$ and 0.5 nm, where the vertical axis denotes the drain current density divided by the ribbon width W, which was calculated by $W = (N-1) \times \sqrt{3}a/2$. The off current density I_{off} was set as $I_{off} = 0.01$ μA/μm.

First of all, it is found that the A-GNR-FET exhibits the largest drain current density, while the A-SiNR and A-GeNR-FETs have almost the same drain current densities, as expected from the band structures shown in Figure 7.30. We may add that the largest drain current flowing through a piece of NR is also obtained with the A-GNR-FET, because of its larger ribbon width $W_{GNR}=1.722\,nm$, compared with $W_{SiNR}=0.974\,nm$ and $W_{GeNR}=1.026\,nm$. However, in Figure 7.31, note that the difference between the A-GNR-FET and the A-SiNR or A-GeNR-FET clearly decreases as T_{ox} reduces to 0.5 nm. This is, again, due to the influence of the quantum capacitance (C_Q) of the channels, as presented below.

7.5.2 Quantum Capacitances

Figure 7.32 shows the gate voltage dependences of C_Q calculated for the three A-NR-FETs.

The horizontal dashed lines represent the oxide capacitances calculated from $C_{ox}=\varepsilon_{ox}/T_{ox}$, where ε_{ox} is the permittivity of the gate oxide. C_Q is basically proportional to the DOS for

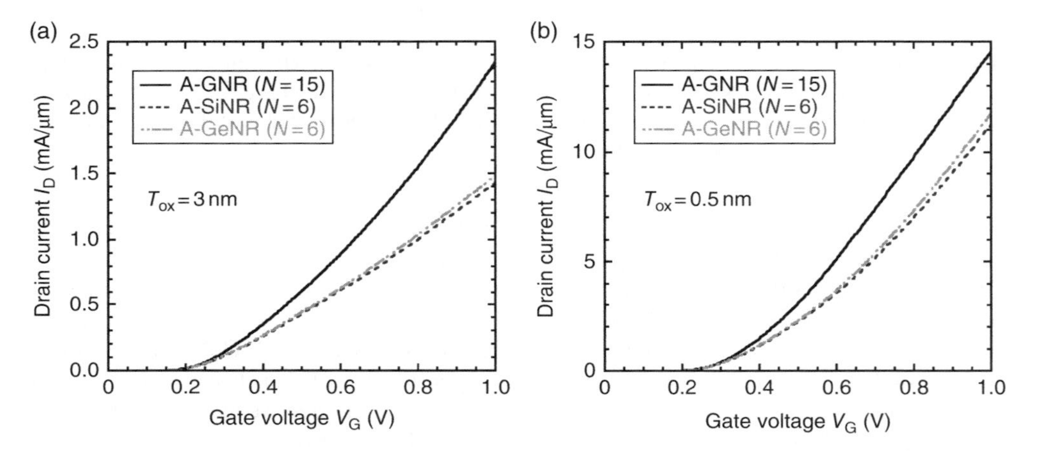

Figure 7.31 $I_D - V_G$ characteristics computed for T_{ox} = (a) 3; and (b) 0.5 nm. The vertical axis denotes the drain current density divided by the ribbon width. $V_D=0.4\,V$, and $I_{off}=0.01\,\mu A/\mu m$.

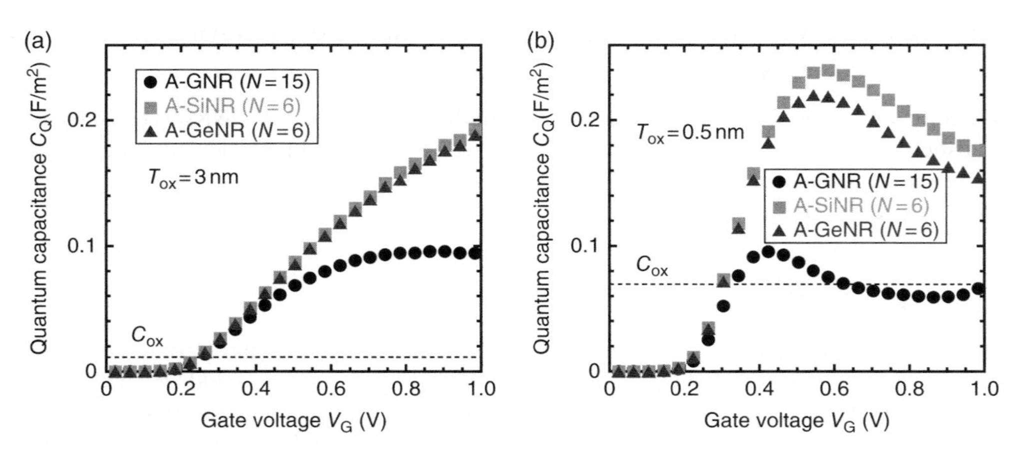

Figure 7.32 Gate voltage dependences of quantum capacitance C_Q calculated for the three A-NR-FETs. T_{ox} = (a) 3; and (b) 0.5 nm.

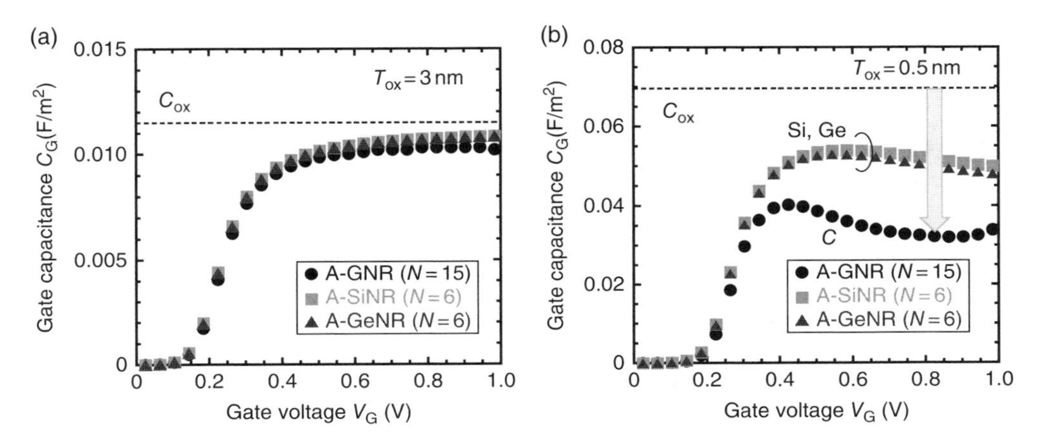

Figure 7.33 Gate voltage dependences of total gate capacitance C_G calculated for T_{ox} = (a) 3 nm; and (b) 0.5 nm.

carriers near the Fermi energy, which has a close relationship with the effective mass of carriers m^* – that is, DOS $\propto \sqrt{m^*}$. Consequently, in the case of T_{ox}=0.5 nm, shown in Figure 7.32(b), only the A-GNR-FET exhibits $C_Q \approx C_{ox}$, because of its lighter effective mass. This indicates that since the total gate capacitance C_G is represented as $C_G = C_{ox}C_Q/(C_{ox}+C_Q)$ in the form of a series-connection of C_Q and $C_{ox,}$ C_G becomes approximately $C_{ox}/2$ if $C_Q \approx C_{ox}$. Therefore, C_G substantially decreases from C_{ox} in the case of the A-GNR-FET with T_{ox} = 0.5 nm, as shown in Figure 7.33, where the actually calculated C_Gs were plotted for T_{ox} = 3 nm and 0.5 nm. On the other hand, the influence of C_Q is found to be negligible in the case of T_{ox} = 3 nm, because C_Q becomes sufficiently larger than $C_{ox,}$ as shown in Figure 7.32(a), which is called a classical capacitance limit (CCL).

7.5.3 Channel Charge Density and Average Electron Velocity

Due to the involvement of the quantum capacitance mentioned above, the channel electron density must be influenced by the decreased C_G in extremely thin T_{ox} as 0.5 nm. Accordingly, we next calculated the channel sheet electron density and the average electron velocity as a function of the gate voltage, as shown in Figure 7.34, where T_{ox} = (a) 3 nm and (b) 0.5 nm. The average electron velocity v was calculated by the equation, $I_D = qNv$.

It is found that for T_{ox} = 3 nm, the electron densities are nearly the same among the three A-NR-FETs (i.e. they are in CCL), and therefore the larger drain current observed for the A-GNR-FET in Figure 7.31 (a) is simply due to the increased electron velocity, as shown in the right column of Figure 7.34(a). As a result, the A-GNR-FET produces the substantially larger drain current for a thick gate oxide such as T_{ox} = 3 nm. On the other hand, for T_{ox} = 0.5 nm the electron density of the A-GNR-FET drastically decreases as compared to the other two A-NR-FETs as shown in the left column of Figure 7.34(b). This is the effect from the decreased C_G owing to the quantum capacitance described in the previous paragraph, and is the cause of the relative decrease in the higher current drivability of the A-GNR-FET with T_{ox} = 0.5 nm.

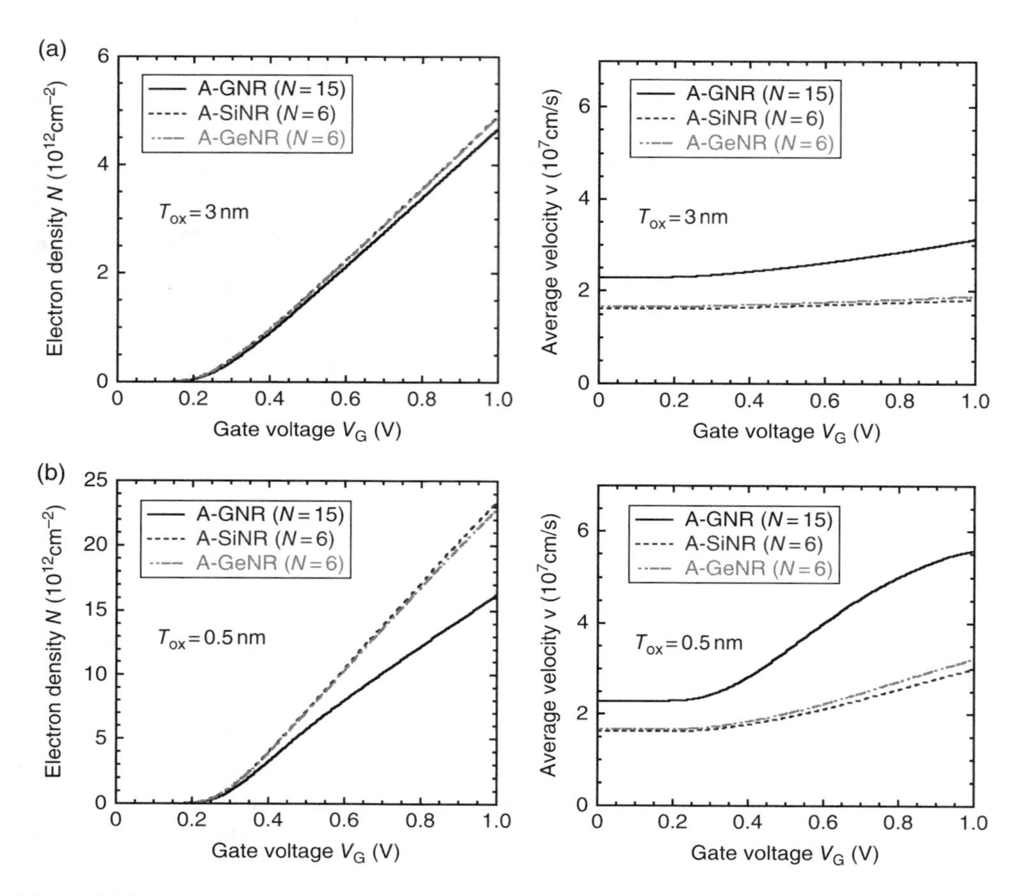

Figure 7.34 (Left column) Channel sheet electron densities and (Right column) average electron velocities as a function of gate voltage computed for T_{ox} = (a) 3 nm; and (b) 0.5 nm.

Again, looking at Figure 7.31(b), A-SiNR and A-GeNR-FETs have the potential to provide high current drivability beyond 2 mA/μm at V_{DD} = 0.5 V under the ballistic transport. This current level meets the requirement for the on-current value specified in the International Technology Roadmap for Semiconductors (ITRS) for 2024 [7.64, 7.67]. Therefore, A-SiNR and A-GeNR are also attractive channel materials for high performance FETs, though A-GNR still maintains the advantage over A-SiNR and A-GeNR under the ideal transport conditions.

7.5.4 Source-drain Direct Tunneling (SDT)

The smaller effective mass in A-GNR may lead to larger off-state leakage current due to SDT, as discussed in Section 5.4. Non-equilibrium Green function (NEGF) simulations of A-SiNR, A-GeNR, and A-GNR-FETs can assess an impact of SDT on their device performances [7.68]. Figure 7.35 shows the $I_D - V_G$ characteristics computed using a NEGF method [7.69] within the nearest-neighbor TB approximation at T = 300 K, where the channel length is 10 nm, the SiO_2 gate oxide thickness T_{ox} = 2 nm, V_D = 0.05 V and the ballistic transport is assumed.

It is found that the A-GNR-FET exhibits the largest drain current, which agrees with the result in Figure 7.31. On the other hand, the off-current of the A-GNR-FET is significantly larger than those of the A-SiNR and A-GeNR-FETs, due to the large tunneling current between the source and the drain in the A-GNR-FET. Figure 7.36 shows the conduction band

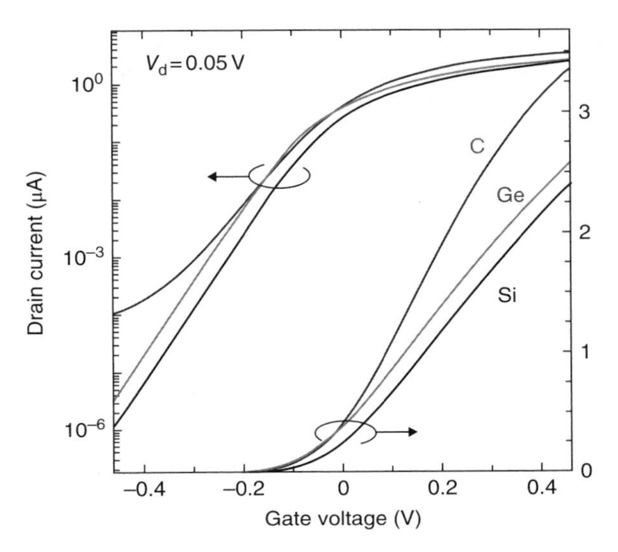

Figure 7.35 $I_D - V_G$ characteristics computed using NEGF method at T=300 K, where the channel length = 10 nm, the SiO_2 gate oxide thickness T_{ox} = 2 nm, V_D = 0.05 V and the ballistic transport is assumed [7.68].

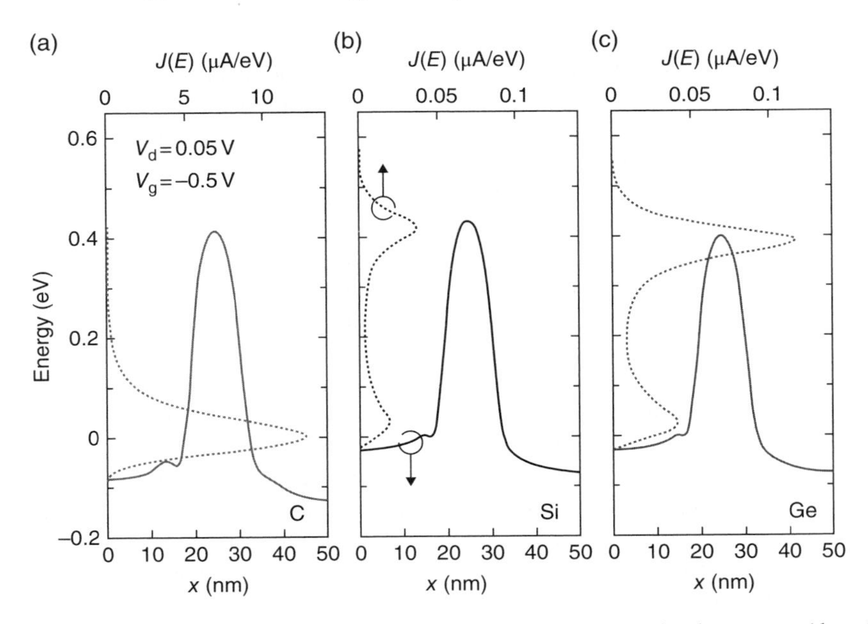

Figure 7.36 Conduction band energy profiles (solid lines) and current density spectra (dotted lines) computed for (a) A-GNR; (b) A-SiNR; and (c) A-GeNR-FETs [7.68].

energy profiles and the current density spectra computed for A-GNR, A-SiNR and A-GeNR-FETs. It can be seen that the tunneling current component at $E \approx 0$ eV dominates in the A-GNR-FET. Note that the fact that the smaller effective mass leads to larger SDT current, and its impact on device performance has been extensively discussed in the context of conventional semiconductor FET scaling [5.25–5.28].

In summary, a performance comparison between A-SiNR, A-GeNR and A-GNR-FETs under ballistic transport was performed. It was found that, comparing at the same bandgap of ≈ 0.5 eV, A-GNR-FET exhibits the largest drain current, even though its advantage over A-SiNR or A-GeNR-FETs reduces in the quantum capacitance limit. It was also pointed out that A-GNR maintains the advantage under the ideal transport situation, whereas A-SiNR and A-GeNR are also attractive channel materials for realizing high performance FETs. Furthermore, the consideration of the off-state leakage current due to SDT was shown to be indispensable for accurately predicting the device performance of ultrashort-channel A-NR-FETs. In the future, we need to consider carrier scattering effects in NR channels.

7.6 Electron Mobility Calculation for Graphene on Substrates

It is well known that graphene has an extremely high electron mobility. As described in Section 7.1, electron mobility as high as $\approx 120\ 000\,\mathrm{cm^2/Vs}$ at 240 K has been experimentally measured in suspended graphene [7.2], and a higher electron mobility, exceeding $10^7\,\mathrm{cm^2/Vs}$ at temperatures up to 50 K, has been demonstrated using graphene layers decoupled from bulk graphite [7.3]. A full-band Monte Carlo (MC) simulation considering the electron-phonon interaction in graphene described by density-functional perturbation theory has estimated an intrinsic electron mobility of $\approx 5 \times 10^6\,\mathrm{cm^2/Vs}$ at 50 K [7.70], which reasonably agrees with the latter experimental result mentioned above. On the other hand, at $T = 300$ K, the simulation predicts a mobility of $9.5 \times 10^5\,\mathrm{cm^2/(V{\cdot}s)}$, which is much higher than expected.

Although practical applications of a graphene field-effect transistor (FET) require a reliable substrate, the mobility in graphene on SiO_2 substrates is limited to $25\,000\,\mathrm{cm^2/(V{\cdot}s)}$ [7.71–7.73]. The reason for this mobility reduction on SiO_2 substrates is the additional scattering mechanisms induced by the substrate, including charged impurities [7.74–7.76], polar and non-polar surface optical phonons in the SiO_2 [7.75, 7.77, 7.78], and substrate surface roughness [7.79. 7.80].

Recently, a drastic improvement of the mobility to $140\,000\,\mathrm{cm^2/(V{\cdot}s)}$ near the charge neutrality point was achieved using a hexagonal boron nitride (h-BN) substrate [7.81]. That substrate has an atomically smooth surface with minimal dangling bonds and charge traps. It also has a lattice constant similar to that of graphite, and a large optical phonon energy and bandgap. Recently it has been reported that h-BN exhibits potential fluctuations, due to charged impurities that are one or two orders of magnitude lower than in SiO_2 [7.82]. Based on these results, h-BN substrate is a suitable substrate for graphene devices.

In this section, the electron mobility in graphene on various substrates is calculated by considering surface optical phonon (OP) and charged impurity (CI) scattering induced by the substrates, as well as the intrinsic acoustic phonon (AP) and OP scattering in pristine graphene. The substrates considered are SiO_2, HfO_2, and h-BN, which are technologically important substrates and gate insulators. Surface AP scattering, which plays an important role in piezoelectric substrates, is not considered in this study. The electron mobility is calculated using a semiclassical Monte Carlo approach [3.12, 7.83], in which both the linear dispersion relation

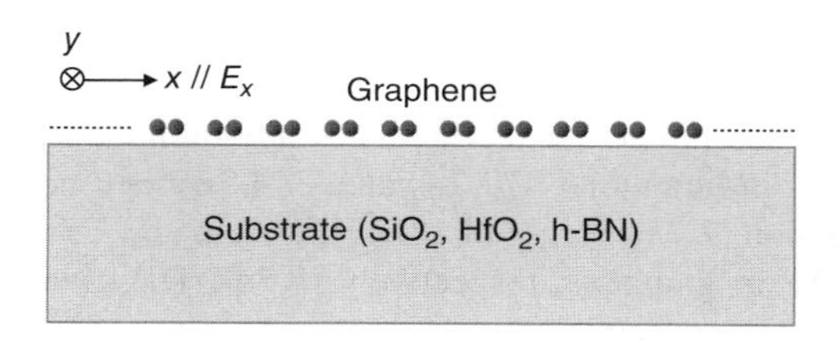

Figure 7.37 Schematic of the graphene-on-substrate system considered in this study. A uniform electric field E_x is applied along the x-direction. Three kinds of substrates (SiO_2, HfO_2, h-BN) are investigated.

of the graphene and the Pauli exclusion principle are taken into account. A schematic of the graphene-on-substrate system considered in this study is shown in Figure 7.37. A uniform electric field E_x is applied along the x-direction. Electron transport in a monolayer graphene on the three substrates (SiO_2, HfO_2, and h-BN) is analyzed using a Monte Carlo simulator for uniform transport, as described below [7.84].

7.6.1 Band Structure

As presented in Section 7.1.1, graphene has a gapless and linear dispersion relation around the K_1 and K_2 points and thus we express it approximately by the following equation.

$$E_{\mathbf{k}} = \hbar v_F |\mathbf{k}| \tag{7.71}$$

v_F is the Fermi velocity, which was taken as $v_F = 10^8$ cm/s in the present calculation, and $|\mathbf{k}|$ is the magnitude of the 2D wavevector relative to the K_1 and K_2 points in **k**-space.

7.6.2 Scattering Mechanisms

The AP, OP, and CI scattering mechanisms are assumed to be isotropic [7.85]. The elastic AP and inelastic OP (with $\hbar\omega_{op} = 164$ meV) cause an intravalley transition $K_{1(2)} \rightarrow K_{1(2)}$, whereas the inelastic AP (with $\hbar\omega_{ac} = 124$ meV) results in an intervalley transition $K_{1(2)} \rightarrow K_{2(1)}$ in intrinsic graphene [7.70]. These are called the intrinsic AP and OP in graphene, and their scattering rates are given by [7.85–7.87]:

$$S_{ac}^{elas} = \frac{E_D^2 k_B T}{4\hbar^3 v_F^2 v_s^2 \rho_s} E \tag{7.72}$$

for elastic AP scattering, where a mass density of $\rho_s = 7.6 \times 10^{-8}$ g/cm^2 and a sound velocity of $v_s = 2 \times 10^6$ cm/s are assumed here, and by:

$$S_{op(ac)}^{inelas} = \frac{D_f^2}{\hbar^2 \rho_s \omega_{op(ac)} v_F^2} \left[\left(E - \hbar\omega_{op(ac)}\right)\left(N_{\hbar\omega_{op(ac)}} + 1\right)\Theta\left(E - \hbar\omega_{op(ac)}\right) + \left(E + \hbar\omega_{op(ac)}\right) N_{\hbar\omega_{op(ac)}} \right] \tag{7.73}$$

for inelastic OP and AP scattering. $\Theta(x)$ is the Heaviside function. In other words, $\Theta = 1$ for $x \geq 1$ and zero otherwise. The function $\Theta(x)$ ensures that the energy is sufficient to emit a phonon. Parameters for the phonon scattering rates are summarized in Table 7.3. For the deformation potential E_D in Equation (7.72), a value of 4.5 eV is used, whereas for the deformation field D_f in Equation (7.73), values of 1.0×10^9 eV/cm and 3.5×10^8 eV/cm are used for the intrinsic OP and AP in graphene, respectively [7.70]. These values are determined by approximating the *ab initio* electron-phonon scattering rates using the simple analytical formulae of Equations (7.72) and (7.73), where E_D and D_f are treated as effective quantities to determine the contribution of all the relevant phonon scatterings.

The OP and CI scatterings induced by the substrates are introduced as follows. First, in semiclassical Monte Carlo studies, it is convenient to use the deformation potential approximation in estimating the scattering rates of various mechanisms [7.85–7.87]. The scattering rates of the polar and non-polar surface OPs in the substrates are comparable to or larger than those of the intrinsic AP and OP, and thus they may reduce the effects from the intrinsic phonon scattering [7.87]. These scattering mechanisms are approximately modeled using Equation (7.73) with the polar and non-polar surface OPs merged. In fact, the approach has successfully reproduced experimental electron velocity versus electric field curves by adjusting the deformation field D_f in Equation (7.73), as reported in Reference [7.87]. Therefore, the resultant D_f for the surface OP scatterings represents an effective interaction strength between the electrons in graphene and the surface OPs, including the non-polar one, so as to reproduce the experimental results. This approach is simple and efficient, and hence it has been widely used to simulate electron transport of graphene as in References [7.85–7.87] and [7.89], where velocity saturation in zero-bandgap graphene FETs has been discussed, based on the surface OP energy of substrates [7.90, 7.91].

On the other hand, a more rigorous approach, where the polar surface OP scattering was modeled using a long-range polarization field created at the graphene and substrate interface, has been also proposed [7.78, 7.92, 7.93]. As far as we know, essentially the same electron density dependency of the electron mobility has been obtained by using the two approaches [7.85, 7.86, 7.89, 7.92]. Thus, we deem the present approach to be applicable to the analysis of the electron mobility for graphene on the polar substrates.

Table 7.3 Parameters of the phonon scattering rates and values of the charged impurity densities.

Electron-phonon interaction parameters of graphene [7.70]			
Optical phonon energy $\hbar\omega_{op}$ (meV)		164	
Intervalley acoustic phonon energy $\hbar\omega_{ac}$ (meV)		124	
Deformation potential of acoustic phonon E_D (eV)		4.5	
Deformation field of optical phonon D_f (eV/cm)		1.0×10^9	
Deformation field of intervalley acoustic phonon D_f (eV/cm)		3.5×10^8	
Parameters of the surface optical phonon and of the charged impurity in substrates			
—	SiO_2	HfO_2	h-BN
Optical phonon energy $\hbar\omega_{op}$ (meV)	55 [7.88]	12.4 [7.88]	200 [7.89]
D_f of optical phonon (eV/cm)	5.14×10^7 [7.87]	1.29×10^9 [7.87]	1.0×10^6 to 1.29×10^9
Charged impurity density n_{imp} (cm^{-2})	2.5×10^{11} [7.82]	—	2.5×10^{10} [7.82]

The parameters of the surface OP scattering used in the simulation are listed in the lower part of Table 7.3. The surface OP energies of SiO_2, HfO_2, and h-BN are 55, 12.4, and 200 meV, respectively [7.88, 7.89], and the deformation fields for SiO_2 and HfO_2 are given by 5.14×10^7 eV/cm and 1.29×10^9 eV/cm, respectively, which give the best match with the experimental electron velocity versus electric field curves [7.87], as mentioned above. However, the deformation field for h-BN is unknown, so it is varied from 1×10^6 eV/cm to 1.29×10^9 eV/cm, spanning the range of the deformation fields for SiO_2 and HfO_2. The results indicate that the electron mobility for h-BN is independent of the value of the deformation fields. Therefore, results will be shown for the h-BN substrate obtained using the largest deformation field, 1.29×10^9 eV/cm. Scattering due to charged impurities in the substrates is taken into account by using the equation [7.85–7.87]:

$$S_{\text{imp}} = \frac{h v_F^2 n_{\text{imp}}}{20E} \tag{7.74}$$

assuming that the impurities are homogeneously distributed throughout the substrates with a fixed sheet density n_{imp}. The 2-D charge density in SiO_2 substrate has been reported [7.82] to range from 0.24 to 2.7×10^{11} cm^{-2}, whereas h-BN exhibits potential fluctuations, due to charged impurities that are one to two orders of magnitude lower than in SiO_2. Accordingly, n_{imp} is taken to be 2.5×10^{11} cm^{-2} and 2.5×10^{10} cm^{-2} for SiO_2 and h-BN, respectively. Graphene on HfO_2 with charged impurities is not simulated, for reasons given in the Section 7.6.4.

In the meantime, some other experimental electron mobilities in graphene on substrates [7.94–7.96] were reported to be significantly smaller than the values of the present study. Hence, scattering mechanisms such as the surface piezoelectric AP scattering may need to be considered to reproduce such lower electron mobilities of graphene-on-substrate samples.

7.6.3 Carrier Degeneracy

The Pauli exclusion principle for the scattering final state is treated using a rejection technique [3.13] described below. Following a scattering event, the new state is accepted or rejected with a probability that depends on the carrier distribution function in energy. The low-field electron mobility is examined, so that the carrier distribution can be approximated by the equilibrium Fermi-Dirac function with lattice temperature T and Fermi energy E_F. The relationship between the electron density and Fermi energy is needed to determine the electron density dependence of the mobility. Using the Fermi-Dirac function and the graphene density-of-states for carriers, given by:

$$g(E) = \frac{2E}{\pi \hbar^2 v_F^2} \tag{7.75}$$

the Fermi energies are calculated for various electron densities relevant to FET operation at $T = 300$ K, as listed in Table 7.4. See Appendix A for the derivation of Equation (7.75). It is seen that graphene has a Fermi energy exceeding a few hundred meV at reasonable densities and, therefore, consideration of the carrier degeneracy is necessary.

The general semiclassical Monte Carlo approach described in References [3.12] and [7.83] is employed. Various parameters, such as the average velocity and average energy, are computed by taking an ensemble average over the particles involved in the transport. Here, an ensemble of 10^5 particles is considered. The electric field E_x is 0.01 kV/cm – sufficiently small to maintain an equilibrium state of the electrons.

7.6.4 Electron Mobility Considering Surface Optical Phonon Scattering of Substrates

Figure 7.38 shows the electron mobility computed as a function of the electron density n at $T = 300$ K, where CI scattering was not included. The "intrinsic" case only considers intrinsic AP and OP scatterings. The inset is a magnified view of the results for graphene on HfO_2.

Table 7.4 Fermi energies for various electron densities at $T = 300$ K.

Density (cm^{-2})	Fermi energy (meV)
5×10^{11}	69
1×10^{12}	107
2×10^{12}	159
5×10^{12}	257
8×10^{12}	327
1×10^{13}	366

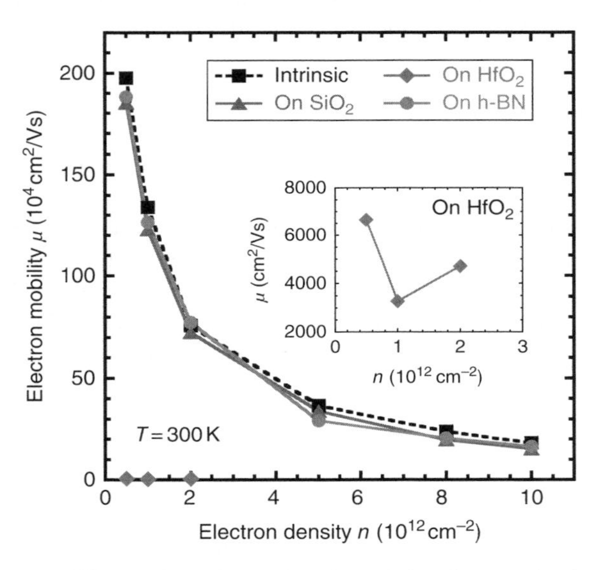

Figure 7.38 Electron mobility computed as a function of the electron density n at $T = 300$ K, not including charged impurity scattering. The "intrinsic" case only considers intrinsic AP and OP scatterings. The inset plots a magnified view of the results for graphene on HfO_2.

The largest deformation field ($D_f = 1.29 \times 10^9$ eV/cm) is used for the surface OP scattering of h-BN. Graphene on SiO_2 and h-BN exhibits slightly decreased electron mobilities, compared with the intrinsic case, whereas graphene on HfO_2 indicates a mobility that is about three orders of magnitude smaller than the others. The drastic reduction in the mobility on the HfO_2 substrate is due to its small surface OP energy of 12.4 meV and its large deformation field, as listed in Table 7.3. When the surface OP energy is smaller than the thermal energy $k_BT \approx 26$ meV at $T = 300$ K, the electrons distributed around the Fermi energy can be scattered by the surface OP, because the Pauli exclusion principle has limited influence.

The slight reduction in the electron mobility for the SiO_2 and h-BN substrates is explained as follows. The h-BN substrate has a large surface OP energy of 200 meV. Accordingly, the Pauli exclusion principle suppresses most of the surface OP scattering, thereby leading to minimal reduction in the electron mobility. On the other hand, the surface OP energy of the SiO_2 substrate is 55 meV, which is not large enough to suppress the surface OP scattering by the Pauli exclusion principle at $T = 300$ K. The rates for each phonon scattering mechanism on a SiO_2 substrate are plotted in Figure 7.39, where the scattering rates for surface OP scattering are plotted by the solid curve with solid triangles (for the surface OP emission process) and by the dashed curve with open triangles (for the surface OP absorption process).

These rates are significantly smaller than the intrinsic AP and OP emission rates in the low- and high-energy regions, respectively. Consequently, electrons in graphene on a SiO_2 substrate are inconsiderably affected by a surface OP, explaining the slight reduction in the electron mobility. As already mentioned, a surface OP in a SiO_2 or h-BN substrate has little impact on the electron mobility in graphene. However, graphene on HfO_2 suffers from significant surface OP scattering, so that HfO_2 is unsuitable as a substrate of graphene. Hereafter, analysis of only the SiO_2 and h-BN substrates is pursued.

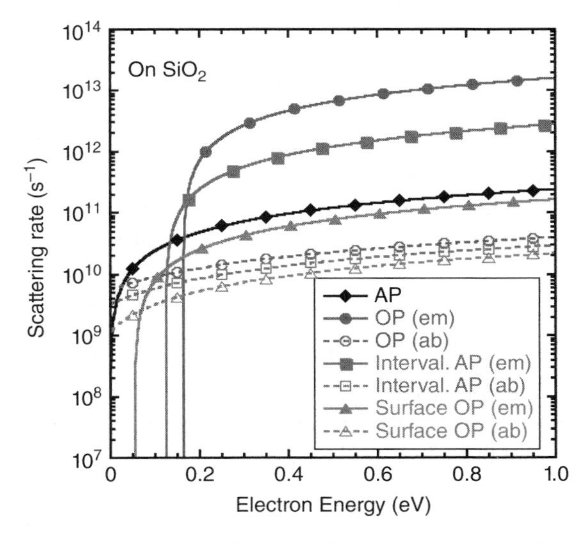

Figure 7.39 Scattering rates computed as a function of electron energy for all phonon scattering mechanisms on a SiO_2 substrate. The rates for surface OP scattering are plotted as the solid curve with solid triangles (for the surface OP emission process) and as the dashed curve with open triangles (for the surface OP absorption process).

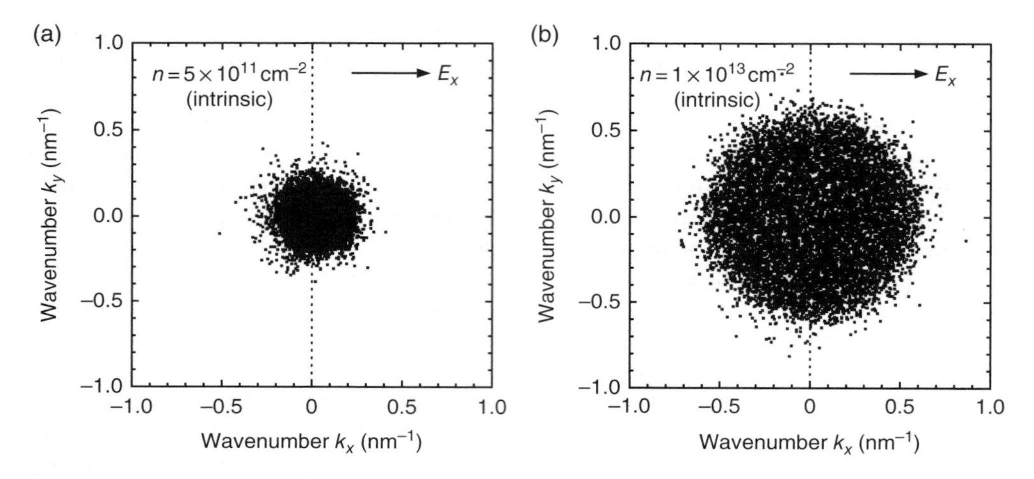

Figure 7.40 Particle distributions in 2-D **k**-space obtained by Monte Carlo simulations for (a) $n = 5 \times 10^{11}\,\text{cm}^{-2}$; and (b) $n = 10^{13}\,\text{cm}^{-2}$, accounting only for the intrinsic AP and OP in graphene. The point $k_x = k_y = 0$ is the Dirac point. The electric field is applied along the x-direction.

As seen in Figure 7.38, the Monte Carlo simulation predicts a large intrinsic mobility of approximately $2 \times 10^6\,\text{cm}^2/(\text{V}\cdot\text{s})$ for $n < 10^{12}\,\text{cm}^{-2}$, in good agreement with Reference [7.70]. However, the mobility drastically decreases by about an order of magnitude as n increases to $10^{13}\,\text{cm}^{-2}$. Similar results have been reported in the literature [7.85, 7.86, 7.89, 7.92], and have been explained using the Pauli exclusion principle.

To demonstrate the effect using a practical carrier distribution, the particle distributions are plotted in 2-D **k**-space in Figure 7.40 for densities of $5 \times 10^{11}\,\text{cm}^{-2}$ and $10^{13}\,\text{cm}^{-2}$, where only the intrinsic AP and OP in graphene are considered. The point $k_x = k_y = 0$ represents the Dirac point. The electric field is applied in the x-direction.

As n increases, the Fermi energy is pushed above the Dirac point, due to the Pauli exclusion principle, as shown in Figure 7.40(b). Since the mobility is mainly determined by electronic states close to the Fermi energy, and raised Fermi energy increases the corresponding scattering rates, the mobility decreases with an increase in the electron density. This mobility decrease with increasing electron density may lead to a performance degradation of graphene FETs in the on-state.

7.6.5 Electron Mobility Considering Charged Impurity Scattering

From the results of the previous section, surface OP scattering for SiO_2 and h-BN has limited impact on the electron mobility in graphene. Hence, it is ignored in the following calculations. Instead, CI scattering from the substrates is included.

The strength of the CI scattering depends on the 2D charge density n_{imp} in the substrates. Based on the experimental analysis of Reference [7.82], in which h-BN exhibits potential fluctuations due to charged impurities that are one or two orders of magnitude lower than for SiO_2, the value of n_{imp} is taken to be $2.5 \times 10^{11}\,\text{cm}^{-2}$ and $2.5 \times 10^{10}\,\text{cm}^{-2}$ for SiO_2 and h-BN substrates, respectively. Figure 7.41 shows the computed electron mobility when CI scattering is considered, along with the results for the "intrinsic" case for comparison.

The mobility for both substrates is considerably decreased by CI scattering. In particular, the mobility on a SiO_2 substrate decreases to 20000 cm²/(V·s), in good agreement with experimental measurements in References [7.71] to [7.73], and it is nearly independent of the electron density. Charged impurity scattering has an increasing influence on the electrons as their kinetic energy decreases, as shown in Figure 7.42, where the CI scattering rates for $n_{imp} = 2.5 \times 10^{11}\,cm^{-2}$

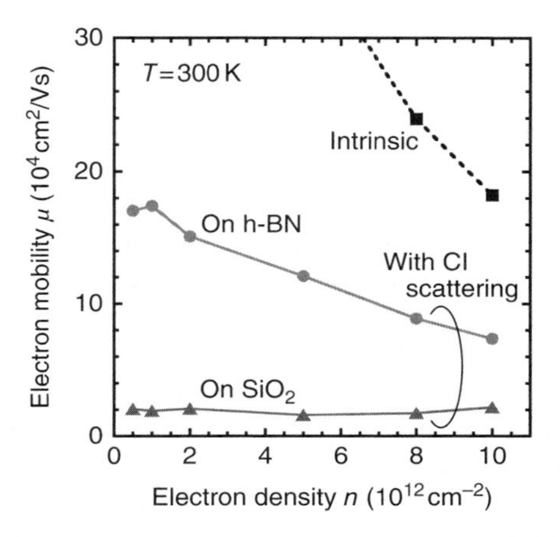

Figure 7.41 Electron mobility computed by considering charged impurity scattering and ignoring surface OP scattering of the substrates. Here n_{imp} is taken to be $2.5 \times 10^{11}\,cm^{-2}$ and $2.5 \times 10^{10}\,cm^{-2}$ for the SiO_2 and h-BN substrates, respectively. The results for the "intrinsic" case are plotted as the dashed line for comparison.

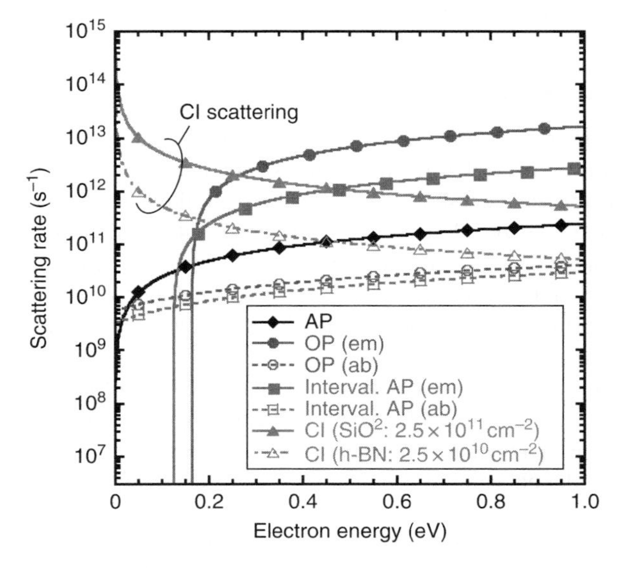

Figure 7.42 Charged impurity scattering rates for $n_{imp} = 2.5 \times 10^{11}\,cm^{-2}$ (on SiO_2) and $2.5 \times 10^{10}\,cm^{-2}$ (on h-BN), along with those for intrinsic phonon scattering. The charged impurity scattering has increasing influence on the electrons as their kinetic energy becomes smaller.

(on SiO_2) and 2.5×10^{10} cm^{-2} (on h-BN) are plotted, together with those for intrinsic phonon scattering. When the electron density is small, electrons in graphene are heavily scattered by charged impurities, because the Fermi energy is then small (*cf.* Table 7.4). As a result, the electron mobility drastically decreases with decreasing electron density and, thus, no electron density dependence of the mobility is observed in a SiO_2 substrate with charged impurities.

The mobility on a h-BN substrate is 170 000 cm^2/(V·s) for $n < 10^{12}$ cm^{-2}, which agrees with Reference [7.81]. This mobility arises from the smaller value of n_{imp} in the h-BN substrate, as confirmed by the smaller CI scattering rate in Figure 7.42. Therefore, the recently observed high electron mobility of 140 000 cm^2/(V·s) for a h-BN substrate [7.82] can be confirmed to be due to its small charged impurity density. A h-BN substrate is thus an appealing choice for graphene devices.

In summary, the electron mobility in graphene on three substrates – SiO_2, HfO_2, and h-BN – has been calculated by considering the extrinsic scattering due to the surface OP and charged impurities in the substrates, as well as the intrinsic phonon scattering in the graphene, based on a semiclassical Monte Carlo method. By investigating the influence of the surface OP scattering, HfO_2 was shown to be unsuitable as a substrate for graphene. On the other hand, the surface OP in the SiO_2 and h-BN substrates was found to have limited impact on the electron mobility in graphene. Therefore, these two substrates were further examined by considering scattering due to charged impurities in the substrates. The mobility for both substrates decreases by CI scattering, but the mobility for the h-BN substrate maintains a relatively high value of 170 000 cm^2/(V·s) for electron densities below 10^{12} cm^{-2}.

This result was confirmed as being due to the smaller charged impurity density in the h-BN substrate. Good agreement is found between the calculated results and the experimental measurements. For the SiO_2 and h-BN substrates, the main scattering mechanism leading to mobility reduction is charged impurities in the substrates and, in that context, h-BN substrate is the best choice for graphene devices.

7.7 Germanane MOSFETs

Germanane is a hydrogen-terminated Ge monolayer in which every Ge atom is terminated with hydrogen atoms above or below the layer, as shown in Figure 7.5(c). Bianco *et al.* have synthesized millimeter-scale crystals of germanane from the topochemical deintercalation of $CaGe_2$, and have characterized their long-term resistance to oxidation and their thermal stability [7.18]. They also have shown that it has a direct bandgap larger than 1.5 eV without a nanoribbon structure and an electron mobility approximately five times higher than that of bulk Ge, using first-principles calculation. Therefore, germanane has great potential for a wide range of optoelectronic and sensing applications and for a high-speed FET channel material.

As shown in Figure 7.5 (c), germanane is a 2-D material similar to graphene, but it maintains sp^3-like orbitals rather than sp^2 orbitals, owing to the hydrogen termination above or below the layer [7.18]. As a consequence, germanane can be thought of as hydrogen-terminated isolated (111) sheets of Ge and, conveniently, no dangling bonds are generated at the surfaces [7.15].

Since ultra-scaled germanane transistors have not yet been fabricated, device simulation plays an alternative important role in exploring the characteristics of future germanane transistors and assessing their potential in advance. So far, several theoretical studies have focused on this research topic. However, the potential of germanane in ultimate transistor scaling technology is poorly understood, and systematic study of germanane transistors is crucially lacking. In this section, the potential of germanane as the FET channel material is evaluated using the atomistic simulation techniques explained in Section 6.1, which combines the electron band structure calculation based on the $sp^3d^5s^*$ TB approach, and the phonon full-band structure calculation using the Keating potential approach, with the semiclassical Boltzmann transport equation [6.8].

7.7.1 Atomic Model for Germanane Nanoribbon Structure

In view of the application to FET channel, channel materials have a finite width which is almost comparable to the channel length in the state-of-the-art technologies. Thus, we adopted a germanane nanoribbon structure with a finite width, as shown in Figure 7.43(a), where the electron confinement due to the monolayer and nanoribbon structures occurs in the $\langle 11\bar{1}\rangle$ and $\langle \bar{1}10\rangle$ directions, respectively, and electrons transport along the <112>direction.

Note that the infinitely-long germanane nanoribbon model is considered here, because this study is intended to clarify the band structure and the electron mobility of germanane nanoribbon channels. As described later, the ribbon width considered here, W, extends from a few nanometers to a few ten nanometers, which is larger than the values chosen for A-SiNR, A-GeNR and A-GNR discussed in Section 7.4. The edge configuration was assumed to be an armchair-edge structure, as shown in Figure 7.43(b). In particular, as found from Figure 7.43(c), germanane have an analogous geometry to a Ge (111) surface in which every Ge atom is terminated with hydrogen atoms above or below the layer, and thus no dangling bonds are generated at the surfaces. Chair-like morphology is employed here, because it is energically more stable than other morphology, such as boat-like one [7.97].

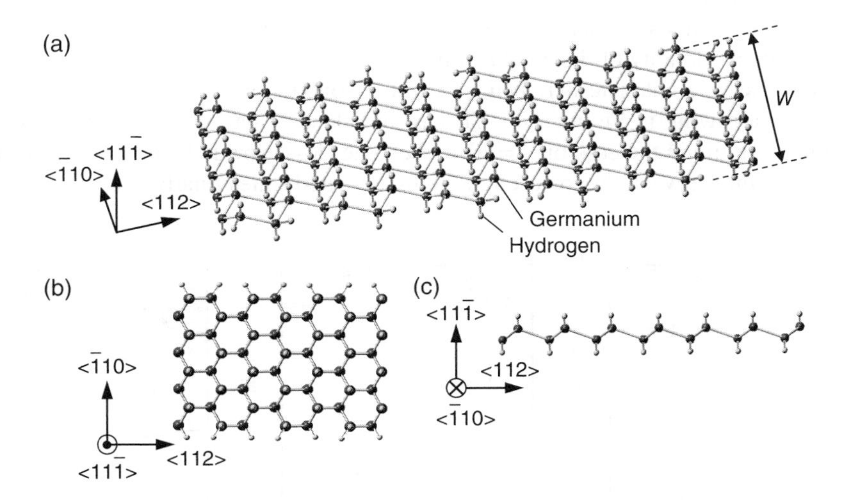

Figure 7.43 (a) Overall view of atomic model for germanane nanoribbon structure used in the simulation. (b) Atomic model looking from the $\langle 11\bar{1}\rangle$ direction. The edge configuration at the ends of the width direction was assumed to be armchair. (c) Atomic model looking from the $\langle 1 1\bar{0}\rangle$ direction. Germanane is confirmed to have an analogous geometry to a Ge (111) surface in which every Ge atom is terminated with hydrogen atoms above or below the layer. Chair-like morphology is employed here.

In addition, since it was reported that the sp^3 bonding network is well maintained, and there is no notable structure distortion in the germanane nanoribbons after relaxation [7.98], we adopted the lattice parameters of bulk Ge in the present simulations: Specifically, the lattice constant $a = 5.66$ Å, the buckling distance $\Delta z = 0.816$ Å, and Ge-Ge distance = 2.449 Å.

7.7.2 Band Structure and Electron Effective Mass

Figure 7.44 shows the band structures computed for germanane nanoribbons with $W = 4$, 11.6 and 20 nm. As in the first-principles calculation [7.98], the present TB calculation suggests that germanane nanoribbons are a direct bandgap material. This is due to the k-space projection of bulk L valleys into the $\langle 11\bar{1} \rangle$ plane, and they has a smaller subband energy than the Γ and X valleys similarly projected into the $\langle 11\bar{1} \rangle$ plane.

Figure 7.45 shows the ribbon width dependence of the bandgap energy of germanane nanoribbons. It is clearly found that the bandgap drastically increases if W becomes narrower than about 10 nm, and this is due to the quantum confinement effect in the width direction. Furthermore, since the bandgap decreases to converge on about 1.4 eV with increasing W, the germanane nanoribbons are found to be always semiconducting [7.98].

Next, we evaluated the electron effective masses of the germanane nanoribbons. Figure 7.46 shows the conduction band structures for germanane nanoribbons with $W = 4$, 11.6, and 20 nm, where the effective mass in the lowest subband is indicated in each figure.

It is found that the germanane nanoribbons have lighter effective mass than that of bulk Ge ($0.082 m_0$) [7.97, 7.99]. This may be partially due to a residual biaxial tensile strain to the germanane nanoribbons [7.97] because of the employment of the bulk lattice parameters. As W decreases, the effective mass increases, similarly to the bandgap result. The detailed ribbon width dependence of the effective mass is shown in Figure 7.47, which reconfirms that the effective mass increases when $W <\approx 10$ nm. The increase in the effective mass may be due to the band nonparabolicity of the conduction band L valleys.

Here, we observe a characteristic behavior in the conduction band structure owing to the ribbon width narrowing in Figure 7.46. That is, the closely-spaced dispersion curves at the conduction minimum in the case of $W = 20$ nm becomes less crowded with decreasing W, and the energy spacing between the lowest and the first excited states finally becomes larger than

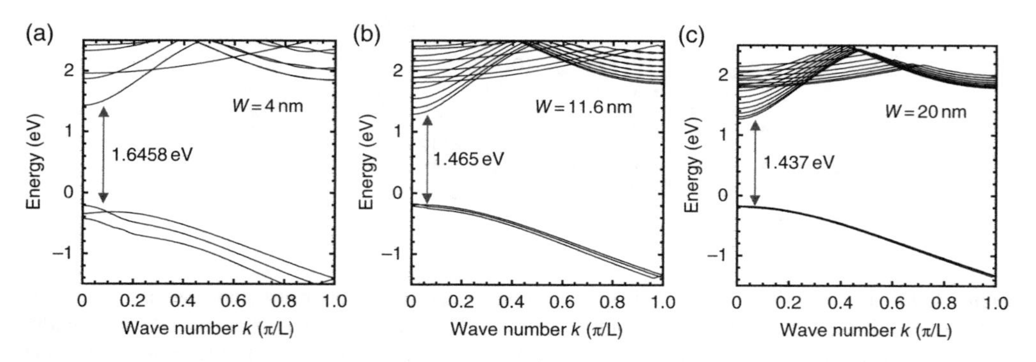

Figure 7.44 Band structures computed for germanane nanoribbons with $W =$ (a) 4; (b) 11.6; and (c) 20 nm. Note that germanane nanoribbons are a direct bandgap material.

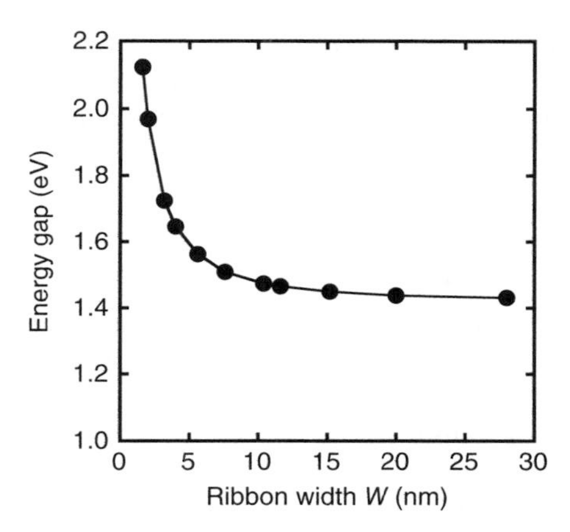

Figure 7.45 Ribbon width dependence of band gap energy of germanane nanoribbons.

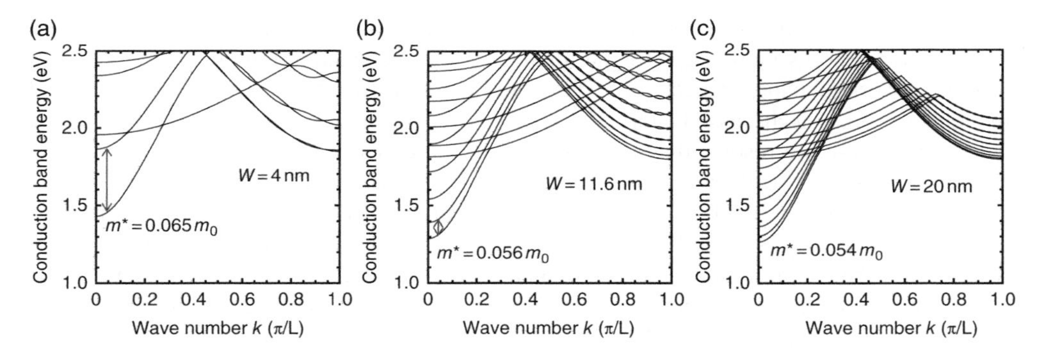

Figure 7.46 Conduction band structures for germanane nanoribbons with W=(a) 4; (b) 11.6; and (c) 20 nm, where the effective mass in the lowest subband is indicated in each figure.

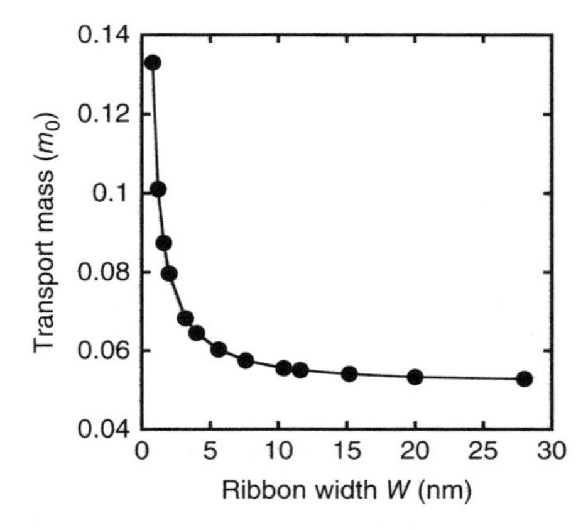

Figure 7.47 Ribbon width dependence of electron effective mass of germanane nanoribbons.

severalfold thermal energy at room temperature for $W<11.6$ nm. This means that the intersubband phonon scattering decreases as W decreases and, therefore, we can expect a mobility enhancement due to the ribbon width narrowing. This speculation will be examined in the next section.

7.7.3 Electron Mobility

Finally, we computed the electron mobility of the germanane nanoribbons. Before showing the result, the phonon band structures in a low energy regime are presented in Figure 7.48 for $W=4$ nm, 11.6 nm, and 20 nm.

There are many modes exhibiting $\omega \propto q^2$ dispersions, which represents a mixed state of transverse and longitudinal acoustic modes [6.8]. Aside from those mixed modes, pure transverse and longitudinal acoustic modes with $\omega \propto q$ dispersion also exist in a long wavelength regime (i.e., $q \approx 0$). Considering the interactions between those phonon modes and the electron modes shown in Figure 7.46, we obtained the electron mobility of the germanane nanoribbons as a function of W, as shown in Figure 7.49(a). The temperature is 300 K. Here, note that the solid and the dashed lines correspond to the results obtained considering multisubbands with energy up to 0.2 eV higher than the lowest subband minimum and considering only the lowest subband, respectively. This is to highlight the influence of the intersubband phonon scattering, as discussed later.

From the solid line results, the electron mobility increases with decreasing W from 28 to 11.6 nm, which agrees with our speculation described in the previous section. In Figure 7.49(b), (c), and (d), the conduction band structures corresponding to $W=28$, 11.6, and 4 nm are plotted, and the number of subbands, *Nsub*, locating within the energy range of about $3k_BT$ from the lowest subband minimum are also indicated in each figure. In accordance with our expectation, the electron mobility increases with decreasing *Nsub*. On the other hand, when W further becomes smaller than 11.6 nm, the electron mobility turns to decline, as shown in Figure 7.49(a). This is due to the fact that the electron-phonon interaction becomes stronger with decreasing W [6.8], and also the effective mass increases in the sub-10 nm nanoribbons, as shown in Figure 7.47.

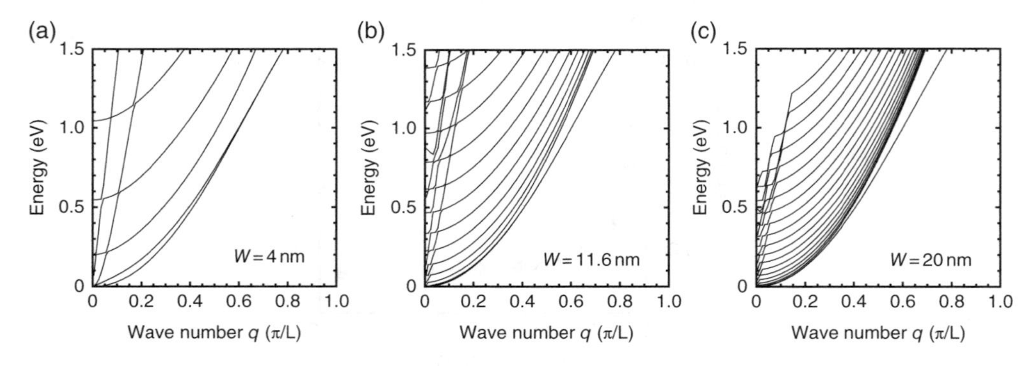

Figure 7.48 Phonon band structures computed for germanane nanoribbons with $W=$(a) 4; (b) 11.6; and (c) 20 nm.

It is interesting to compare between the solid and the dashed lines in Figure 7.49(a), which reveals that when $W > 11.6$ nm (i.e., when the multisubbands are involved), the electron mobility substantially decreases from the results considering only the lowest subband, as shown in the solid line. This suggests that the intersubband phonon scattering significantly decreases the electron mobility and, as a result, it produces the characteristic ribbon width dependence of the electron mobility, especially producing the maximum electron mobility at $W = 11.6$ nm. The maximum electron mobility is 1332 $cm^2/V \cdot s$ in the present simulation for germanane nanoribbons, which is much smaller than the value reported for 2-D germanane in [7.18]. However, it is about 10 or 100 times larger than that of MoS_2 monolayer transistors

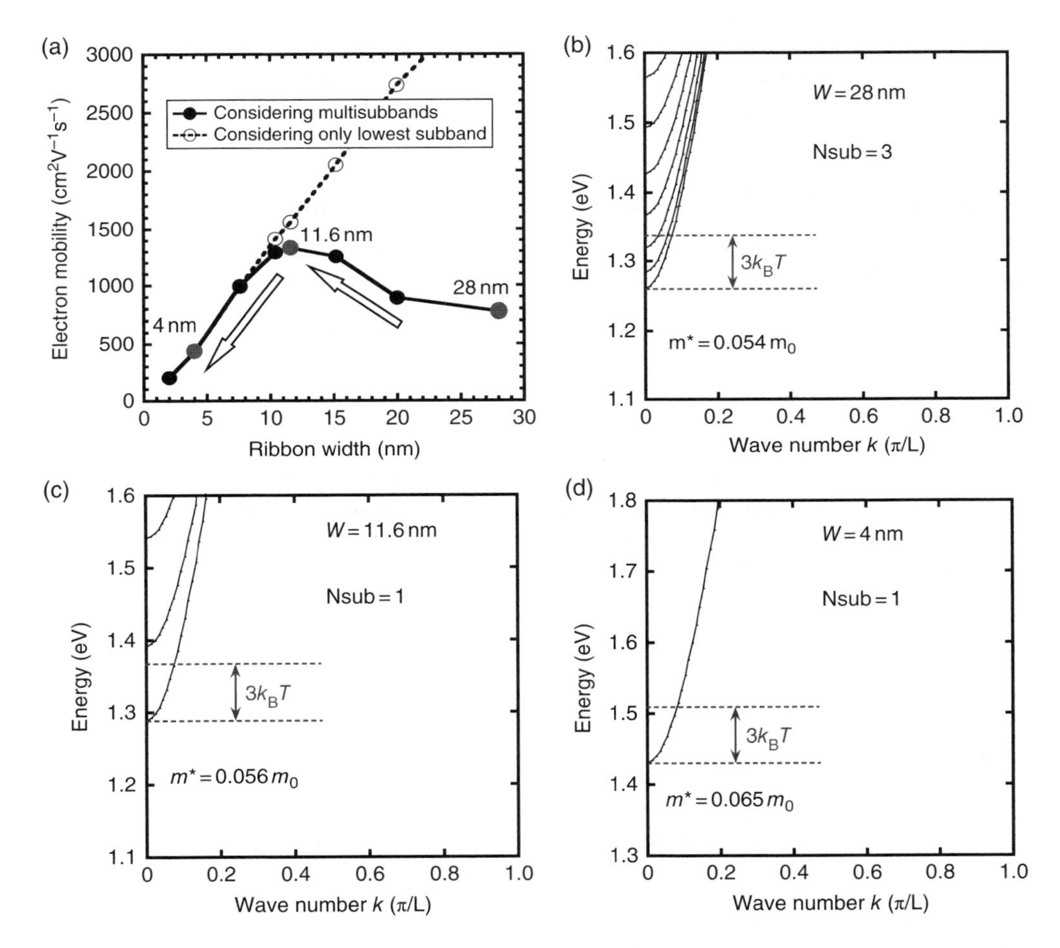

Figure 7.49 (a) Computed electron mobility of germanane nanoribbons as a function of W. Note that the solid and the dashed lines correspond to the results obtained considering multisubbands with energy up to 0.2 eV higher than the lowest subband minimum and considering only the lowest subband, respectively. (b) Conduction band structures with $W = 28$ nm; (c) $W = 11.6$ nm; and (d) $W = 4$ nm, also indicated with the number of subbands, *Nsub*, locating within the energy range of about $3k_BT$ from the lowest subband minimum.

[7.19, 7.20]. Accordingly, germanane channels are found to have the potential to outperform the transition metal dichalcogenide 2-D materials.

In this section, the electronic properties of germanane nanoribbons were examined using the atomistic simulator. First, it was shown that the germanane nanoribbons are a direct bandgap semiconductor with a bandgap energy larger than about 1.4 eV. Next, the germanane nanoribbons were shown to have a lighter effective mass than that of bulk Ge but, when the ribbon width decreases to less than about 10 nm, the effective mass increases along with the bandgap. Finally, the electron mobility of germanane nanoribbons was computed, and it was shown to be significantly decreased by the intersubband phonon scattering. As a result, the electron mobility strongly depends on the ribbon width (i.e. the number of subbands), which produced the maximum electron mobility at $W = 11.6$ nm. It is important to note that this ribbon width is approximately one order of magnitude greater than those for A-SiNR, A-GeNR and A-GNR with a reasonable bandgap. Therefore, practical fabrication will become far less difficult in germanane FETs.

Appendix A: Density-of-states for Carriers in Graphene

Considering that graphene has two full Dirac cones inside the first BZ and also has the spin degeneracy of two, total number of states from 0 to E in energy, $N(E)$, is represented as:

$$N(E) = \frac{\pi k^2}{\left(\frac{2\pi}{L}\right)^2} \times 2 \times 2 = \frac{k^2}{\pi} = \frac{SE^2}{\pi \hbar^2 v_F^2} \tag{A.1}$$

where $S = L^2$ denotes the area of graphene, and the linear dispersion relation in Equation (7.71) is used in the second equal sign. Accordingly, the density-of-states for carriers per unit area and unit energy is obtained by differentiating Equation (A.1) with respect to the energy and further dividing it by S, as follows.:

$$g(E) = \frac{1}{S} \frac{\partial N(E)}{\partial E} = \frac{2E}{\pi \hbar^2 v_F^2} \tag{A.2}$$

References

[7.1] K.S. Novoselov, A.K. Geim, S.V. Morozov, D. Jiang, Y. Zhang, S.V. Dubonos, I.V. Grigorieva and A.A. Firsov (Oct. 2004). Electric field effect in atomically thin carbon films. *Science* **306**, 666–669.

[7.2] K.I. Bolotin, K.J. Sikes, Z. Jiang, M. Klima, G. Fudenberg, J. Hone, P. Kim and H.L. Stormer (Mar. 2008). Ultrahigh electron mobility in suspended graphene. *Solid State Communications* **146**, 351–355.

[7.3] P. Neugebauer, M. Orlita, C. Faugeras, A.-L. Barra and M. Potemski (Sep. 2009). How perfect can graphene be? *Physical Review Letters* **103**(13), p. 136403.

[7.4] K. Takeda and K. Shiraishi (Nov. 1994). Theoretical possibility of stage corrugation in Si and Ge analogs of graphite. *Physical Review B* **50**(20), 14916–14922.

[7.5] A. Fleurence, R. Friedlein, T. Ozaki, H. Kawai, Y. Wang and Y. Yamada-Takamura (June 2012). Experimental evidence for epitaxial silicene on diboride thin films. *Physical Review Letters* **108**(24), p. 245501.

[7.6] A.H. Castro Neto, F. Guinea, N.M.R. Peres, K.S. Novoselov and A.K. Geim (Jan.–Mar. 2009). The electronic properties of graphene. *Reviews of Modern Physics* **81**(1), 109–162.

[7.7] P.O. Löwdin (Mar. 1950). On the non-orthogonality problem connected with the use of atomic wave functions in the theory of molecules and crystals. *Journal of Chemical Physics* **18**(3), 365–375.
[7.8] H. Min, B. Sahu, S.K. Banerjee and A.H. MacDonald (Apr. 2007). *Ab initio* theory of gate induced gaps in graphene bilayers. *Physical Review B* **75**(15), p. 155115.
[7.9] P.R. Wallace (May 1947). The band theory of graphite. *Physical Review* **71**(9), 622–634.
[7.10] S. Cahangirov, M. Topsakal, E. Aktürk, H. Şahin and S. Ciraci (June 2009). Two- and one-dimensional honeycomb structures of silicon and germanium. *Physical Review Letters* **102**(23), p. 236804.
[7.11] J.-A. Yan, R. Stein, D.M. Schaefer, X.-Q. Wang and M.Y. Chou (Sep. 2013). Electron-phonon coupling in two-dimensional silicene and germanene. *Physical Review B* **88**(12), p. 121403(R).
[7.12] G.L. Lay, B. Aufray, C. Léandri, H. Oughaddou, J.-P. Biberian, P.D. Padova, M.E. Dávila, B. Ealet and A. Kara (Aug. 2009). Physics and chemistry of silicene nano-ribbons. *Applied Surface Science* **256**, 524–529.
[7.13] S. Cahangirov, M. Topsakal and S. Ciraci (May 2010). Armchair nanoribbons of silicon and germanium honeycomb structures. *Physical Review B* **81**(19), p. 195120.
[7.14] Y.-W. Son, M.L. Cohen and S.G. Louie (Nov. 2006). Energy gaps in graphene nanoribbons. *Physical Review Letters* **97**(21), p. 216803.
[7.15] H. Tsuchiya, S. Kaneko, N. Mori and H. Hirai (May 2015). Simulation of electron transport in atomic monolayer semiconductor FETs. *Journal of Advanced Simulation in Science and Engineering* **2**(1), 127–152.
[7.16] M.Y. Han, B. Özyilmaz, Y. Zhang and P. Kim (May 2007). Energy band-gap engineering of graphene nanoribbons. *Physical Review Letters* **98**(20), p. 206805.
[7.17] X. Li, X. Wang, L. Zhang, S. Lee and H. Dai (Feb. 2008). Chemically derived, ultrasmooth graphene nanoribbon semiconductors. *Science* **319**, 1229–1232.
[7.18] E. Bianco, S. Butler, S. Jiang, O.D. Restrepo, W. Windl and J.E. Goldberger (Mar. 2013). Stability and exfoliation of germanane: A germanium graphane analogue. *ACS Nano* **7**(5), 4414–4421.
[7.19] B. Radisavljevic, A. Radenovic, J. Brivio, V. Giacometti and A. Kis (Mar. 2011). Single-layer MoS_2 transistors. *Nature Nanotechnology* **6**, 147–150.
[7.20] M.S. Fuhrer and J. Hone (Mar. 2013). Measurement of mobility in dual-gated MoS_2 transistors. *Nature Nanotechnology* **8**, 146–147.
[7.21] J. Kang, W. Liu and K. Banerjee (Mar. 2014). High-performance MoS2 transistors with low-resistance molybdenum contacts. *Applied Physics Letters* **104**(9), p. 093106.
[7.22] S.P. Koenig, R.A. Doganov, H. Schmidt, A.H. Castro Neto and B. Özyilmaz (Mar. 2014). Electric field effect in ultrathin black phosphorus. *Applied Physics Letters* **104**(10), p. 103106.
[7.23] A.N. Rudenko and M.I. Katsnelson (May 2014). Quasiparticle band structure and tight-binding model for single- and bilayer black phosphorus. *Physical Review B* **89**(20), p. 201408(R).
[7.24] T. Ohta, A. Bostwick, T. Seyller, K. Horn and E. Rotenberg (Aug. 2006). Controlling the electronic structure of bilayer graphene. *Science* **313**, 951–954.
[7.25] S. Zhou, G.-H. Gweon, A. Fedorov, P. First, W. de Heer, D.-H. Lee, F. Guinea, A. Castro Neto and A. Lanzara (Oct. 2007). Substrate-induced bandgap opening in epitaxial graphene. *Nature Materials* **6**, 770–775.
[7.26] G. Giovannetti, P. Khomyakov, G. Brockes, P. Kelly and J. van den Brink (Aug. 2007). Substrate-induced band gap in graphene on hexagonal boron nitride: *Ab initio* density functional calculations. *Physical Review B* **76**(7), p. 073103.
[7.27] E. McCann (Oct. 2006). Asymmetry gap in the electronic band structure of bilayer graphene. *Physical Review B* **74**(16), p. 161403(R).
[7.28] N. Harada, M. Ohfuti and Y. Awano (Feb. 2008). Performance estimation of graphene field-effect transistors using semiclassical Monte Carlo simulation. *Applied Physics Express* **1**, p. 024002.
[7.29] H. Hosokawa, R. Sako, H. Ando and H. Tsuchiya (Nov. 2010). Performance comparisons of bilayer graphene and graphene nanoribbon field-effect transistors under ballistic transport. *Japanese Journal of Applied Physics* **49**, p. 110207.
[7.30] G. Liang, N. Neophytou, D.E. Nikonov and M.S. Lundstrom (Apr. 2007). Performance projections for ballistic graphene nanoribbon field-effect transistors. *IEEE Transactions on Electron Devices* **54**(4), 677–682.
[7.31] D. Gunlycke and C.T. White, Tight-binding energy dispersions of armchair-edge graphene nanostrips. *Physical Review B* **77**(11), p. 115116, Mar. 2008.
[7.32] H. Raza and E.C. Kan (June 2008). Armchair graphene nanoribbons: Electronic structure and electric-field modulation. *Physical Review B* **77**(24), p. 245434.
[7.33] R. Sako, H. Hosokawa and H. Tsuchiya (Jan. 2011). Computational study of edge configuration and quantum confinement effects on graphene nanoribbon transport. *IEEE Electron Device Letters* **32**(1), 6–8.

[7.34] M. Fujita, K. Wakabayashi, K. Nakada and K. Kusakabe (July 1996). Peculiar localized state at zigzag graphite edge. *Journal of the Physical Society of Japan* **65**(7), 1920–1923.

[7.35] B. Obradovic, R. Kotlyar, F. Heinz, P. Matagne, T. Rakshit, D. Nikonov, M.D. Giles and M.A. Stettler (Apr. 2006). Analysis of graphene nanoribbons as a channel material for field-effect transistors. *Applied Physics Letters* **88**(14), p. 142102.

[7.36] K. Wakabayashi (Sep. 2001). Electronic transport properties of nanographite ribbon junctions. *Physical Review B* **64**(12), p. 125428.

[7.37] K. Nakada, M. Fujita, G. Dresselhaus and M. Dresselhaus (June 1996). Edge state in graphene ribbons: Nanometer size effect and edge shape dependence. *Physical Review B* **54**(24), 17954–17961.

[7.38] C. Berger, Z. Song, X. Li, X. Wu, N. Brown, C. Naud, D. Mayou, T. Li, J. Hass, A.N. Marchenkov, E.H. Conrad, P.N. First and W.A. de Heer (May 2006). Electronic confinement and coherence in patterned epitaxial graphene. *Science* **312**, 1191–1196.

[7.39] Y. Zhang, J.P. Small, W.V. Ponyius and P. Kim (Feb. 2005). Fabrication and electric field-dependent transport measurements of mesoscopic graphite. *Applied Physics Letters* **86**(7), p. 073104.

[7.40] J. Cai, P. Ruffieux, R. Jaafar, M. Bieri, T. Braun, S. Blankenburg, M. Muoth, A.P. Seitsonen, M. Saleh, X. Feng, K. Müllen and R. Fasel (July 2010). Atomically precise bottom-up fabrication of graphene nanoribbons. *Nature* **466**, 470–473.

[7.41] H. Miyazaki, K. Tsukagoshi, A. Kanda, M. Otani and S. Okada (Aug. 2010). Influence of disorder on conductance in bilayer graphene under perpendicular electric field. *Nano Letters* **10**, 3888–3892.

[7.42] P. Zhao, J. Chauhan and J. Guo (2009). Computational study of tunneling transistor based on graphene nanoribbon. *Nano Letters* **9**(2), 684–688.

[7.43] S. Reich, J. Maultzsch and C. Thomsen (July 2002). Tight-binding description of graphene. *Physical Review B* **66**(3), p. 035412.

[7.44] N. Hasegawa, K. Shimoida, H. Tsuchiya, Y. Kamakura, N. Mori and M. Ogawa (Sep. 2013). *Performance comparison of graphene nanoribbon, Si nanowire and InAs nanowire FETs in the ballistic transport limit.* Extended Abstracts of International Conference on Solid State Devices and Materials (SSDM), Fukuoka, 664–665.

[7.45] Y. Ouyang, Y. Yoon and J. Guo (2008). Edge chemistry engineering of graphene nanoribbon transistors: A computational study. *IEDM Technical Digest*, 517–520.

[7.46] J. Guo, A. Javey, H. Dai and M. Lundstrom (2004). Performance analysis and design optimization of near ballistic carbon nanotube field-effect transistors. *IEDM Technical Digest*, 703–706.

[7.47] Y.M. Banadaki and A. Srivastava (June 2015). Investigation of the width-dependent static characteristics of graphene nanoribbon field effect transistors using non-parabolic quantum-based model. *Solid-State Electronics* **111**, 80–90.

[7.48] R. Sako, H. Tsuchiya and M. Ogawa (Oct. 2011). Influence of band-gap opening on ballistic electron transport in bilayer graphene and graphene nanoribbon FETs. *IEEE Transactions on Electron Devices* **58**(10) 3300–3306.

[7.49] H. Tsuchiya, H. Hosokawa, R. Sako, N. Hasegawa and M. Ogawa (Apr. 2012). Theoretical evaluation of ballistic electron transport in field-effect transistors with semiconducting graphene channels. *Japanese Journal of Applied Physics* **51**, p. 055103.

[7.50] E. Sano and T. Otsuji (Apr. 2009). Theoretical evaluation of channel structure in graphene field-effect transistors. *Japanese Journal of Applied Physics* **48**, p. 041202.

[7.51] H. Hosokawa, R. Sako, H. Ando and H. Tsuchiya (Nov. 2010). Performance comparisons of bilayer graphene and graphene nanoribbon field-effect transistors under ballistic transport. *Japanese Journal of Applied Physics* **49**, p. 110207.

[7.52] G. Fiori and G. Iannaccone (Aug. 2007). Simulation of graphene nanoribbon field-effect transistors. *IEEE Electron Device Letters* **28**(8), 760–762.

[7.53] T. Fang, A. Konar, H. Xing and D. Jena (Nov. 2008). Mobility in semiconducting graphene nanoribbons: Phonon, impurity and edge roughness scattering. *Physical Review B* **78**(20), p. 205403.

[7.54] Y. Yang and R. Murali (Mar. 2010). Impact of size effect on graphene nanoribbon transport. *IEEE Electron Device Letters* **31**(3), 237–239.

[7.55] A. Yazdanpanah, M. Pourfath, M. Fathipour, H. Kosina and S. Selberherr (Feb. 2012). A numerical study of line-edge roughness scattering in graphene nanoribbons. *IEEE Transactions on Electron Devices* **59**(2), 433–440.

[7.56] D. Basu, M. J. Gilbert, L.F. Register and S.K. Banerjee (Jan. 2008). Effect of edge roughness on electronic transport in graphene nanoribbon channel metal-oxide-semiconductor field-effect transistors. *Applied Physics Letters* **92**(4), p. 042114.

[7.57] J. Bai, X. Zhong, S. Jiang, Y. Huang and X. Duan (Mar. 2010). Graphene nanomesh. *Nature Nanotechnology* **5**, 190–194.

[7.58] X. Liang, Y.-S. Jung, S. Wu, A. Ismach, D.L. Olynick, S. Cabrini and J. Bokor (June 2010). Formation of bandgap and subbands in graphene nanomeshes with sub-10 nm ribbon width fabricated via nanoimprint lithography. *Nano Letters* **10**, 2454–2460.

[7.59] R. Sako, N. Hasegawa, H. Tsuchiya and M. Ogawa (Apr. 2013). Computational study on band structure engineering using graphene nanomeshes. *Journal of Applied Physics* **113**(14), p. 143702.

[7.60] E.V. Castro, K. S. Novoselov, S.V. Morozov, N.M.R. Peres, J.M.B. Lopes dos Santos, J. Nilsson, F. Guinea, A.K. Geim and A.H. Castro Neto (Nov. 2007). Biased bilayer graphene: Semiconductor with a gap tunable by the electric field effect. *Physical Review Letters* **99**, p. 216802.

[7.61] G. Fiori and G. Iannaccone (Mar. 2009). On the possibility of tunable-gap bilayer graphene FET. *IEEE Electron Device Letters* **30**(3), 261–264.

[7.62] G. Liang, N. Neophytou, M.S. Lundstrom and D.E. Nikonov (Sep. 2007). Ballistic graphene nanoribbon metal-oxide-semiconductor field-effect transistors: A full real-space quantum transport simulation. *Journal of Applied Physics* **102**(5), p. 054307.

[7.63] G. Liang, N. Neophytou, M.S. Lundstrom and D.E. Nikonov (June 2008). Contact effects in graphene nanoribbon transistors. *Nano Letters* **8**(7), 1819–1824.

[7.64] N. Harada, S. Sato and N. Yokoyama (Aug. 2013). Theoretical investigation of graphene nanoribbon field-effect transistors designed for digital applications. *Japanese Journal of Applied Physics* **52**, p. 094301.

[7.65] K. Nagashio, T. Nishimura, K. Kita and A. Toriumi (May 2010). Systematic investigation of the intrinsic channel properties and contact resistance of monolayer and multilayer graphene field-effect transistor. *Japanese Journal of Applied Physics* **49**, p. 051304.

[7.66] S. Kaneko, H. Tsuchiya, Y. Kamakura, N. Mori and M. Ogawa (Mar. 2014). Theoretical performance estimation of silicene, germanene and graphene nanoribbon field-effect transistors under ballistic transport. *Applied Physics Express* **7**, p. 035102.

[7.67] International Technology Roadmap for Semiconductors (2013) [http://www.itrs.net/].

[7.68] C. Clendennen, N. Mori and H. Tsuchiya (May 2015). Non-equilibrium Green function simulations of graphene, silicene and germanene nanoribbon field-effect transistors. *Journal of Advanced Simulation in Science and Engineering* **2**(1), 171–177.

[7.69] S. Datta (1995). *Electronic Transport in Mesoscopic Systems*. Cambridge University Press, Cambridge.

[7.70] K.M. Borysenko, J.T. Mullen, E.A. Barry, S. Paul, Y.G. Semenov, J.M. Zavada, M. Buongiorno Nardelli and K.W. Kim (Mar. 2010). First-principles analysis of electron-phonon interactions in graphene. *Physical Review B* **81**(12), p. 121412(R).

[7.71] Y.-W. Tan, Y. Zhang, K. Bolotin, Y. Zhao, S. Adam, E.H. Hwang, S. Das Sarma, H.L. Stormer and P. Kim (Dec. 2007). Measurement of scattering rate and minimum conductivity in graphene. *Physical Review Letters* **99**(24), p. 246803.

[7.72] S. Cho and M.S. Fuhrer (Feb. 2008). Charge transport and inhomogeneity near the minimum conductivity point in graphene. *Physical Review B* **77**(8), p. 081402(R).

[7.73] J. Yan and M.S. Fuhrer (Nov. 2011). Correlated charged impurity scattering in graphene. *Physical Review Letters* **107**(20), p. 206601.

[7.74] E. Rossi, S. Adam and S. Das Sarma (June 2009). Effective medium theory for disordered two-dimensional graphene. *Physical Review B* **79**(24), p. 245423.

[7.75] J.-H. Chen, C. Jang, S. Xiao, M. Ishigami and M.S. Fuhrer (Apr. 2008). Intrinsic and extrinsic performance limits of graphene devices on SiO_2. *Nature Nanotechnology* **3**, 206–209.

[7.76] J. Martin, N. Akerman, G. Ulbricht, T. Lohmann, J.H. Smet, K. von Klitzing and A. Yacoby (Feb. 2008). Observation of electron-hole puddles in graphene using a scanning single-electron transistor. *Nature Physics* **4**, 144–148.

[7.77] E.H. Hwang and S. Das Sarma (Mar. 2008). Acoustic phonon scattering limited carrier mobility in two-dimensional extrinsic graphene. *Physical Review B* **77**(11), p. 115449.

[7.78] S. Fratini and F. Guinea (May 2008). Substrate-limited electron dynamics in graphene. *Physical Review B* **77**(19), p. 195415.

[7.79] M. Ishigami, J.H. Chen, W.G. Cullen, M.S. Fuhrer and E.D. Williams (June 2007). Atomic structure of graphene on SiO_2. *Nano Letters* **7**(6), 1643–1648.

[7.80] S.V. Morozov, K.S. Noveselov, M.I. Katsnelson, F. Schedin, D.C. Elias, J.A. Jaszczak and A.K. Geim (Jan. 2008). Giant intrinsic carrier mobilities in graphene and its bilayer. *Physical Review Letters* **100**(1), p. 016602.

[7.81] C.R. Dean, A.F. Young, I. Meric, C. Lee, L. Wang, S. Sorgenfrei, K. Watanabe, T. Taniguchi, P. Kim, K.L. Shepard and J. Hone (Oct. 2010). Boron nitride substrates for high-quality graphene electronics. *Nature Nanotechnology* **5**, 722–726.

[7.82] K.M. Burson, W.G. Cullen, S. Adam, C.R. Dean, K. Watanabe, T. Taniguchi, P. Kim and M.S. Fuhrer (July 2013). Direct imaging of charged impurity density in common graphene substrates. *Nano Letters* **13**, 3576–3580.

[7.83] N. Harada, Y. Awano, S. Sato and N. Yokoyama (May 2011). Monte Carlo simulation of electron transport in a graphene diode with a linear energy and dispersion. *Journal of Applied Physics* **109**(10), p. 104509.

[7.84] H. Hirai, H. Tsuchiya, Y. Kamakura, N. Mori and M. Ogawa (Aug. 2014). Electron mobility calculation for graphene on substrates. *Journal of Applied Physics* **116**(8), p. 083703.

[7.85] J. Chauhan and J. Guo (July 2009). High-field transport and velocity saturation in graphene. *Applied Physics Letters* **95**(2), p. 023120.

[7.86] R.S. Shishir and D.K. Ferry (July 2009). Velocity saturation in intrinsic graphene. *Journal of Physics: Condensed Matter* **21**, p. 344201.

[7.87] J.K. David, L.F. Register and S.K. Banerjee (Apr. 2012). Semiclassical Monte Carlo analysis of graphene FETs. *IEEE Transactions on Electron Devices* **59**(4), 976–982.

[7.88] T. O'Regan and M. Fischetti (May 2007). Remote phonon scattering in Si and Ge with SiO_2 and HfO_2 insulators: Does the electron mobility determine short channel performance?. *Japanese Journal of Applied Physics* **46**(5B), 3265–3272.

[7.89] D.K. Ferry (2012). *Transport in graphene on BN and SiC*. 12th IEEE International Conference on Nanotechnology (IEEE-NANO), Birmingham, 20–23 August.

[7.90] I. Meric, C.R. Dean, A.F. Young, N. Baklitskaya, N.J. Tremblay, C. Nuckolls, P. Kim and K.L. Shepard (Jan. 2011). Channel length scaling in graphene field-effect transistors studied with pulsed current-voltage measurements. *Nano Letters* **11**, 1093–1097.

[7.91] I. Meric, M.Y. Han, A.F. Young, B. Ozyilmaz, P. Kim and K.L. Shepard (Nov. 2008). Current saturation in zero-bandgap, top-gated graphene field-effect transistors. *Nature Nanotechnology* **3**, 654–659.

[7.92] M. Bresciani, P. Palestri, D. Esseni, L. Selmi, B. Szafranek and D. Neumaier (Sep. 2013). Interpretation of graphene mobility data by means of a semiclassical Monte Carlo transport model. *Solid-State Electronics* **89**, 161–166.

[7.93] V. Perebeinos and P. Avouris (May 2010). Inelastic scattering and current saturation in graphene. *Physical Review B* **81**(19), p. 195442.

[7.94] L. Liao, J Bai, Y. Qu, Y. Huang and X. Duan (Nov. 2010). Single-layer graphene on Al_2O_3/Si substrate: better contrast and higher performance of graphene transistors, *Nanotechnology* **21**, p. 015705.

[7.95] I. Meric, C. Dean, A. Young, J. Hone, P. Kim and K.L. Shepard (2010). Graphene field-effect transistors based on boron nitride gate dielectrics. *IEDM Technical Digest*, 556–559.

[7.96] A.Y. Serov, Z.-Y. Ong, M.V. Fischetti and E. Pop (July 2014). Theoretical analysis of high-field transport in graphene on a substrate. *Journal of Applied Physics* **116**(3), p. 034507.

[7.97] R.K. Ghosh, M. Brahma and S. Mahapatra (July 2014). Germanane: a low effective mass and high bandgap 2-D channel material for future FETs. *IEEE Transactions on Electron Devices* **61**(7), 2309–2315, and references cited there.

[7.98] S. Dong and C.-Q (June 2015). Chen, Stability, elastic properties and electronic structure of germanane nanoribbons. *Journal of Physics: Condensed Matter* **27**, p. 245303.

[7.99] K.L. Low, W. Huang, Y.-C. Yeo and G. Liang (May 2014). Ballistic transport performance of silicane and germanane transistors. *IEEE Transactions on Electron Devices* **61**(5), 1590–1598.

Index

Carrier Transport in Nanoscale MOS Transistors, First Edition. Hideaki Tsuchiya and Yoshinari Kamakura.